AUTOMATIC CHEMICAL ANALYSIS

ELLIS HORWOOD SERIES IN ANALYTICAL CHEMISTRY

General Editor: Dr. R. A. Chalmers, University of Aberdeen

Founded as an international library of fundamental books on important and growing subject areas in analytical chemistry, this series will serve chemists in industrial work and research, and in teaching or advanced study

Published or in active preparation:

HANDBOOK OF PROCESS STREAM ANALYSIS
K. J. Clevett, Crest Engineering (U.K.) Inc., London

THEORETICAL FOUNDATIONS OF CHEMICAL ELECTROANALYSIS
Z. Galus, Warsaw University

ELECTROANALYTICAL CHEMISTRY
G. F. Reynolds, University of Reading

ANALYSIS OF ORGANIC SOLVENTS
V. Sedivec, J. Flek } Institute of Hygiene and Epidemiology, Prague

METHODS OF CATALYTIC ANALYSIS
G. Svehla, Queen's University of Belfast
H. Thompson, University of New York

ORGANIC REAGENTS IN INORGANIC ANALYSIS
Z. Holzbecher et al., Institute of Chemical Technology, Prague

ANALYSIS OF SYNTHETIC POLYMERS
J. Urbanski et al.

COLORIMETRIC DETERMINATION OF THE ELEMENTS
Z. Marczenko, Warsaw Technical University

PARTICLE SIZE ANALYSIS
Z. Jelinek, Organic Synthesis Research Institute, Pardubice

ANALYTICAL APPLICATIONS OF COMPLEX EQUILIBRIA
J. Inczedy, University of Chemical Engineering, Veszprem

GRADIENT LIQUID CHROMATOGRAPHY
S. Liteanu and S. Gocan, University of Cluj

STRIPPING VOLTAMETRIC ANALYSIS
F. Vydra, Heyrovsky Polarographic Institute and E. Julakova and K. Stublik both of Charles University, Prague

AUTOMATIC CHEMICAL ANALYSIS

JAMES K. FOREMAN and
PETER B. STOCKWELL

The Laboratory of
The Government Chemist, London

ELLIS HORWOOD LIMITED
Publisher Chichester

Halsted Press: a division of
JOHN WILEY & SONS Inc.
New York · London · Sydney · Toronto

English Edition first published in 1975
by
ELLIS HORWOOD LIMITED
Coll House, Westergate, Chichester, Sussex, England
The Publisher's colophon is reproduced from
James Gillison's drawing of the ancient Market Cross,
Chichester

Distributed in:

Australia, New Zealand, South-east Asia by

JOHN WILEY & SONS AUSTRALASIA PTY LIMITED
110 Alexander Street, Crow's Nest, N.S.W., Australia

Europe, Africa by

JOHN WILEY & SONS INC.
605 Third Avenue, New York, N.Y. 10016, U.S.A.

ISBN 0 85312 009 9 (Ellis Horwood Limited)
Library of Congress Catalog No. 74–14671

(Ellis Horwood series in analytical chemistry)
Includes bibliographies.
1. Chemistry, Analytic – Automation. I. Stockwell,
Peter B., joint author. II. Title.
QD75.4.A8F67 1974 543 74–14671
ISBN 0–470–26619–8

Filmset and Printed Offset Litho in Great Britain by
Cox & Wyman Ltd., London, Fakenham and Reading

Contents

Preface

Automatic methods of chemical analysis, virtually unknown a quarter of a century ago, are now sufficiently commonplace that they are in use to a greater or lesser extent in most analytical laboratories. This growth reflects the increasing role of chemical analysis in support of objectives as diverse as production industries on the one hand and social well-being, as illustrated by medical care and environmental quality, on the other. Although many factors have contributed to the overall stimulus to develop automated methods, two of these have proved to be particularly significant. Firstly, where the volume of analytical information required is large the requirement to produce it as economically as possible becomes paramount. Secondly, there are many instances, notably analytical services to industrial processes, where the value of such information is greatly enhanced when when it can be produced and evaluated rapidly.

To date the documentation of advances in automating analytical procedures is largely confined to research papers in the scientific literature and manufacturers' literature describing specific instruments which have emerged as commercial products, together with a number of review-type papers formulating the demands and scope for automatic analysis in particular scientific areas. In addition Technicon Corporation, manufacturers of the 'AutoAnalyzer' have sponsored a number of symposia on automatic analysis, the proceedings of which have been published (Mediad Press Ltd.). Consolidation and co-ordination of progress in book form is confined to a few volumes dwelling on specific aspects of the subject, for example, *Automatic Methods in Volumetric Analysis* by D. C. M. Squirrell, Hilger & Watts, London (1964) and *Automatic Titrations* by J. P. Phillips, Academic Press, New York and London (1959). *Practical Automation for the Clinical Laboratory* by W. L. White, M. M. Erickson and S. C. Stevens, C. V. Mosby Co., St. Louis (1968) is a detailed laboratory manual covering several commercial automatic analysers of special relevance to clinical analysis. The volume *Continuous Analysis of Chemical Process Systems* by S. Siggia, Wiley, New York (1959) also relates to certain aspects of automatic analysis. Now that almost every analytical measurement and separation technique has been studied with a view to producing automatic methods, it is opportune to review progress across the whole field of laboratory methods; the present volume attempts to do this.

In planning the book the authors had in mind the ever-increasing number of analytical chemists who will be confronted with requirements to automate, partially or completely, procedures which have hitherto been performed manually. They will need to develop a competence in mechanical, pneumatic and electronic techniques or alternatively to work alongside fellow-scientists with such expertise and to appreciate the power and limitations of their disciplines. The emphasis of the book is therefore on technique. We have not attempted to be comprehensive or to compile a manual of methods; rather we have tried to produce a broad backcloth indicating the real advances made, against which new problems and demands can be seen in perspective. Automatic analysis is a subject in which scientific requirements and performance cannot be divorced from economic factors and the importance of matching the solution to the problem in terms of cost and instrumental complexity is stressed throughout. We have limited our coverage to laboratory analysis and have in consequence given no more than passing consideration to the role of automatic analysis in the control of industrial processes. This is an area in which automatic analysis has brought, and will continue to bring, major benefits and is of sufficient scope and importance to merit a separate volume devoted to it. Nevertheless, much that appears in these pages is of direct applicability to, or can offer partial solutions to, 'on-line' analytical problems.

Whenever a laboratory chemical method is automated the final procedure represents a blend of chemistry, handling techniques and data processing and this has determined the structure of the book. An introductory chapter considers the broad philosophies of automating analytical procedures and also the scientific and economic factors which bear on the choice of approach. Thereafter progress in the automation of the principal methods of measurement is discussed followed by separation procedures. Finally a chapter on the application of computer techniques presents a digest of the diverse ways in which computers can contribute to solving problems of data handling, processing and presentation which arise in larger scale automatic procedures.

But what is an 'automatic' method? In the context of this volume it implies the elimination, partially or fully, of human intervention in a chemical method. The Commission on Analytical Nomenclature of the Analytical Chemistry Division of the International Union of Pure and

Applied Chemistry (IUPAC) prefers a more rigorous terminology. Automation is defined as 'the use of combinations of mechanical and instrumental devices to replace, refine, extend or supplement human effort and facilities in the performance of a given process, in which at least one major operation is controlled, without human intervention, by a feed-back system', and mechanization as 'the use of mechanical devices to replace, refine, extend or supplement human effort'. In these terms there are, as yet, relatively few completely automatic methods in laboratory use; and yet the overwhelming majority of the procedures discussed in this book represent mechanical and instrumental advances which are essential components of any ultimate automatic method. In IUPAC terminology, therefore, this book should be titled 'Mechanization and Automation of Laboratory Methods'. Nevertheless our definition of 'automatic' reflects the interpretation of the vast majority of workers upon whose studies we have drawn and because we have reviewed the historical as well as the more recent aspects of the subject we have preferred not to change it in the light of the IUPAC recommendation.

It is a pleasure to acknowledge the help of those whose contributions have been so valuable in the preparation of this volume. We thank particularly Mr. W. H. Scates for the illustrations and Mrs. K. M. Bradley, Mrs. M. Stuart, Miss S. Waldron, Mrs. P. Cordroy, Mrs. B. Malone and Mrs. M. Evans for typing the manuscript and the many revisions it underwent. We are indebted to Mr. H. Baxter for his comprehensive literature searching and for compiling the index. We are fortunate to work at a laboratory where a number of our colleagues are rich in experience of automatic analyses, both as designers and users. To all of them who have stimulated us by advice, opinion and argument we offer our thanks.

Chapter 1

Introduction

This volume describes the major areas of the progress which has been made towards the partial and complete automation of analytical methods. Almost all of this progress dates from the past three decades and the majority of it, in particular the rapidly expanding commercial availability of automatic analysers, relates to the last 10–15 years. In this context the introduction, in 1957, of the Technicon 'AutoAnalyzer' has proved to be the dominating influence as judged by publications in the scientific literature. It is pertinent at the outset to examine the motivations which have stimulated research in what is now one of the fastest growing areas of analytical chemistry. As with almost every area of technological growth, the pattern can be related to considerations of demand and supply. 'Demand' in the present context requires little elaboration; the essential role which analytical chemistry fulfils in providing fundamental information and quality control data in major commitments such as chemical and manufacturing industry, public health, environmental quality and international trade, is becoming increasingly more complex and exacting. In all such instances chemical analysis represents a service cost and there are obvious incentives therefore to improve the quality, quantity and economic efficiency of the service. Automation and mechanization of analytical procedures are consistent with all three requirements. They constitute the 'supply' position and in order to satisfy the demand suitable techniques for handling and processing samples and of presenting and evaluating data must be, and are being, devised. But technique development is not necessarily the limiting factor in the automation of an analytical procedure. Automation frequently implies long-term operation of an analytical facility and in consequence places demands on materials of construction which, especially in the case of moving parts, should ideally yield reproducible long-term performance with minimal downtime. Because automation of analytical methods is a relatively new subject, it is not surprising that technological research and achievements are moving more rapidly than published systematic data on the performance, both in total and in detail, of equipment in regular use. Where such information is available it is emphasized in the appropriate chapters but it must be remembered that, in the majority of instances, performance data relate only to the particular circumstances under which the equipment is operated. To quote but one general example, the design criteria for handling samples or reagents which are toxic, corrosive or inherently unstable are likely to be more demanding, both in terms of handling technique and constructional materials, than for analyses where these properties are not encountered. There is, indeed, every incentive for authors to publish accumulated performance data for automated analytical systems; ideally the information should relate to component parts as well as the overall system. A valid rationale of automatic analysers, upon which predictive assessments of performance in new circumstances can be based, is ultimately dependent upon the availability of proven performance characteristics for methods which have been extensively tested.

Now that commercial instrument manufacturers are becoming more active in marketing automatic analysers the availability of performance characteristics has a twofold importance; it provides an independent assessment of the present generation of commercially produced analysers, and, in the case of analysers developed in users' laboratories, it assists the manufacturers in evaluating their potential commercial viability for more general application.

An important feature in the proving of analytical methods of wide-ranging importance is the establishment of their performance in defined circumstances by collaborative testing between laboratories involved in the analysis under consideration. This is typified by the studies sponsored by the Association of Official Analytical Chemists in the U.S.A. and by the Analytical Methods Committee of the Society for Analytical Chemistry in the U.K. Procedures involving a degree of automation or mechanization have already been implicated in such collaborative exercises in a very limited way. The evaluation of automated procedures by collaborative testing at a more extensive level than hitherto will clearly contribute to the status of the subject, but the difficulties and limitations, particularly in terms of finding sufficient laboratories equipped with identical or similar automatic procedures, are obvious.

1.1 Economics of Automatic Analysis

Automatic analysers, like other analytical instruments, must be purchased or designed and built within the laboratory. In either case money is involved and it is standard practice to seek to justify the investment in terms of the advantages, scientific or economic, which will be derived from it.

Although scientific advantages may result from the installation of automatic analytical equipment, and these are discussed below, the main incentive for most laboratories to convert manual analyses to full or partial automation is economic. The basic cost of an analysis performed manually (C_M) may be expressed as

$$C_M = F_M (T_S A_S + T_P A_P + T_M A_M + T_C A_C) \quad (1)$$

where T_S, T_P, T_M and T_C represent the time taken for sampling, pretreatment (physical or chemical), measurement and computation respectively, and the corresponding A terms are the fraction of T for which the operator is directly involved at each stage. Thus A would approximate to unity for operations such as weighing, titration or solvent extraction but would be significantly less for ashing, drying, incubation, etc. F_M is a tariff factor to convert time into money and therefore includes salaries and overhead expenses. The cost (C_A) of undertaking the same analysis under conditions of full automation is

$$C_A = F_A (T_A A_A + T_M A_M) \quad (2)$$

where the suffixes A and M refer to automatic analysis and equipment maintenance respectively. If the method is only partially automated then equation (2) must include the terms from equation (1) for the remaining manual stages.

For the automatic method to be preferable on purely economic grounds the cost of automatic analyses must be less than the manual cost by at least an amount equal to the cost of the automatic equipment amortized over, say, three years. The appropriate costs are obtained by multiplying both C_M and C_A by the number of samples to be analysed over the chosen amortization period.

This simple algebraic treatment is clearly only approximate; it takes no cognizance of factors such as differences in reagent costs, power requirements and supervisory costs between the two methods. Nevertheless, in most cases these are of second order magnitude relative to analysis time and operator costs, and in any event for the economic case to be sound it should not need to be dependent on marginal issues of this type. The treatment is limited to the economics of the analytical laboratory only; this is valid where samples are received from an outside source but it is unlikely to be so where the laboratory is an adjunct to a processing plant and performing a quality control function. In the latter case laboratory costs tend to be small compared with production costs and if the introduction of automatic analysis reduces the cost of product hold-up while analytical clearance is awaited, then the additional laboratory cost of the automatic equipment is likely to be of small overall significance. The installation of automatic analysers in the production line represents the ultimate ideal for quality control; there have been several accounts of economic benefits deriving from in-line monitoring and control[1, 2] but discussion of these is outside the scope of a volume dealing with laboratory automation.

Two additional features are evident from equations (1) and (2). Where the automatic method is more rapid than the manual procedure, i.e. T_A is less than T_M, then the capacity of the laboratory to perform the analysis is increased, and if the laboratory is under pressure to increase its sample throughput, then the costing becomes proportionately more favourable to the automatic method. Alternatively, the automatic method will enable a given sample load to be analysed more rapidly and thereby release staff for other duties. Secondly, although most automatic methods are more rapid than the manual counterpart, equations (1) and (2) indicate that economic benefits can accrue even when this is not so. A reduction of the A values is equivalent to saving operator time and provided the reduction is sufficiently large, the operator will be free to carry out other laboratory work. It is not uncommon for a manual method requiring full-time operator attention to be automated in such a way that A_A in equation (2) is very small, say 0·1. In many laboratories increased operator availability represents a benefit as valuable as increased sample throughput and capacity.

A potentially large economic advantage offered by fully automatic analysis is the capability to perform long runs of repetitive analyses during non-working hours. Depending on the laboratory requirements the additional analytical freedom so gained can be utilized to achieve a more rapid turn-round of samples or to increase the overall analytical capacity.

Unattended equipment operation implies a high order of reliability and safety. Generally speaking, extensive pre-operational trials are necessary before committing the analyser to unattended use, as an instrumental fault developing during 'silent hours' may invalidate, perhaps irrevocably, the results for a large number of samples. Major faults such as power failure, cessation of reagent or sample feed, etc. are readily recognized from inspection of the output data, or lack of it, and fail-safe devices can be incorporated to minimize the complications that can result. It is preferable that failures of this type should shut down the analyser rather than allow samples to be incorrectly analysed. Transient or intermittent effects may be more troublesome and may pass undetected by an inexperienced operator. They can arise not only in the instrument but also in the chemistry of the system; also, in many laboratories the ambient temperature falls appreciably at night and unless adequate temperature compensation is provided, many analytical reactions can be adversely affected. Where this type of effect can occur it is essential to include periodic standards with the samples; if temperature control is inadequate a systematic variation in the standard results is observed. Again, voltage variations in the mains power supply may affect the lamp intensity in a colorimetric analyser; this may be of no consequence in an instrument providing compensation facilities as in a double-beam analyser, but in a single-beam system results could be affected. Many automatic analysers now display results in digital form, but in an analyser intended for unattended use the supporting evidence of a recorder trace can be invaluable as a means of monitoring performance.

An approach to assessing the economic merits of automatic analysis similar to that outlined above has been described by Kehoe[1] whose paper contains hypothetical worked examples with typical time, labour and capital costs inserted. Although such detailed calculations offer a useful guideline, and indeed are amenable to some measure of extrapolation, individual cases differ so widely in relative cost emphasis for the various stages of the procedure that each situation must be considered on its merits.

The ramifications of installing automatic analysers may extend beyond purely economic effects. If the equipment is highly complex it may well be necessary to examine carefully the type of scientist best suited to operate it. Although the operational regime for the equipment may be within the compass of a junior analyst, the successful long-term performance is likely to demand the presence of a more highly skilled scientist. He should have some understanding in depth of the system, be able to detect and correct faulty operation, to undertake modifications and adjustments to meet new circumstances such as significant changes in sample composition, and to deal with unusual situations. Such variations in staffing will pose little problem to a laboratory endowed with people having expertise in a range of disciplines, but the smaller, monofunctional type of analytical laboratory may well meet difficulties, particularly in terms of equipment maintenance. The higher the capital cost of the automatic equipment, the more important it is to maintain a high utilization rate. Once again considerations of this nature are highly individual in relation to the laboratory situation and more detailed generalizations are not meaningful.

The approach to training in automatic analysis needs careful consideration in the light of the projected level of involvement of the laboratory. Where the commitment is limited to the purchase of a particular instrument which, without modification, can be used to automate a manual method, then training by the manufacturer may well be adequate provided that efficient servicing facilities are available. In many instances the situation is likely to be more complex. The economic case for automation may well be compelling but the scientific solutions to the problems involved in the transition may not be immediately evident, or may require an appraisal in depth of several alternatives. At the authors' laboratory, where automation of methods is economically justifiable in several subject areas, a small research and development team has been set up to enable the problems, and most important the philosophy, of automation, to be studied in depth[3]. For each problem the following possibilities must initially be considered:

(*a*) is a commercial instrument available to perform the analysis?
(*b*) can any commercial instrument be modified to perform the analysis?
(*c*) if neither (*a*) nor (*b*) applies, can the design and development costs involved in automation be justified in cost-benefit terms?

An affirmative answer to (*c*) generates an applied research project to design, build, and test an automatic analyser. Research in automatic analysis ideally requires a multi-disciplinary team including specialists in chemistry, physics and electronics, in addition to an adequate degree of engineering support. Although the team should function in a closely integrated manner to exploit the contributions from each discipline, it should guard against being insular. Throughout the project it should work in continuous unison with the laboratory on behalf of which the analyser is being made. For preference the user laboratory should participate fully in the design, construction and testing stages. The user laboratory must specify the requirements and ensure that they are understood and respected by the research team. The relationship between the research and user laboratories is critical; loss of confidence on the part of the user can be damaging to the whole project and therefore good liaison between the two brings psychological as well as scientific benefits.

Although the transition from manual to automatic analysis would appear to centre largely on equipment development and testing, the role of the chemist remains a fundamental one. The nature and complexity of the analyser depend upon the chemical method to be automated. The success, both in scientific and economic terms, of an automatic method is likely to be related to the optimization of the chemical approach. Preliminary studies may well be necessary to decide between a proven method which involves a separation stage and a less well-developed one which offers the possibility of being specific for the analyte[4]. The chemist must also provide background information on the 'ruggedness' of the method, the extent to which changes in conditions such as temperature, pH, concentration of extraneous material in the sample, can be tolerated without prejudicing the desired accuracy and precision. In particular he must draw attention to any parameters for which close control is imperative. An adequate understanding of the 'ruggedness' should ensure that the analyser is designed to the right level of performance. Overdesign can be unnecessarily costly in terms of development, construction and maintenance, and underdesign frustrating in that the desired performance is not achieved.

1.2 Advantages of Automatic Analysis

The economic advantages of automated analysis have been considered in the previous section; there are, however, many other scientific or managerial advantages which may or may not have economic connotations.

A well designed and constructed automatic analyser can operate reproducibly over long periods. Consequently it may be expected that for the analysis of a large batch of samples its analytical precision will be superior to that of manual analysis. It does not follow that in absolute terms the automatic method is more precise than the highest performance achievable by a skilled analyst, but the automatic method eliminates human error and fatigue, both of which are likely to become more prevalent as the sample batch size increases. Robinson[2] demonstrated that whereas the precision of analyses performed manually or with simple mechanical aids worsened over the course of a working day, that obtained with an automatic analyser remained consistently good.

Of especial significance is the reproducibility of the timing sequences. This enables reaction conditions to be controlled, hence improved performance can be achieved where the method involves stages where reactions do not proceed to completion or where the parameter being measured is unstable. Separation techniques such as dialysis and solvent extraction, in which recovery of the desired species is frequently incomplete, can give highly reproducible performance in an automatic analyser and it is possible to employ colorimetric reactions where the colour stability would be considered inadequate for a manual method. Indeed, a feature of instruments having closely controlled sequencing, such as the 'AutoAnalyzer', is that reproducible results can be obtained even when the analytical reaction is incomplete, provided standards are processed in exactly the same way as the samples. This approach can increase the range of potentially useful analytical reactions provided that suitable standards are available. A further extension becomes possible where the automatic analysis is carried out in a closed system, in that materials which are toxic or unstable in air can be more conveniently processed than by a manual method.

The increased work capacity or sample throughput offered by an automatic analyser has already been referred to; for the normal analytical laboratory this has the effect of either reducing the delay in completing an analysis or, alternatively, of enabling the laboratory to analyse a greater number of samples. On a larger scale, however, these facilities can have far-reaching effects. The most intensive single use of automatic analysis has been in the field of clinical medicine, and multichannel analysers capable of determining, usually in blood or urine, as many as twelve clinically significant parameters simultaneously are now well established tools. Typically up to 300 measurements per channel per hour can be made. With this level of analytical capacity it becomes possible to contemplate the extension of clinical analysis beyond diagnostic analysis on patients to general health screening of the populace. Several medical schools, mostly in the U.S.A., are sufficiently well equipped to offer this type of service, which could represent a significant advance in the area of preventive medicine. Similar considerations apply in other fields such as environmental monitoring and process control.

As the analytical capacity of automatic analysers is increased, whether by speeding up the manual procedure or by the use of multichannel analysers, the full utilization of this capacity requires similar improvements in the handling and evaluation of results if the full economic benefits are to be secured. Advantage may be taken of the recent extensive advances in data processing techniques. In particular, conversion of the analytical signal into digital form (if it is not initially digital) enables a choice from the now considerable range of digital computing techniques to be made. Such is the design flexibility of automatic data processing techniques that they can be adapted to the size and range of the analytical requirement; indeed it is economically important that a careful assessment of data processing needs should be performed in the early stages of design or purchase of an automatic analyser. Overprovisioning can incur unnecessary expense and underprovisioning can hinder full exploitation of the chemical system. Computer-based data processing of results from automatic analysers is assuming ever-increasing importance and Chapter 11 is devoted to it.

1.3 Limitations of Automatic Analysis

The automatic analysers currently available are designed to process batches of samples of similar composition or samples where expected variations in composition do not affect the measurement stage. The analyser performs functions which are essentially fixed by design, although in some instances the programming unit may be capable of performing minor variations, for example in volume of reagent addition. Usually such variations are predetermined by the operator rather than by response to variations in sample composition. Thus it must be accepted that automatic analysis at the present time is applicable only to runs of similar samples. The value of this can hardly be over-estimated, but it must nevertheless be borne in mind that it is generally not applicable to samples having wide variations in composition.

Although day-to-day instrument operation of automatic analysers can be performed by junior staff of limited experience and qualifications, in all but the simplest instruments fairly close supervision by a qualified and experienced scientist is necessary to validate results and to diagnose malfunction or malperformance. Once a run of samples has been loaded into the automatic analyser, the degree of control of the equipment available to the operator is minimal; little can be accomplished other than to switch off the instrument when its performance deviates from that expected. Minor deviations, due to, say, unexpected changes in sample composition or marginal equipment malfunction may well pass unnoticed by inexperienced or unqualified operators.

1.4 Automatic Analysis by Discrete and Continuous Methods

All automated analyses of samples in the liquid state are performed by one of two methods, discrete or continuous, occasionally by a combination of both[5]. In the discrete method each sample is maintained as a separate entity, it is placed in a separate receptacle and the analytical stages of dilution, reagent addition and mixing are performed separately by mechanically transporting the sample to dispensing units where controlled additions are made to each sample individually. Likewise each treated sample is presented in turn to the measuring unit (colorimeter, electrodes, etc.). In continuous analysis the sample is converted into a flowing stream by a pumping system and the necessary reagent additions are made by continuous pumping and merging of the sample and reagent streams. Ultimately the treated sample is pumped to a flow-through measuring unit and thence to waste.

In designing an automated analytical instrument, or selecting the most appropriate commercially available type, the choice between the discrete or continuous approach is frequently the fundamental one. Both types are considered in detail under the various technique headings but the broad relative merits of the two approaches are set out here for general reference.

The discrete method has the advantage that samples can be processed at a high rate. For example, commercial colorimetric analysers are capable of yielding between 100 and 300 measurements per hour, whereas for continuous analysers a processing rate of 20–80 samples per hour is normal. However, the high-throughput discrete analysers are appreciably more expensive than the continuous analysers.

Because the discrete approach retains the sample as an entity, cross-contamination between samples is almost entirely eliminated. Since the sample retains its identity throughout the analysis, its fate at any time is known and there is little chance of confusing one sample with another at the measurement and recording stage. In continuous analysis the identity of the sample is lost once the processing is under way and in a normal continuous analyser several samples are being processed at any one time between the sampling and measurement stages. Because of the regular timing intervals between stages, as controlled by the pumping rate and length of delay lines, there is usually no difficulty in associating each recorder response with a particular sample, and indeed it is becoming increasingly common to include a sample identification system on the recorder trace. Nevertheless, in simpler instruments devoid of this facility, problems can arise if a succession of samples have zero responses. Where this situation is suspected the insertion of frequent standards in the sample sequence affords regular datum points. More significantly, unless precautions are taken, interaction between successive samples can occur in a continuous system thereby causing overlap and loss of discrimination at the recording stage. Interaction between samples is discussed in greater depth in Chapter 3 on Automatic Colorimetry. In general, interaction can be minimized by optimizing the design of the timing sequence between samples and by reducing the processing rate and, uneconomically, by inserting water between each sample. The higher the sample throughput the greater is the interaction between samples and this accounts for the restriction on sample processing rate in continuous analysers.

Continuous analysers have the merit of being mechanically simpler than the discrete ones. Sample transport and reagent addition require only a suitable pump. Peristaltic pumps are almost always used in commercial continuous analysers and the ready availability of multichannel peristaltic pumps enables a single motor to control the entire sequence of events. Discrete analysers require a number of moving parts for sample transport, and valves or automatic syringes for reagent addition. In general, therefore, discrete analysers are more demanding on maintenance than the continuous type and regular attention to the moving parts is essential. Clearly this requirement can be minimized by good design and optimum selection of materials of construction.

In the discrete method each analytical operation requires a separate group of moving parts and in general discrete analysers are best suited to heavy loads of chemically straightforward analyses. For this reason their major field of application has so far been in hospitals for clinical analysis of blood and urine, using a multichemical system capable of several simultaneous simple analyses where the stages involved are dilution and addition of one or two reagents followed by measurements. Discrete analysers offering any form of chemical or physical separation such as precipitation, distillation or solvent extraction are rare. Facilities for deproteinization are offered in a few commercial analysers. Discrete solvent extraction units have been designed and proved but await commercial exploitation. Continuous analysers are favoured where chemical pretreatment stages are necessary. The design problems involved in incorporating pretreatment stages in a flowing system are less formidable than with discrete operation and many satisfactory techniques are now available. Consequently continuous analysis is likely for some to come to be the method of choice where intermediate separation stages are required between sampling and final measurement.

Continuous analysis requires flexible tubes which are not attacked by the materials under examination, and this places certain limitations on the scope of the method. Certain reactive and corrosive materials cannot be satisfactorily pumped although advances have been made in the development of inert plastics and other synthetic materials. Displacement pumping with the aid of a liquid compatible with sample and reagents provides an alternative, though generally inconvenient, approach. No such limitations arise in discrete analysers because there is no restriction on the choice of materials for sample and reagent containment.

The comments above regarding the relative merits and limitations of the discrete and continuous methods are intended as broad generalizations only, though they will frequently be of value in deciding the approach to a new problem. The two approaches are not mutually exclusive. For example, a number of commercially available automatic amino-acid analysers utilize continuous ion-exchange separation followed by discrete analysis of individual fractions of column eluate. As emphasized earlier, it is the chemistry of the method which determines the instrumental approach; only in relatively few cases, of which clinical analysis is the principal example, has the choice of instrumental approach been optimized. In other instances the chemist is likely to find that, even though many papers have been written on automatic analysis, the subject is still in its infancy. This is one of the main reasons why a multidisciplinary approach to automation commends itself as being the most profitable means of answering the first and most vital question, how should a new problem in automatic analysis be approached?

References

1. Kehoe, T. J. *Treatise on Analytical Chemistry* Part 3, Vol. 1, eds. I. M. Kolthoff, P. J. Elving and F. M. Stross, pp. 159–186, Interscience, New York (1967).
2. Robinson, R. *Technicon Symp. 3rd, Paris, 1966*, **2**, 211.
3. Foreman, J. K. and Stockwell, P. B. *Lab. Equipment Digest*, 1971, **9**, 129.
4. Sawyer, R. and Stockwell, P. B. *Lab. Equipment Digest*, 1972, **10**, 51.
5. Tucker, K. B. E., Sawyer, R. and Stockwell, P. B. *Analyst*, 1970, **95**, 730.

Chapter 2

Electrochemical Methods

The field of electrochemical analysis embraces a wide range of techniques including potentiometry, polarography, amperometry, conductimetry, coulometry, chronopotentiometry and ion-selective electrodes, and by suitable choice of technique and experimental conditions a high degree of analytical sensitivity and specificity can be achieved. It is not surprising, therefore, that electrochemical techniques have found considerable favour in the design of continuous and automatic methods, especially where trace components of a sample are to be determined and where their selectivity can reduce or eliminate the need for pretreatment stages, thereby simplifying the design of the automatic equipment. The range of electrochemical methods which are amenable to automation is extremely diverse, incorporating the determination of organic and inorganic constituents in aqueous, non-aqueous, gaseous and molten salt media. A particular advantage of electrochemical methods is their independence of sample colour. Also in many instances the response of the sensing electrodes is a linear function of the concentration of the species being determined. However, electrode systems can sometimes display a marked sensitivity to extraneous influences such as deposition of particulate matter from the flowing sample at electrode surfaces and, under highly oxidizing potential, difficulties due to irreversible reactions at the electrode surface can be encountered and such factors must be taken into account in designing systems for continuous or automatic operation.

In the following sections progress in designing automatic equipment is discussed with respect to each of the principal electrochemical techniques.

2.1 **Polarography**

The first accounts of the application of polarography to flowing systems date back to the early 1940s and involved the determination of dissolved oxygen in lake water[1], activated sewage sludge[2,3], water used for metabolic studies[4], and of oxygen in gases[5,6,7]. These studies were essentially empirical and designed for the specific application described. For example, the polarographic diffusion current is affected by stirring and the flow-rate of the sample across the electrode system. An understanding of the effect of these and other variables is clearly an essential prerequisite to the logical development of polarography for analysis of flowing streams.

Wilson and Smith[8] made a definitive study of the effect of sample flow-rate and cell design on the response of a dropping mercury electrode to a 0·004*M* solution of Cd^{2+}. Horizontal flow in cells of 25 mm and 12·3 mm bore and both upward and downward flow through a cell of 22 mm bore were examined and the variation of diffusion current as a function of volume flow-rate and linear flow-rate was derived. Figures 2.1 and 2.2 summarize the results; for each flow system there is a threshold flow-rate below which the diffusion current is not distorted; for vertical flow, distortion is less in the downward direction where the mercury drop grows in the direction of sample flow. For similar linear flow-rates, ascending flow yields greater diffusion currents than descending flow. Figures 2.1 and 2.2 demonstrate that undistorted diffusion currents can be produced over a fairly wide range of flow-rates and that experimental

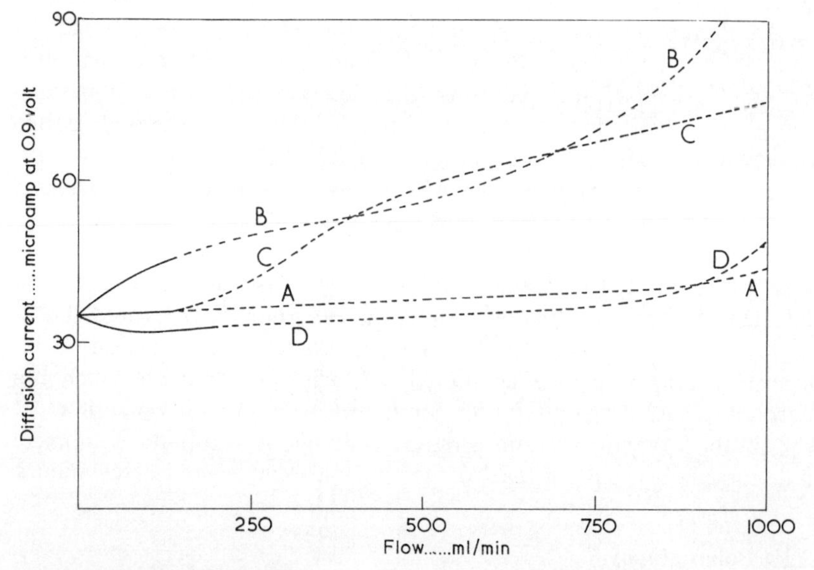

Fig. 2.1 Effect of Volume Flow Rate on Diffusion Current
A. Large horizontal cell; B. Small horizontal cell; C. Vertical cell, upward flow; D. Vertical cell, downward flow.
——— Normal diffusion current
----- Distorted diffusion current
Reproduced with permission from Wilson and Smith[8] and American Chemical Society.

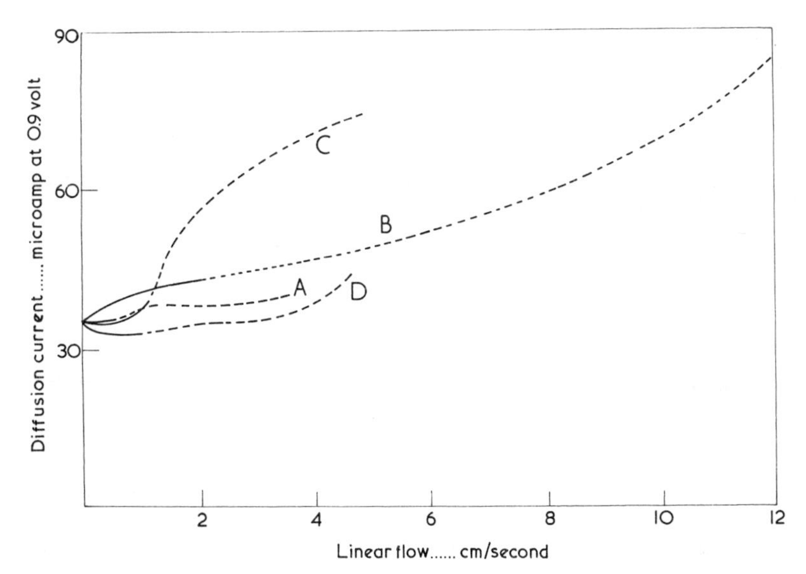

Fig. 2.2 Effect of Linear Flow Rate on Diffusion Current
A. Large horizontal cell; B. Small horizontal cell; C. Vertical cell, upward flow; D. Vertical cell, downward flow.
——— Normal diffusion current
----- Distorted diffusion current
Reproduced with permission from Wilson and Smith[8] and American Chemical Society.

conditions can be selected such that the effects of small variations in flow-rate are minimized.

Solid microelectrodes, especially of platinum, are of importance in polarographic studies of reactions occurring at high positive potentials which are beyond the working potential range of the dropping mercury electrode. Their application to flowing systems was first studied in detail by Müller[9] who examined the current–voltage relationships for quinone–hydroquinone and several inorganic couples at a platinum microelectrode. Reproducible responses resulted, provided that the following conditions were held constant: rate of flow of sample solution past the electrode, rate and direction of change of applied voltage, temperature, pretreatment of the electrode, and nature and concentration of the supporting electrolyte. The electrode behaved reversibly to those redox systems where both oxidized and reduced forms exist as free ions, but the quinone–hydroquinone couple did not respond reversibly although it does so under static conditions. It is likely that in weak acid–weak base couples of this type the rates of dissociation and association are inadequate to maintain equilibrium during the course of the electrode reaction. Nevertheless if the experimental conditions listed above are kept constant, the measured limiting current is a linear function of the concentration of the reacting substance. It is also logarithmically related to the rate of flow of the sample stream. Müller also drew attention to the fact that for a flowing stream, even if neither the flow-rate nor the concentration of electroactive species is known, the polarographic response will still be an indication of the rate at which the species reaches the electrode and that this type of measurement can be of much value in physiological studies such as rate of transfer of substances between blood and tissues.

2.1.1 The Dropping Mercury Electrode (DME)

The dropping mercury electrode is by far the most widely used sensing system in static polarography and to date it holds similar prominence for continuous or automatic systems. The development of highly selective separation methods involving ion-exchange chromatographic columns has produced a requirement for a fast-response, repetitive technique to monitor column effluents continuously or intermittently and so eliminate the time-consuming procedure of manual analysis of individual fractions. Polarography using the dropping mercury electrode has been examined in detail for this purpose.

The cell shown in Fig. 2.3 was used[10] to follow the elution of lead at a concentration of 0·01m*M* from a cation-exchange column (Dowex 50) with 1·05*M* nitric acid. The cell is constructed of borosilicate glass and

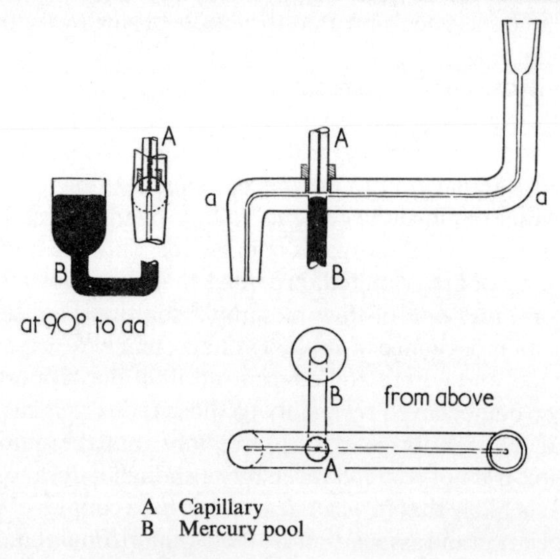

A Capillary
B Mercury pool

Fig. 2.3 Polarographic cell for monitoring ion exchange eluates. *Reproduced with permission from Mann[10] and American Chemical Society.*

is connected to the base of the ion-exchange column by a standard taper-joint. The capillary tube carrying the column effluent is fitted with a port to admit the dropping mercury electrode and, directly opposite, a side-arm to hold the mercury pool electrode. The body of the cell is reamed out to accommodate the electrode capillary. A rubber gasket seated in a glass flange provides an adequate water-tight seal. The cell was used in conjunction with a standard commercial polarograph. To follow the elution of lead from the column the polarograph was operated in the amperometric mode by applying a constant potential (−0·92 V *vs.* the mercury pool) on the plateau of the lead reduction wave. Flow-rate of the column effluent through the cell was held constant. The hold-up volume of the cell was small and the response time of the electrode system was acceptably fast; a typical recorded elution curve is shown in Fig. 2.4. For a given

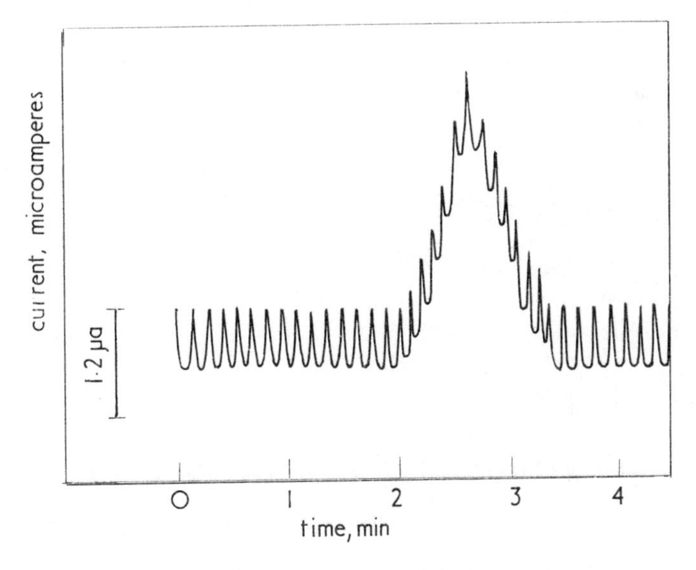

Fig. 2.4 Typical elution curve using the cell shown in Fig. 3.
Reproduced with permission from Mann[10] and American Chemical Society.

concentration of lead the measured current is greater than in a static system because solution flow augments the normal diffusion process. Diffusion current response became irregular at flow-rates above 4 ml/min; over the range 0–2 ml/min the measured diffusion current increased by about 5%. In this case the flow-rate restriction is imposed by the size of the cell; higher rates could be tolerated in a higher volume cell, possibly at the expense of some increase in response time.

In the study above, dissolved oxygen was not removed from the column effluent before making the polarographic measurements, and reduction of oxygen accounts for the high observed background currents and also limits the sensitivity of the method. Nevertheless, where relatively high concentrations of electroreducible species are to be measured, removal of oxygen is unnecessary unless the highest precision is required and both Drake[11] and Kemula[12] satisfactorily monitored millimolar quantities of amino-acids separated on ion-exchange columns. For greater sensitivity deaeration of the sample stream is essential. A highly satisfactory system for this purpose, and one which has been evaluated in some detail, was described by Blaedel and Todd[13]. The apparatus is shown diagrammatically in Fig. 2.5. The column effluent is first deaerated with nitrogen, the nitrogen is vented to atmosphere, and the solution passed through the polarographic cell and thence to waste or a collecting vessel. The column is attached by a spherical joint at *A*, and stopcocks *B* and *E* enable the solution either to pass to the measuring cell or to be diverted to permit measurements on standard solutions. The nitrogen stream enters at *F* to the deaerator *G* which is a length of 2-mm bore capillary tubing.

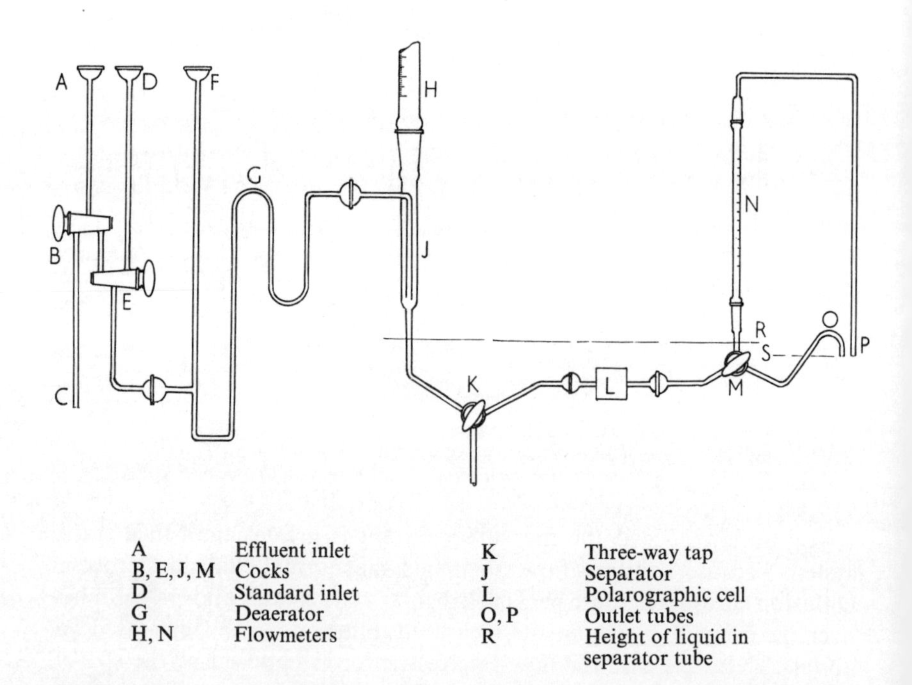

Fig. 2.5 Polarographic system, incorporating deaeration, for monitoring ion exchange eluates.
Reproduced with permission from Blaedel and Todd[13] and American Chemical Society.

The nitrogen carries the solution in a state of turbulence to the separator *J* where the nitrogen exits via flowmeter *H*. The sample solution re-forms at the base of the separator and passes to the polarographic cell *L*. The three-way stopcock *M* provides for sample outlet either through flowmeter *N* and exit tube *P*, or direct to waste or subsequent collection at *O*. Flowmeter *N* monitors the sample flow-rate which is manually controlled by the setting of stopcocks *B* and *E*. To minimize sample hold-up the internal dimensions of the apparatus are kept small, 1 mm bore to the deaerator and 2 mm bore thereafter. With these dimensions sample flow-rates of about 0·25–4 ml/min can be employed with a nitrogen flow of 1–3 l/min. The deaerator removes 99·9% of the dissolved oxygen, a steady state being reached in less than 5 min after commencement of an experiment. The level *S*, at which the outlet tube terminates, determines the height *R* at which solution stands in the separator. It is not critical but requires some measure of control to avoid spray formation in the separator (if *R*–*S* is too great), or formation of gas slugs in the solution passing to the cell (if *R*–*S* is too small). The polarographic cell (Fig. 2.6) has a total internal volume of about 1 ml. To prevent mixing between solutions in the cell and calomel reference electrode compartments, these are separated by a salt-bridge fitted with sintered glass discs. The bridge is filled with ground glass, and deaerated half-saturated KCl solution, fed from a constant-level storage vessel, flows through it at a rate sufficient to yield a slight leakage. This renders deaeration of the reference electrode solution unnecessary.

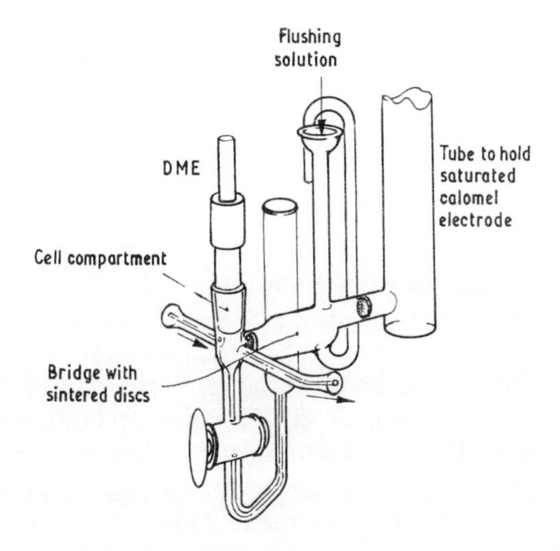

Fig. 2.6 Cell for continuous polarographic analysis.
Reproduced with permission from Blaedel and Todd[13] *and American Chemical Society.*

The response time of the equipment was derived by flowing pure 1*M* KCl through it and allowing the response at an applied voltage of −0·8 V *vs.* a saturated calomel electrode (SCE) to become steady and then abruptly changing the influent solution to $10^{-2}M$ cupric chloride in 1*M* HCl, the latter being admitted at *D* (Fig. 2.5). Decay times were similarly measured by reverting abruptly to 1*M* HCl. The satisfactory response of the equipment can be seen from Table 2.1 which lists response times for three different flow-rates.

Table 2.1

Solution flow-rate (ml/min)	1	2	4
Dead-time (min)	1·5	0·7	0·3
Build-up time (min) to reach			
90% of steady state	3·5	2·4	0·6
99% of steady state	5·0	3·0	1·0
Decay time (min) to reach			
1% of original current	4·0	2·0	1·0
0·1% of original current	6·0	3·0	1·5
0·01% of original current	12·0	6·0	2·0

The system was evaluated for two analytical separations, the cation-exchange separation of Cd^{2+}, Cu^{2+} and Pb^{2+} with hydrochloric and perchloric acids as eluents, and the anion-exchange separation of maleic and fumaric acids with hydrochloric acid as eluent. In each case the limiting current was a linear function of the concentration of each reducible ion over the concentration range 2×10^{-5}–$10^{-3}M$. Provided that an adequate number of standards was run during the course of the separation, concentrations at the upper end of the range could be measured with a relative standard error of about 2%.

Subsequently Blaedel and Todd[14] demonstrated that the same equipment could be satisfactorily used to monitor the ion-exchange separation of α-amino-acids. The method involved passing the column eluate through a bed of copper phosphate to form a copper–amino-acid complex, converting the latter into the copper–EDTA complex, and determining the copper polarographically.

The cell arrangement shown in Fig. 2.7 was also developed for monitoring the progress of ion-exchange separations[15]. Its performance was demonstrated for the anion-exchange separation of Ni^{2+}, Mn^{2+}, Co^{2+} and Cu^{2+} with eluents of various concentrations of hydrochloric acid as described by Kraus and Moore[16]. The flow of eluent through the column was controlled with a Sigmamotor pump; a bubble of nitrogen resting on the top of the column served to damp out flow pulses generated by the pump. Supporting electrolyte (NH_4OH/NH_4Cl) and column eluate converged and were thoroughly mixed by passing through a column of

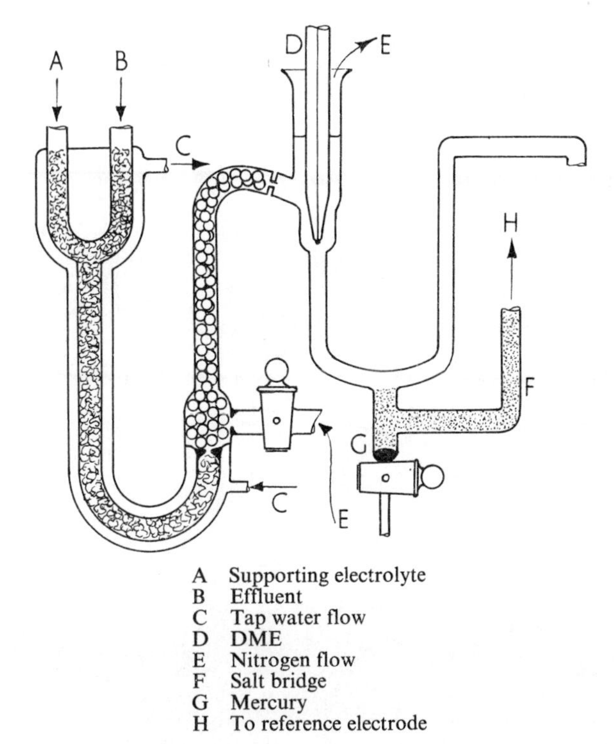

A Supporting electrolyte
B Effluent
C Tap water flow
D DME
E Nitrogen flow
F Salt bridge
G Mercury
H To reference electrode

Fig. 2.7 Cell for polarographic analysis of ion-exchange eluates.
Reproduced with permission from Rebertus, Cappell and Bond[15] *and American Chemical Society.*

glass beads immediately before flowing into the electrode compartment. The flow-rate of supporting electrolyte was controlled by nitrogen pressure on a 4-litre reservoir and was maintained at ten times the flow-rate of column eluate. The cell has a considerably larger volume than that of Blaedel and Todd[13] and consequently response times are longer, 20 min being required to attain a steady-state reading. On the other hand higher flow-rates can be tolerated without distortion of the response. Deaeration was accomplished in two stages: the supporting electrolyte was prepurged to remove the dissolved oxygen and about 95% of that subsequently introduced by the column eluate was removed by the nitrogen flow. Performance of the cell was tested by calibration experiments involving column separation of 2·0, 0·2 and 0·02 mg each of the four elements. Relative standard deviations on mean values varied from 0·6 to 3·0% at the 2·0-mg level, 1·8 to 8·7% at 0·2 mg and 8.6 to 19% at 0·02 mg.

Several other successful designs of cell have been reported. Blaedel and Strohl[17] devised the unit shown in Fig. 2.8 for use in conjunction with

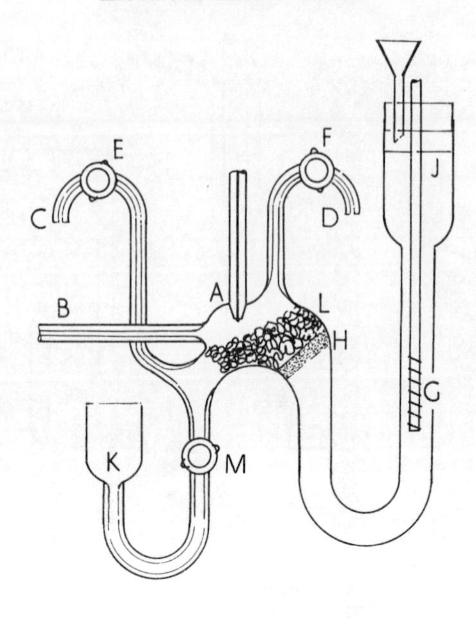

A	DME	H	Glass frit
B	Inlet	J	Reference solution
C, D	Outlets	K	Mercury reservoir
E, F	Cocks	L	Crumpled Pt wire
G	Reference electrode	M	Stopcock

Fig. 2.8 Cell for polarographic monitoring of flowing streams.
Reproduced with permission from Blaedel and Strohl[17] *and American Chemical Society.*

the deaerator described above[13]. The principal attributes of the cell, which has an internal volume of 2 ml, are fast response time and low ohmic resistance. The DME (*A*) is ground roughly into the cell body and subsequently sealed with epoxy resin. The flowing stream enters at *B* and two outlets *C* and *D* are provided with terminations at approximately the same height. By adjustment of the stopcocks *E* and *F* most of the stream is diverted through the upper outlet *D*, thereby continuously removing any entrained gas bubbles. An Ag/AgCl reference electrode *G*, immersed in $\sim 2M$ NaCl is connected to the cell through a fine sintered frit *H*. A small leakage of sodium chloride solution (0·1 ml/min) into the cell is provided by maintaining the reference solution level *J* above outlets *C* and *D*. This leakage ensures that the reference electrode solution is not contaminated with, or diluted by, the sample solution. Should the leakage be inadvertently reversed, the U-tube connection provides a resistance to sample solution reaching the reference electrode. To avoid contact between the DME and the leaking reference solution, the latter should be more dense than the sample solution so that it flows smoothly down the face of the frit to the bottom outlet tube *C*. The lower part of the cell is loosely packed with 26-gauge platinum wire *L* which serves to reduce turbulence and cycling in the sample solution, and also provides an additional means of isolating the DME from the leaking reference solution. Spent mercury drops collect at the bottom of the cell and, if the stopcock *M* is kept open during operation of the cell, find their own level in reservoir *K*. The response time of the cell, derived from measurements on $10^{-3}M$ Cd^{2+} solutions, is such that flushing volumes of approximately 2 and 3 ml are required for the diffusion current to reach 90% and 99% respectively of the steady-state reading. The effect of flow-rate on the diffusion current is small; at 4 ml/min the response is only about 5% greater than the static value, again with respect to $10^{-3}M$ Cd^{2+} solution after flushing volumes of 0·5 and 0·75 ml respectively. Subsequently Blaedel and Strohl[18] developed a cell of smaller internal volume and hence more rapid response, current values of 90% and 99% of steady state being obtained on passing 0·5 and 0·75 ml respectively of sample solution. The design, which is self-explanatory, is shown in Fig. 2.9. Construction of the cell involves some

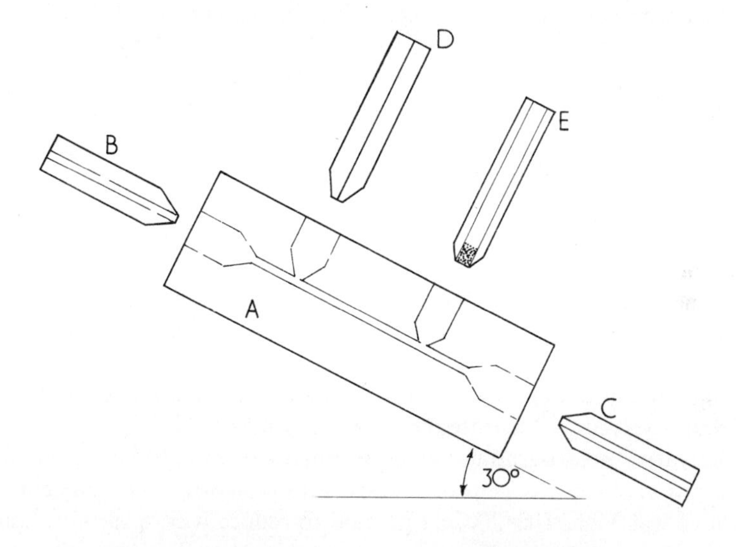

A Teflon cylinder with $\frac{1}{16}$ in hole bored axially
B Inlet, 2 mm i.d.
C Outlet, 2 mm i.d.
D DME capillary
E Bridge to reference electrode, 3 mm i.d. with glass frit

Fig. 2.9 Low-volume, fast-response cell for continuous polarographic analysis. *Reproduced with permission from Blaedel and Strohl[18] and American Chemical Society.*

exacting machining and finishing, details of which are given in the paper. Because of the small diameter of the axial hole, any gas bubbles entering the cell are not trapped but are flushed rapidly through with a momentary interruption of the polarographic response. Cells of the same design, but with axial holes of diameter greater than $\frac{3}{32}$ in. do tend to trap gas bubbles.

Jura[19] used polarography at a shielded dropping mercury electrode to determine the cyanide content of flowing streams. Over the concentration range of interest (10^{-2}–$10^{-3}M$) it was demonstrated that the measurements could, under alkaline conditions, be made in the presence of dissolved oxygen. Sample solution flowed upwards through the cell, having first been continuously diluted tenfold with supporting electrolyte by using a Zenith metering pump. The effective cell volume was about 3 ml and at flow-rates of 95 ml/min the diffusion current reached 98% of the steady state value in 2 min. Current remained diffusion-controlled at flow-rates up to 120 ml/min and polarograms were identical in all respects with corresponding static measurements.

Continuous polarographic determination of trace quantities of cadmium in the presence of up to 1000-fold excess of bismuth has been described[20]. Two dropping mercury electrodes were used differentially, the mercury drop-rates from each cathode being synchronized by a valve-operated relay, a modification of the method described by Airey and Smales[21]. The applied voltages on the two electrodes were set to correspond to the plateaux of the bismuth and cadmium waves respectively; the compensating effect of the two cells eliminates the bismuth wave and only the cadmium wave is recorded. This approach should be of general applicability to the elimination of the effect of ions which reduce at more positive potentials than the ion sought. The flowing sample was deaerated by passing nitrogen through a sintered glass frit at the base of the inlet vessel, and overflowed into an annular vessel with two outlets, each carrying the solution to the cell. An additional vessel was interposed before the cell to remove residual gases and smooth out small variations in flow-rate. The equipment described was simple and essentially demonstrational but it served to illustrate the potential utility of the approach.

Derivative polarography has been utilized to monitor the uranium concentration of process feeds of dissolved uranium ore[22]. The concentration range was high, 100–200 mg/ml, and to reduce it to a suitable concentration for polarographic measurement, less than 1 mg/ml, the feed stream was continuously diluted 400-fold with supporting electrolyte. The dilution stage also served to reduce the interference effects from other metals with the exception of vanadium and molybdenum, which, if present, affected the accuracy of the measurement. Derivative polarograms were achieved by using a standard resistance–capacitance circuit[23]. In the supporting electrolyte used (0·1M H_2SO_4, 0·1M Na_2SO_4, $4\times10^{-4}M$ thymol) uranium reduces at $-0{\cdot}5$ V *vs.* the mercury pool and the applied voltage was scanned from $-1{\cdot}0$ to $-0{\cdot}4$ V. By reading the uranium wave during the cycle in the direction of positive potential, oscillation at the top of the derivative peak was minimized. The flow system is shown in Fig. 2.10. The most critical feature was the accurate dilution of the feedstream with supporting electrolyte. Several proportioning methods were examined, including a displacement-type proportioning pump and a constant-flow measuring device, but the most reproducible performance

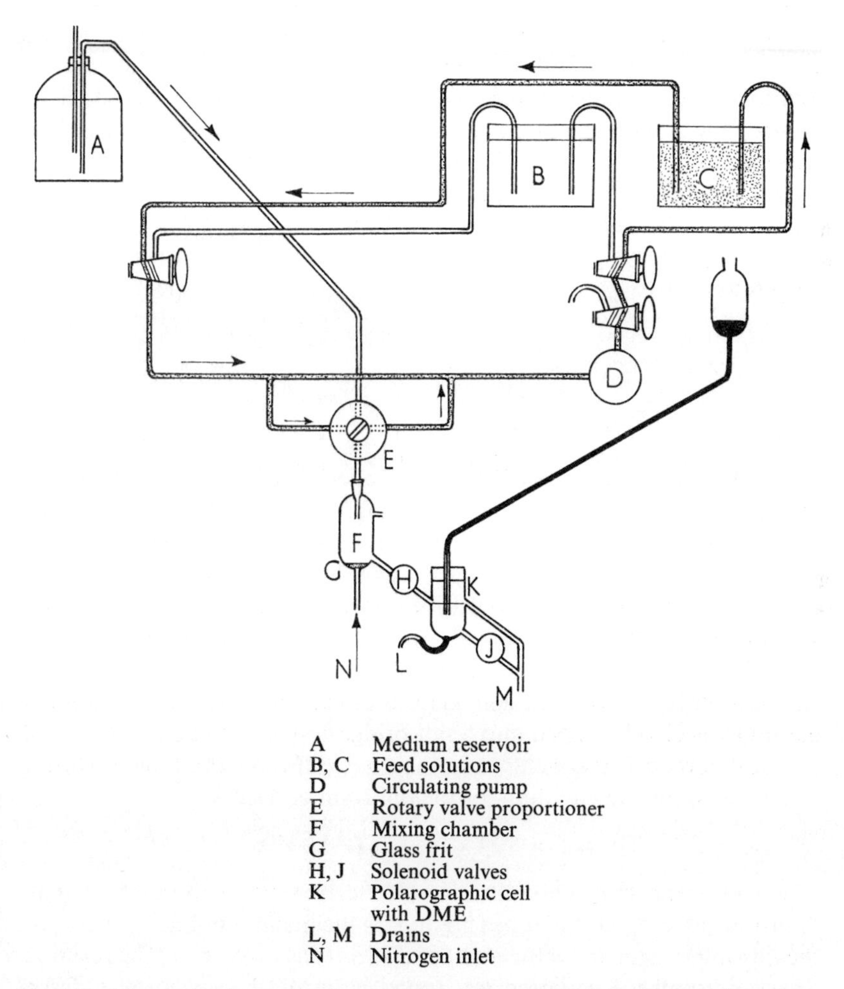

A	Medium reservoir
B, C	Feed solutions
D	Circulating pump
E	Rotary valve proportioner
F	Mixing chamber
G	Glass frit
H, J	Solenoid valves
K	Polarographic cell with DME
L, M	Drains
N	Nitrogen inlet

Fig. 2.10 Flow system for continuous derivative polarographic determination of uranium. *Reproduced with permission from Bertram et al.*[22] *and American Chemical Society.*

was achieved by using a rotary-valve proportioning unit, which yielded dilution factors of up to 400 with an error of less than $\pm 0{\cdot}5\%$. It comprised a rotating plug-valve controlling the flow of the two lines arranged perpendicular to one another; the valve body was stainless steel and the plug was 'Teflon'. On rotation of the valve (at 1 rpm) a small amount of the sample solution is trapped in the plug and subsequently diluted with vertically flowing supporting electrolyte when the valve is rotated a further 90°. The composite solution flows into a mixing vessel and thence into the DME compartment and ultimately to waste. The cycle time for complete analysis is 5 min; during the scan from $-1{\cdot}0$ to $-0{\cdot}4$ V a measurement is made, then the flow is stopped and the cell emptied. On the reverse voltage scan the proportioner is restarted and the cell flushed and refilled in readiness for the next measurement. Two solenoid valves control the flow of solution into and out of the electrode vessel. The relative standard deviation of results on flowing samples was $\pm 0{\cdot}5\%$ and the maximum bias $\pm 5{\cdot}5\%$ except for one sample rich in molybdenum and vanadium, when the bias increased to about 12%. Temperature control of the solutions is desirable because the polarographic response was shown to change by 3% per °C.

2.1.2 Solid Electrode Systems

2.1.2.1 *The Tubular Platinum Electrode* (*TPE*)

Following the work of Müller[9], a number of workers have investigated the performance of solid electrodes for polarography in flowing solutions. Blaedel, Olson and Sharma[24] describe a tubular platinum electrode which, when used in conjunction with a reference electrode such as the SCE, is capable of reproducible and sensitive measurements on a flowing sample stream. The electrode was fabricated from 0·1–1 in. lengths of narrow bore (0·02–0·04 in.) seamless platinum tubing with the ends finished squarely and smoothly, sealed into soft glass tubing. The outlet of the tubular platinum electrode dipped into a salt bridge connected to an SCE.

A theoretical treatment[25] of convective diffusion at a tubular surface yields the following equation for the diffusion current I:

$$I = 5{\cdot}24 \times 10^{5}\, n\, C\, D^{\frac{2}{3}}\, x^{\frac{2}{3}}\, V^{\frac{1}{3}}$$

where C is the bulk concentration of electroactive substance, D is its diffusion current, x is the tube length, V is the maximum linear velocity of the flowing stream and n the number of electrons involved in the reaction. The validity of the equation was tested by using the oxidation of ferrocyanide ion over the electrode dimension range given above and for applied potentials in the diffusion-limiting region, for flow-rates from 2 to 20 ml/min and for concentrations from 10^{-5} to $10^{-4}M$. In all circumstances the equation was obeyed. The sensitivity of the tubular platinum electrode is high and concentrations down to $10^{-8}M$ are detectable. Furthermore, current fluctuations are much smaller than with a dropping mercury electrode. Indeed the tubular platinum electrode offers considerable practical advantages over earlier attempts to use in flowing streams solid electrodes such as the rotating wire[26], disc[25] and conical[27] electrodes.

2.1.2.2 *The Mercury Coated Tubular Platinum Electrode* (*MTPE*)

A limitation to the applicability of the tubular platinum electrode is that it lacks the high hydrogen overvoltage of mercury. A combination of the latter property with the experimental simplicity and high sensitivity of solid electrodes would clearly be advantageous. Marple and Rogers[28] evaluated amalgamated platinum electrodes and demonstrated that stationary electrodes yielded acceptable results in both stirred and unstirred solutions. Amalgamated silver electrodes were recommended by Cooke[29] but they were subsequently shown to be limited in that the calibration curve is time-dependent[28].

Oesterling and Olson[30] developed a mercury coated tubular platinum electrode (Fig. 2.11) in order to exploit the greater hydrogen overvoltage of mercury. To coat the platinum tube with mercury an electrodeposition procedure[31] was used in which the tube (0·5 × 0·06 in.), previously cleaned with hot concentrated nitric acid and distilled water, was first subjected to a potential of −3 V for 15 min with $1M$ perchloric acid flowing through it at 1 ml/min, and then triply-distilled mercury was pumped forward and backward through the tube for 5 cycles. Although the weight of mercury deposited was not constant the performance of the electrode could be readily checked by running a current–voltage curve with an oxygen-free supporting electrolyte of $0{\cdot}1M$ KNO_3 flowing through the cell. A current of less than 0·5 μA over the range from 0 to −1 V *vs.* SCE at a voltage scan rate of 500 mV/min was taken to be proof of adequate performance.

The electrode was evaluated by studying the reduction of a number of metal ions [Bi^{3+}, Cd^{2+}, Cu^{2+}, $Co(NH_3)_6^{3+}$, Pb^{2+}, Tl^{+}] at concentrations ranging from 10^{-4} to $10^{-6}M$. At a flow-rate of 3 ml/min the relationship between diffusion current and ionic concentration was linear for each species over the 100-fold concentration range. The current response was diffusion-limited and independent of the quantity of mercury in the coating. A typical scan for a stream containing $5 \times 10^{-5}M$ each of Cu^{2+} and Pb^{2+} is shown in Fig. 2.12.

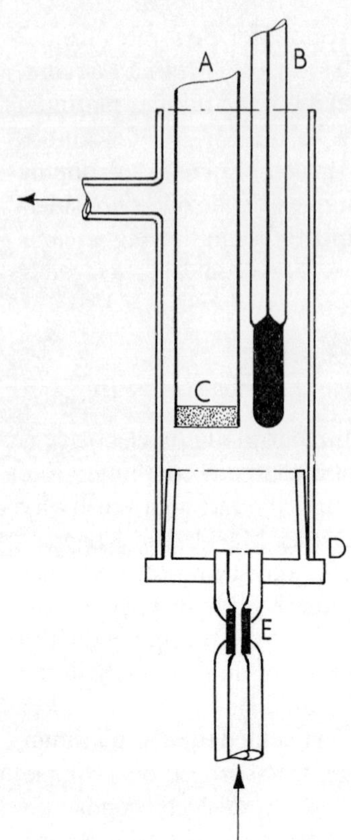

Fig. 2.11 Cell for continuous polarographic analysis using the mercury coated tubular platinum electrode. *Reproduced with permission from Oesterling and Olson*[30] *and American Chemical Society.*

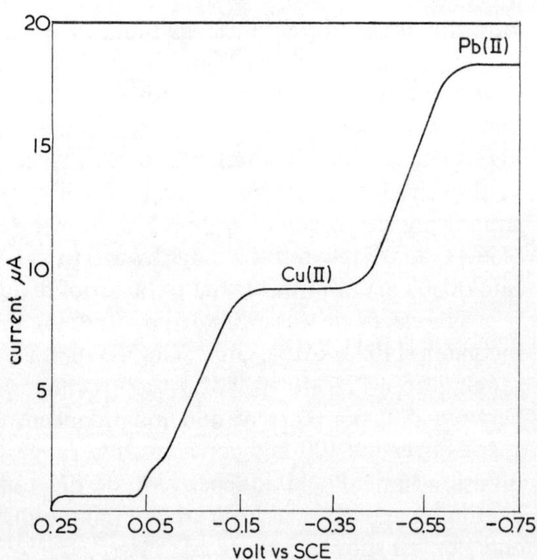

Fig. 2.12 Response of mercury coated tubular platinum electrode to Cu^{2+} and Pb^{2+} at 5×10^{-5} M in a flowing stream. *Reproduced with permission from Oesterling and Olson*[30] *and American Chemical Society.*

In seeking an automatic method for the determination of calcium and magnesium, Booth *et al.*[32] investigated the performance of several voltammetric sensors. The method involves the reactions of calcium and magnesium with both EDTA and EGTA [ethyleneglycol bis(α-amino-ethyl ether)-*N*,*N*,*N*′,*N*′-tetra-acetic acid]. EDTA and EGTA yield anodic waves at the dropping mercury electrode, owing to complex formation with mercury. In the presence of calcium and magnesium the wave height is reduced by an extent proportional to the metal ion concentration. To determine calcium and magnesium in the presence of one another two measurements are necessary; calcium + magnesium is determined from the reduction in height of the anodic EDTA wave at pH 9–10 and calcium alone by reduction of the EGTA wave height at the same pH, the magnesium complex being too weak to yield a significant response. Some distortion of the EGTA wave may occur and pentasodium tripolyphosphate is added as an ancillary masking agent to overcome the effect.

The Technicon 'AutoAnalyzer' forms the basis of the analytical system, as shown in Fig. 2.13. It is necessary to feed the EDTA and EGTA reagents sequentially, and at any one time one line is blanked off by means of the cam-operated taps T_1, T_2, T_3 or T_4. The cam-shaft is driven by the peristaltic pump motor through a gear-train, the reagent line being closed by the action of a push rod operated by the cam. Samples are fed to the manifold from an automatic sample turntable ('AutoAnalyzer' Sampler

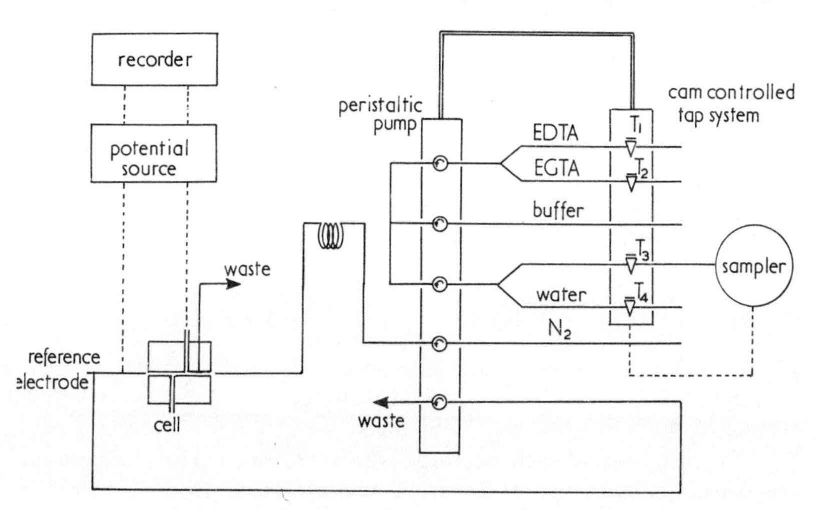

Fig. 2.13 'AutoAnalyzer' manifold for continuous voltammetric determination of calcium and magnesium. Flow rates: complexone and buffer 0·32 ml/min, nitrogen 1·2 ml/min, waste 0·42 ml/min. Double lines and dotted lines represent mechanical and electrical connections respectively.
Reproduced with permission from Booth et al.[32] *and Elsevier Scientific Publishing Company.*

II). To couple the latter to the manifold the microswitch which controls the sampling rate is disconnected and its input leads fixed to a microswitch operated by a cam on the drive shaft of the tap assembly, which is activated after each complete analytical cycle. The mixed reagents and sample are pumped through the electrode assembly and thence to waste. A low-resistance calomel reference electrode was used throughout the study. The voltammetric sensors studied were the dropping mercury, hanging mercury-drop, tubular platinum, mercury coated tubular platinum and pyrolytic graphite electrodes. The flow-through cell for use with the

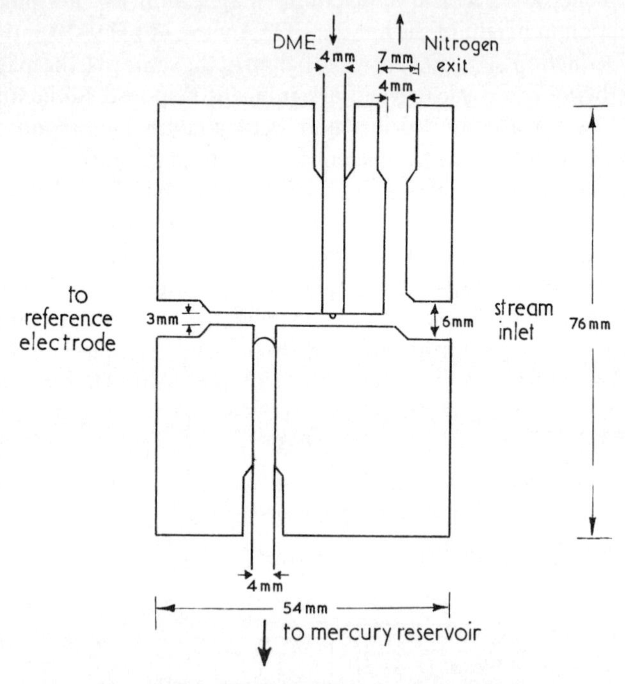

Fig. 2.14 Flow-through polarographic cell design for use with 'AutoAnalyzer'. *Reproduced with permission from Booth et al.*[32] *and Elsevier Scientific Publishing Company.*

dropping mercury electrode is depicted in Fig. 2.14, the assembly being fabricated from 'Perspex' sheet 1 in. thick.

The responses of each electrode pair were recorded for calcium- and magnesium-containing solutions of concentration from 2×10^{-5} to $10^{-3} M$. The most satisfactory responses are obtained with the dropping mercury and hanging mercury-drop electrodes, a typical trace for the former being shown in Fig. 2.15. The latter offers advantages arising from the elimination of capillary noise from the trace and from the decreased

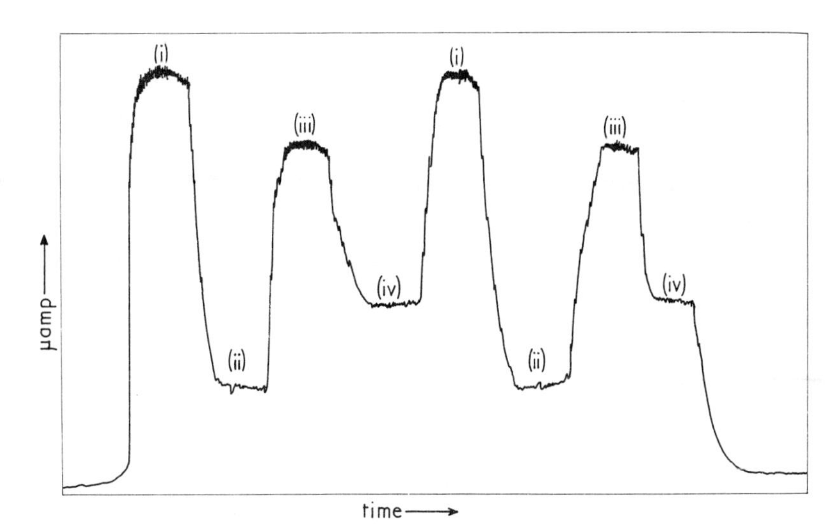

Fig. 2.15 Recorder trace for complete cycle of polarographic analyzer.
(i) 0·5 mM EDTA; (ii) (i)+sample (0·25 mM Ca+0·15 mM Mg); (iii) 0·5 mM EGTA; (iv) (iii)+sample.
Reproduced with permission from Booth et al.[32] *and Elsevier Scientific Publishing Company.*

charging current in comparison to the dropping mercury electrode, and it merits further study for use as a continuous sensor. The tubular platinum electrode, though sensitive, gives irreproducible results and the mercury coated tubular platinum electrode suffers from interference due to oxygen reduction. Even thoroughly deaerated solutions give an unacceptably high residual current. The pyrolytic graphite electrode proved to be of no analytical value because the waves were poorly defined. With the dropping mercury electrode the precision on an individual measurement is ±3% and the long-term stability is good.

Fano and Scalvini[33] describe the electronic circuits for an automatic polarograph using the hanging-drop mercury electrode, either as a single-cell device or with two similar cells used to obtain differential polarograms. The sequencing is controlled by timers (Crouzet), microswitches and relays. These permit preselection of the time for electrodeposition of the active species on to the hanging drop, the applied voltage and the scanning time for the polarogram. The scanning function is provided by a scanning helipot driven by a wide-ratio stepping motor. The scanning linearity is better than 0·25%. The power supply, which has two outputs to provide the differential function, is stabilized with Zener diodes. The power generator comprises a relaxation oscillator with a junction transistor, the generated impulses being fed to a Schmitt trigger which controls a bistable multivibrator. The hanging-drop polarograph is inherently capable of measuring electroactive ions to concentrations as low as 1 ng/l. In the polarograph of Fano and Scalvini the lower current levels are measured with a Keithley picoammeter. A standard XY recorder is used for the higher levels. Available performance data are limited to those connected with circuit performance, no specific analytical measurements being presented.

2.1.2.3 *The Porous Catalytic Silver Electrode*

Fleet, Ho and Tenygl have utilized a porous silver electrode for the determination of oxygen and have demonstrated its usefulness in both the polarographic[34] and coulometric[35] modes. In each case they have applied it to the study of chemical reactions in which oxygen is quantitatively liberated.

The polarographic application involves the determination of sodium hypochlorite and hydrogen peroxide with particular reference to the concentration control of reagents used in textile bleaching processes. For the determination of hypochlorite the reaction with hydrogen peroxide is used and for hydrogen peroxide the reaction with potassium periodate, both being carried out in alkaline solution. Fig. 2.16 is a diagrammatic representation of the polarographic cell. Two porous silver electrode tubes are connected in series and the liberated oxygen is swept through them in a stream of nitrogen from an 'AutoAnalyzer' system, the manifold for which is shown in Fig. 2.17. This permits the mixing of hypochlorite, hydrogen peroxide and nitrogen. The liquid phase is removed by passage through two phase-separators. For the peroxide method the same manifold design is used, with the reagent stream of hydrogen peroxide replaced by potassium periodate. The cell is operated at $-0{\cdot}85$ V *vs.* a mercury/mercuric oxide reference electrode and reduction of oxygen occurs in the annular space of the two porous silver electrodes. The counter-electrode is a platinum wire isolated from the rest of the solution by a sintered glass frit. The porous silver electrode (Type LD 848, Heyrovský Institute of Polarography, Prague) depends for its efficient performance on the maintenance of equilibrium between the gas stream and the electrolyte (25% potassium hydroxide solution) in the pores. The LD 848 electrode has a hydrodynamic resistance of some 10–15 mmHg and to ensure equilibrium the sample gas overpressure must exceed this value. This may be achieved by differential pumping with the waste-lines pumping at a slower rate than the input-lines. An alternative, and preferred, method is to incorporate a pressure-control valve immediately following the cell to regulate automatically the rate of flow to waste. By this means a constant gas overpressure can be maintained at the electrode regardless of variations in

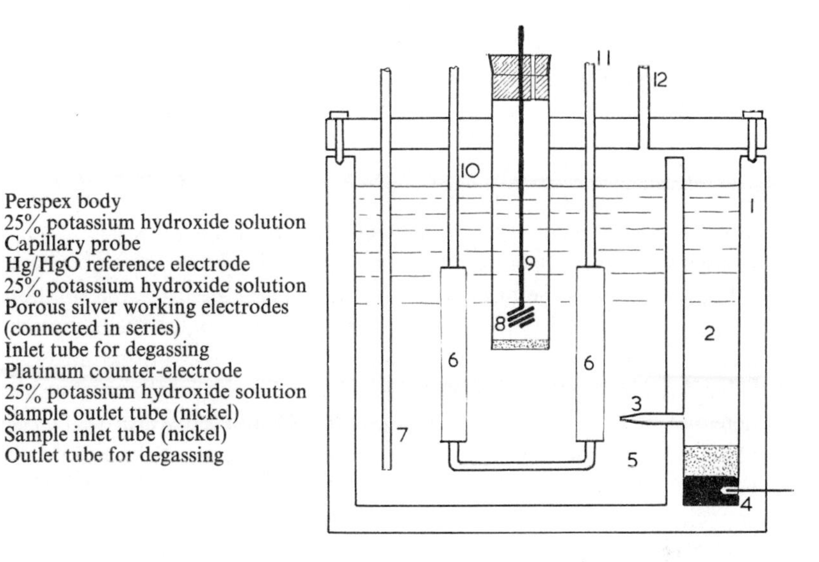

Fig. 2.16 Polarographic cell utilizing the porous catalytic silver electrode.
Reproduced with permission from Fleet, Ho and Tenygl[34] and Elsevier Scientific Publishing Company.

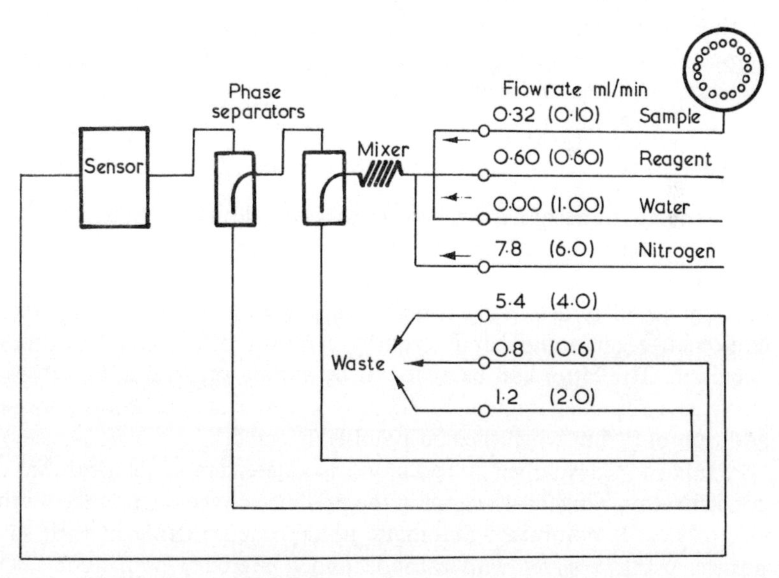

Fig. 2.17 'AutoAnalyzer' manifold for the determination of sodium hypochlorite and hydrogen peroxide in bleaching agent using the porous catalytic silver electrode.
Reproduced with permission from Fleet, Ho and Tenygl[34] and Pergamon Press.

input flow-rates. Use of a pressure regulator obviates the need to pass the waste through the pump and so reduces problems associated with pump pulsing.

The method was tested for sodium hypochlorite by feeding standard solutions in the available chlorine concentration range 0·44–2·2 g/l to the analyser and measuring the electrode response by both integrated and non-integrated techniques. Coefficients of variation were 1·1% and 0·8% respectively, but the integrated method can process samples twice as quickly (20/hr) as the non-integrated one and is the preferred method for treating discrete samples. Used as a continuous monitor the equipment as described has a dead-time of $2\frac{1}{2}$ min and a sensor transition time (95%) of a further 2 min. Thus a variation in hypochlorite concentration can be detected in $2\frac{1}{2}$ min and the full change in $4\frac{1}{2}$ min. However this dynamic response of $4\frac{1}{2}$ min could be shortened by optimizing the manifold design to minimize hold-up and by using a flow-through electrode of smaller capacity. For determinations of 0·2–1·2-vol. hydrogen peroxide the coefficients of variation were 1·5% for the integrated method and 1·0% for the non-integrated one.

2.2 **Amperometry**

Automation of methods based on amperometry has been developed in three main areas, amperometric titration of electroactive materials, amperometric measurement in analysis by reaction rates, and amperometric monitoring of flowing streams.

2.2.1 Automated Amperometric Titrations

Manual amperometric titrations involve adding the titrant stepwise, recording the measured current at each stage and plotting the titration curve to determine the end-point. Automation of such titrations requires first a method of delivering titrant, either from a burette or by internal coulometric generation, and secondly a means of automatic end-point detection. The latter can be achieved by simply presenting the titration curve as a recorder trace or by the use of a mechanism which stops the addition of titrant when the end-point is detected.

Juliard[36] developed a technique for determining chloride in the concentration range 2–40 μg/ml in the presence of gross amounts of other electrolytes. It comprised automatic amperometric titration with silver nitrate, using a silver wire cathode and a mercury pool anode. This titration yields a V-shaped curve of current against volume of titrant, the end-point being at the minimum current value. The titration was performed on 30-ml volumes of solution in a 50-ml glass beaker; a stirrer with two bent blades permitted a very rapid stirring rate (1800 rpm) without cavitation or trapping of air bubbles. The detector for minimum current was a 2-mV full-scale self-balancing recorder. It was provided with a microswitch which was pushed along by the recorder pen as long as the current decreased, the contact being released upon a small increase in measured current. Titrant was delivered by a screw-driven glass syringe lubricated with graphite. The screw was linked by a clutch to the chart-driving shaft of the recorder. The gear ratio was adjusted so that one chart division corresponded exactly to the delivery of 0·010 ml of titrant. The delivered volume of titrant at the end-point was read from a revolution counter attached to the syringe drive mechanism, which was capable of measuring to the nearest 0·001 ml. The accuracy of the method is affected by a small lag in response of the mechanism for detecting the current minimum, and for the best results a correction factor based on titration of standard solutions is necessary.

The rotating platinum electrode is a satisfactory cathode for automatic amperometric titrations, as illustrated by the studies of Murayama[37], who determined mercaptans by titration with silver nitrate, and by Gonzales Barredo[38] who used potassium bromate as the titrant for determining arsenic trioxide. The applicability of a dropping mercury cathode was demonstrated by Myers and Swann[39] in determining sulphate with standard lead solutions. In this work a mercury pool constituted the anode. If distortions of the titration curve are to be avoided the dimensions of the electrode system must be carefully standardized. The design of the anode system is shown in Fig. 2.18. Mercury is supplied from a levelling bulb which is connected by 'Tygon' tubing to a stopcock. Electrical contact is made by means of a piece of platinum wire sealed into the glass tubing at point B. The anode is contained in 1-mm bore capillary tubing. This dimension is critical but the others are not. The end of the electrode is cup-shaped with the top of the pool level with the edge of the cup. The anode surface is readily replenished by adjusting the height of the levelling bulb, but, in fact, contamination of the anode surface does not prove a serious problem and some 10 titrations could be performed using the same mercury surface. In this study titrant was delivered from a 50-ml Luer-type syringe fitted with a constant-rate drive and the automatic recording polarograph used was a Leeds and Northrup Electrochemograph Type E. No performance data are quoted but satisfactory results on samples containing as little as 0·001 mmole of sulphate are claimed.

Automatic methods are clearly advantageous where highly radioactive solutions are to be titrated because handling problems are greatly magnified when hot-cell containment is necessary. Kubota and Surak[40] described

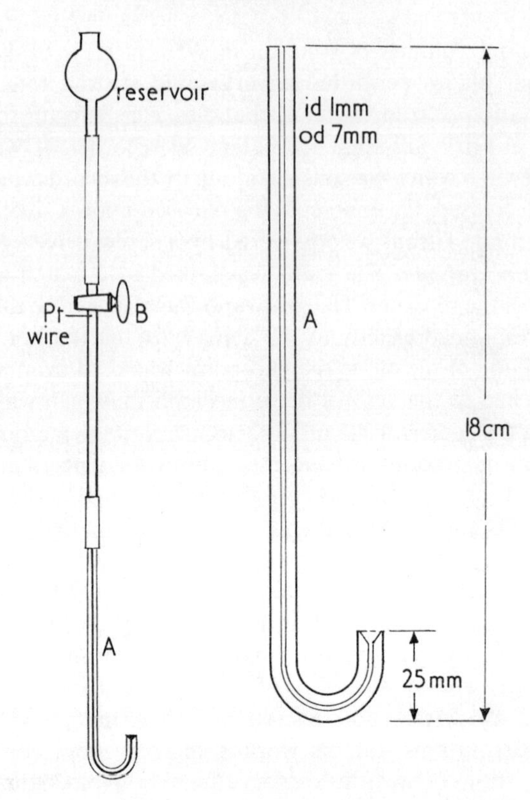

Fig. 2.18 Mercury pool anode for automatic amperometric titration of sulphate. *Reproduced with permission from Myers and Swann*[39] *and Microforms International Marketing Corporation.*

the automatic amperometric titration of zirconium which is present at 9·6 mole per cent in MSRE (Molten Salt Reactor Experiment) salt. The titrating reagent was cupferron and the titration medium 1*N* H_2SO_4, in which very few metals other than zirconium react with cupferron[41], and at this acidity the hydrolysis of zirconium is minimal. Small amounts of fluoride do not interfere and aluminium ion can be used to eliminate the effect of larger amounts. Titrant was delivered from a syringe at a rate of 0·34 ml/min. Constant delivery was achieved by using a portable infusion-withdrawal pump (Harvard Apparatus Company) equipped with a 2-rpm motor. A polythene delivery tube of $\frac{1}{8}$-in. bore drawn to a tip at the end was attached to the syringe. The syringe was also fitted with a small outlet tube at the withdrawal end to permit remote filling by application of vacuum. The titration cell was a 50-ml lipless beaker with a loosely fitting 'Teflon' cap containing entry holes for the burette tip, stirrer, reference electrode, electrical connection to a mercury pool anode and inert gas inlet. Maintenance of a small positive inert gas pressure obviated the need to seal the lid to the beaker. A potential of −0·5 V was applied across the indicator electrode pair (platinum wire and mercury pool) and the titration curve was plotted with a recording polarograph. Aliquots containing about 5 mg of zirconium were taken for the titration, which was completed in about 5 min. Long-term precision studies indicated that a relative standard deviation of 1·5% could be achieved under 'hot-cell' conditions, compared with 1% when titrations were performed on non-radioactive material.

2.2.2 Automatic Amperometric Measurement of Reaction Rates

Analysis by reaction rate methods, particularly for trace quantities, has been extensively developed during the past decade[42]. Electrometric techniques are well suited to automated measurement of reaction rates and both potentiometry and amperometry have been successfully used. Although there are several ways of making the appropriate rate measurement the constant-time and variable-time methods have proved the most popular. In the former all samples in a series are reacted for the same time and the extent of reaction in each sample is determined. In the latter the time taken for a predetermined extent of reaction to occur is measured, and provided that experiments are completed at close to zero reaction time then the concentration of the species sought is inversely proportional to the measured time. Examples have been described of automatic amperometric measurement of reaction rates in both static and flowing systems. The static method is demonstrated by the studies of Pardue[43] on the determination of glucose in serum, plasma and whole blood, by using glucose oxidase. The following reaction sequence is utilized.

$$\text{Glucose} + O_2 \xrightarrow{\text{glucose oxidase}} \text{gluconic acid} + H_2O_2 \qquad (1)$$

$$H_2O_2 + 2H^+ + 2I^- \xrightarrow{\text{molybdate}} I_2 + 2H_2O \qquad (2)$$

Since reaction (1) is very slow compared with reaction (2) the former is the rate-determining step and if all reagent concentrations are kept much larger than that of glucose and hence essentially constant, the overall reaction is pseudo first-order with respect to glucose.

Fig. 2.19 shows in outline the measuring equipment used. The cell volume is a few ml and the analysis is performed on 1 ml of deproteinized sample and 1 ml of a composite solution containing all the reagents. The electrode system is a stationary platinum and rotating platinum pair. Iodide is oxidized at the stationary electrode and simultaneously iodine is

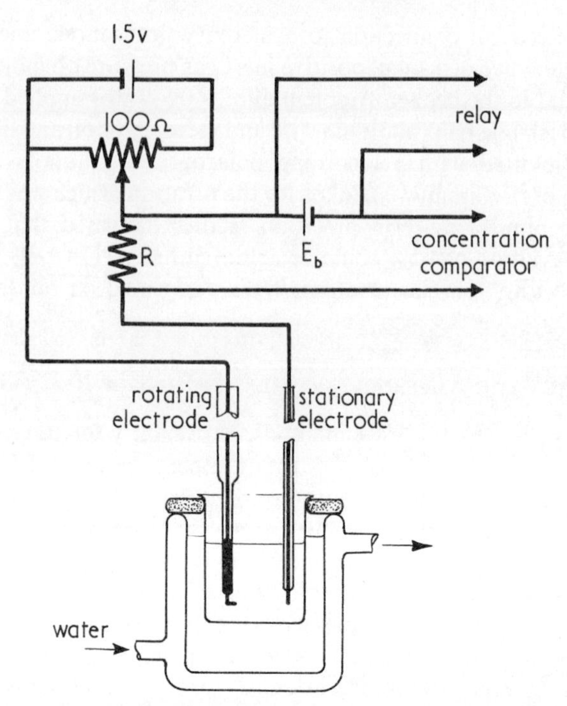

Fig. 2.19 System for automatic amperometric analysis by the reaction rate method. *Reproduced with permission from Pardue*[43] *and American Chemical Society.*

reduced at the rotating one. The total electrolysis current is measured with a Sargent Model Q concentration comparator and the automatic sequence and time measurement is achieved by a relay system. A potential difference of 1·0 V is applied across the electrodes, the desired current preset on the comparator and the reaction sequence started from an auxiliary relay. When the predetermined current is reached the relay stops the comparator timer and the reaction time is read off.

The relative standard deviation of the method over the glucose concentration 50–300 mg/100 ml of serum or plasma is better than 2% and recoveries for standard glucose solutions are in the range $100 \pm 5\%$. The method may be used in the opposite sense to determine the activity of glucose oxidase in the rate-limiting stage[44]. It has also been demonstrated as satisfactory for determining galactose by reaction with galactose oxidase[45].

A continuous automatic amperometric method for glucose, utilizing continuous pumping, is described by Blaedel and Olson[46]. It was designed for use over the concentration range 1–100 μg/ml and is based on the following reaction sequence.

$$\text{Glucose} + O_2 + H_2O \xrightarrow{\text{glucose oxidase}} \text{gluconic acid} + H_2O_2 \quad (1)$$

$$Fe(CN)_6^{4-} + H_2O_2 + 2H^+ + e^- \xrightarrow{\text{peroxidase}} Fe(CN)_6^{3-} + 2H_2O \quad (2)$$

The ferricyanide is determined amperometrically.

The experimental procedure utilizes the constant-time method, the time interval taking the form of a fixed delay period between two measurements of current. The flow path is illustrated in Fig. 2.20. The sample and a second stream containing reagents are both pumped at constant rate and are allowed to mix; the reacting mixture then passes through a delay

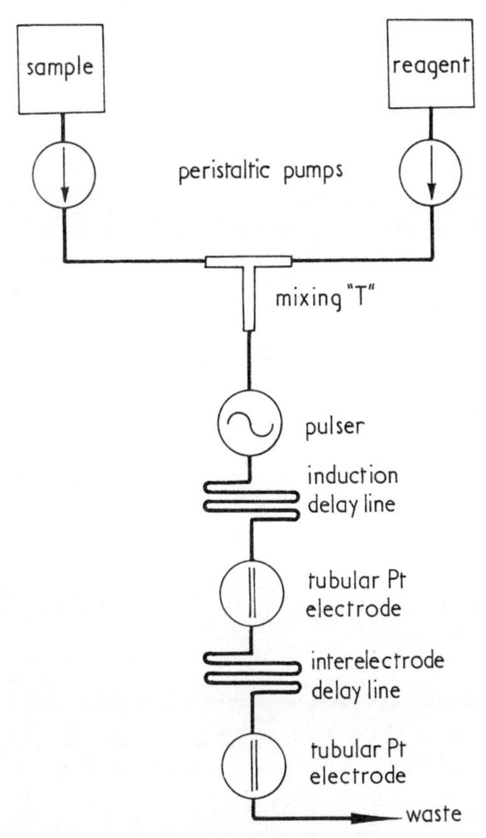

Fig. 2.20 Flow system for continuous amperometric analysis by the reaction rate method using a fixed time delay.
Reproduced with permission from Blaedel and Olson[46] and American Chemical Society.

line to allow for any induction effects and then successively through two tubular platinum electrodes separated by a fixed delay line. The difference in concentration of the electroactive species at the two electrodes is determined by differential amperometry at a fixed applied potential. Because the pumping rate is constant and the electrodes are separated by a fixed distance the measured current is proportional to the concentration difference of the electroactive species at the two electrodes and consequently is proportional to the reaction rate. The high sensitivity of the tubular platinum electrode enables short reaction times to be used, thereby ensuring that measurements are made close to zero reaction time, so preserving the linearity between the measured current and the glucose concentration.

A dual channel variable speed peristaltic pump (Harvard Apparatus Co.) was used, typical flow rates being 1–2 ml/min. Mixing was achieved in a borosilicate T-piece of 2 mm bore with a length of platinum wire inserted in the common arm. The delay lines were lengths of narrow-bore PTFE tubing, the induction delay line being 18 in. long and the inter-electrode delay line 40 in. Both were coiled and embedded in ceresin wax to eliminate temperature fluctuations.

The electrode system consisted of two tubular platinum electrodes (0·5 × 0·04 in.) bridged to a common standard calomel reference electrode. A fixed potential of 0·08 V was applied, the tubular platinum being cathodic with respect to the SCE. The current difference between the two

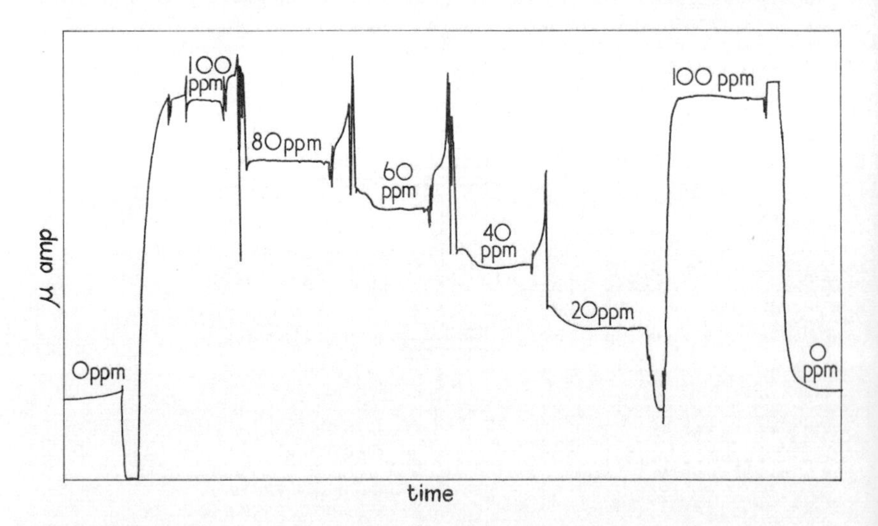

Fig. 2.21 Recorder response for glucose standards using the equipment shown in Fig. 2.20. *Reproduced with permission from Blaedel and Olson[46] and American Chemical Society.*

measuring electrodes was determined by using a simple bridge circuit in which the two tubular platinum electrodes constituted two of the arms, the other two being variable resistors. The currents generated were of the order of μA or less and care was required to ensure adequate electrical shielding and earthing.

The composite reagent contained 2 g of glucose oxidase and 20 mg of peroxidase per litre and was 0·010*M* with respect to potassium ferrocyanide. In common with the glucose sample stream it was buffered to pH 7·5 with 0·1*M* KCl and 0·1*M* Na_2 HPO_4. The composite reagent was deaerated with a continuous stream of nitrogen to prevent aerial oxidation of ferrocyanide.

The sensitivity of the differential amperometric system is such that a detection limit approaching $10^{-9}M$ ferrocyanide is achieved. At $5 \times 10^{-8}M$, recording conditions are steady and essentially free from drift. A typical recorder response for a range of glucose standards is depicted in Fig. 2.21.

2.2.3 Automatic Amperometry in Flowing Streams

An interesting example of continuous amperometry involving a preliminary separation stage is the determination of small amounts of cyanide in Sudan grass (Sorghum vulgare)[47]. Following homogenization at pH 5 in the presence of emulsin (β-glucosidase) to hydrolyse cyanogenic glycosides, and filtration to remove insoluble matter, the filtrate is treated in a continuous-type apparatus in which the cyanide is volatilized, absorbed in alkali and amperometrically determined, by use of a tubular gold electrode in combination with a silver/silver chloride reference electrode.

A block diagram of the continuous analyser is shown in Fig. 2.22. The sample, neutralizing and absorbing solution (0·03*M* NaOH) are transported at 1·5 ml/min by means of peristaltic pumps *A* and *B* (Harvard Apparatus Company); the absorbent flows directly into absorber *F* and the sample together with 0·76*M* $(NH_4)_2SO_4$, pumped at 0·3 ml/min, passes to a volatilization unit *E*. This comprises two coils in tandem immersed in a heating bath containing ethylene glycol maintained at $88 \pm 0{\cdot}5°$ C. Volatilization of the cyanide occurs in two streams of nitrogen. The cyanide-bearing nitrogen passes through 'Tygon' lines into absorber *F* and spent liquid is pumped to waste at *C*, the pumping rate being maintained at slightly greater than the combined inlet rates to avoid subsequent flooding of the gas–liquid separator *G*. The function of the separator is to ensure the absence of gas bubbles from the solution passing through the tubular gold electrode *H*. A single-channel peristaltic pump *D* removes the measured solution to waste. The electrode response

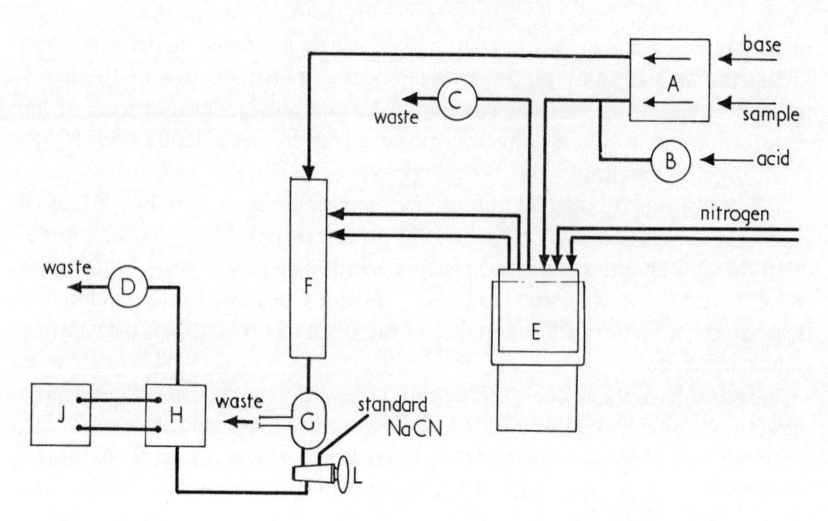

Fig. 2.22 Block diagram of equipment for continuous amperometric determination of cyanide.
Reproduced with permission from Easty, Blaedel and Anderson[47] *and American Chemical Society.*

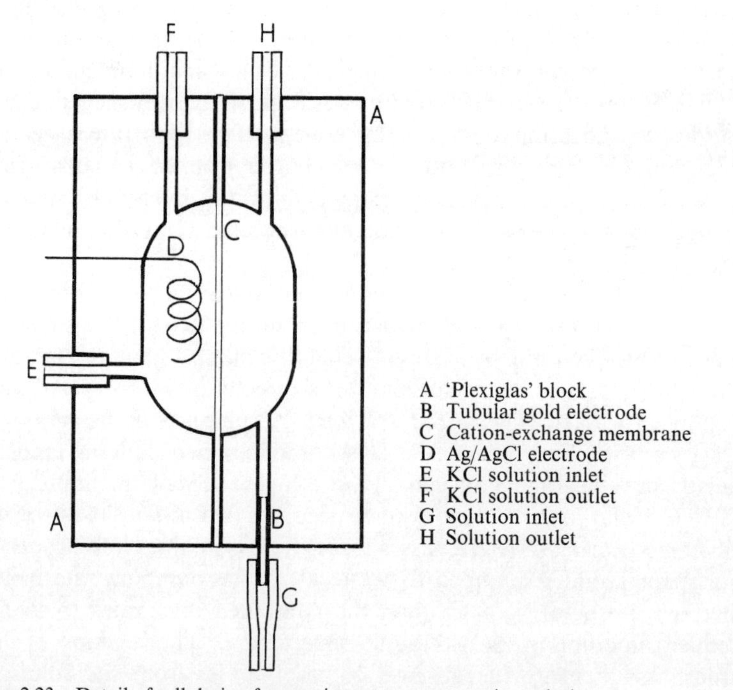

Fig. 2.23 Detail of cell design for continuous amperometric analysis.
Reproduced with permission from Easty, Blaedel and Anderson[47] *and American Chemical Society.*

is measured with a Sargent Model XXI polarograph *J* which supplies the applied voltage to the electrode system. The three-way stopcock *L* enables blank and standard cyanide solutions to be fed directly to the cell for calibration purposes.

Details of the electrode cell are given in Fig. 2.23. It is made from two identically milled 'Plexiglas' blocks subsequently heat-sealed together to form the electrode compartment, which is divided into working and reference segments by means of a sheet of cation-exchange membrane. Before measurements are started the gold electrode is cathodized by applying −1·0 V, thereafter the applied voltage is held at 0±0·15 V. The electrode reaction, which probably involves the formation of $Au(CN)_2^-$ or $Au(CN)_4^-$ is sensitive and linear in response over a cyanide concentration range of 10^{-4}–10^{-6}*M*. A relative standard deviation of ±2% is obtainable in the upper part of this concentration range. A cyanide blank of 2–3 mg/kg was found and tentatively attributed to interference in the amperometric measurement from traces of sulphide arising from the volatilization stage.

Recently Blaedel and Boyer[48] have demonstrated that the tubular platinum electrode (TPE) can be used for amperometric measurement of sub-micromolar quantities of ferricyanide ion. Instead of making measurements successively on sample and blank solutions, a technique which is limited at low concentrations owing to the non-Faradaic background current, a stopped-flow method was used. The cell is fabricated from 'Plexiglas', and the two electrodes, the TPE and an Ag/AgCl reference, are contained in a cavity of about 2 ml hold-up volume. Solutions are fed through the cell by gravity in 2-mm bore glass capillary tubing. The technique comprises measuring the current generated at an appropriate applied potential (−0·1 V) when the solution is flowing at about 3 ml/min and then measuring the blank Faradaic background current with the flow stopped. The difference between the two values is a measure of the concentration of the electroactive species. Over the concentration range 0–1μM ferricyanide an almost linear calibration graph is obtained. Although this technique has not been applied to sequential automatic analyses there is no apparent reason why, in conjunction with a suitable sample feed system, it should not.

Differential amperometry has been utilized in an automatic analyser for determining penicillins[49]. The instrument is based on continuous peristaltic pumping; the manifold is outlined in Fig. 2.24, and 8 samples per hour can be analysed. To provide the differential measurement the sample is split into two streams, one serving as a blank and the other providing the desired analytical reaction. This involves oxidation of the penicillin to penicilloic acid by alkali or penicillinase and iodometric

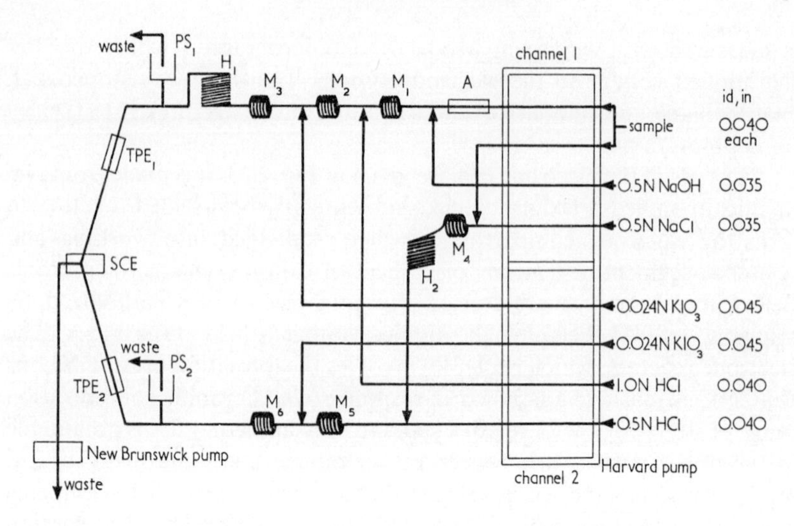

Fig. 2.24 Manifold for continuous amperometric determination of penicillins. *Reproduced with permission from Bomstein et al.*[49] *and American Pharmaceutical Association.*

determination of the penicilloic acid. Thus in practice the only difference in solution composition between the analysis and blank manifolds is the absence of oxidant from the latter. The electrode system comprises a tubular platinum electrode in each of the blank and sample lines, both being bridged to a common reference SCE. Physical separation of the SCE from the working electrodes is provided by a porous glass wall. Saturated potassium chloride/mercurous chloride solution flows continuously past the reference electrode to carry off unwanted impurities which diffuse into the SCE. The differential amperometric response is measured by a simple Wheatstone bridge circuit. The peristaltic pumping action produces pulsing of the flowing solutions, which affects the recorder trace; to eliminate this the pumping rate to waste is adjusted to be marginally lower than the input rate and the excess of liquid is taken up in the vented outlets PS_1 and PS_2 (Fig. 2.24). This difficulty does not arise when the air-segmented solution pumping mode of the 'AutoAnalyzer' is employed.

2.3 Potentiometry

Analyses utilizing potentiometric end-point detection are amenable to automation under both batch and continuous conditions. In the former the full titration curve is recorded or control circuitry is used to terminate the titration at the end-point, whereas in the latter equilibrium end-point conditions are set up and the rate of titrant delivery is measured, since this is then proportional to the sample concentration. In both cases the titrant may be delivered mechanically or generated internally coulometrically.

2.3.1 Automatic Potentiometric Titration

Probably the first automatic potentiometric titrimeter to be described in detail is that of Robinson[50]. It utilized a motor-driven syringe burette to give stepwise delivery of the titrant and a Leeds and Northrup Micromax potentiometric recorder to plot the titration curve. The syringe was mounted vertically and operated by an accurately machined screw driven by a 19-rpm reversible motor fitted with reduction gearing. It was provided with guide rods and suitable clamping to ensure mechanical rigidity and reproducibility of motion. The tip of the syringe terminated in a three-way stopcock to facilitate both delivery of titrant and refilling the syringe with titrant.

Depending upon the titration chemistry the electrode system was either a metal–SCE pair or glass–SCE. There is no obvious limitation on the types of electrode which can be used. The electrode potential was fed to, and recorded on, the Micromax recorder, which, by provision of an additional cam and microswitch system, controlled the movement of the syringe. The circuit arrangement was such that titrant flowed only when the recorder was balanced. The normal recorder balancing cycle was once every 2 sec, but if voltage fluctuations were large, as occurred in the region of the end-point, then titrant addition was delayed for further 2-sec intervals until equilibrium was reached. As a result addition of reagent was rapid in the well-buffered regions of a titration curve but was progressively slowed near the end-point. Quantitative measurement of volume of titrant added was achieved by proportioning the chart travel to the amount of reagent fed.

The titrimeter performance was evaluated for acid–base and redox titrations at approximately decimolar concentrations; 10 ml samples could be titrated in 15 min with a precision of about 1%.

The general principles of the Robinson titrimeter were embodied in the Precision-Dow Recordomatic Titrator, one of the earliest commercial automatic potentiometric titrimeters. They have also served as the basis of automatic titrimeters developed at Oak Ridge National Laboratory for the analysis of radioactive solutions[51]. Inevitably improvements and simplifications in both mechanical and electronic design have been introduced. Thus in the titrimeter described by Kelley *et al.*[51] burette mountings based on universal joints accommodate microburettes and syringe burettes of 1–50 ml capacity. Also the intermittent recorder balancing is achieved with an unsymmetrical multivibrator which operates on the recorder signal after amplification by a gated-beam amplifier.

Solenoid-operated clamps used in conjunction with a standard burette offer an alternative to the motor-driven syringe for delivery of titrant. In the titrimeter described by Shain and Huber[52] the titrant flowed from the burette tip through a short piece of capillary tubing, which controlled the flow rate, and then through the solenoid-operated burette clamp and through the delivery tip. The clamp closed a piece of rubber tubing between the capillary tubing and the delivery tip. This arrangement was shown experimentally to give rapid and reproducible burette closure. Titrations were performed in ordinary beakers, the solution being stirred magnetically. Gold microelectrodes 11 mm in length were used. The relative disposition of the electrodes and burette delivery tip within the solution proved unimportant if analytical precision of 1% was adequate. The error could, however, be reduced to 0·1% by placing the delivery tip and cathode opposite each other near the walls of the beaker. The anode position was not critical. These figures apply to titrations of ferrous ion and thiosulphate with ceric ion, manganous ion with permanganate and chloride ion with silver at concentrations in the range 0·03–0·1*M*. For concentrations below 0·01*M* the error increased considerably. The instrument was designed to operate with a constant current of a few μA applied across the indicator electrodes. Under these circumstances a peaked titration curve usually results and the electronic control circuitry was designed to detect this peak. Indeed, the applicability of the titrimeter is limited to such systems. An advantage of the approach is that knowledge of the end-point potential is not required. The control unit comprised an amplifier, a trigger circuit and relays to operate the solenoid clamp. The electrode potential was amplified and the potential at the plate of a triode valve was arranged to remain constant until the sudden change in electrode potential, representing the end-point, was detected. This was achieved by earthing the plate through a silicon diode which had a high resistance in one direction of polarity only. The trigger circuit, of the Schmitt type, was actuated by the sudden change of potential and operated a locking-type relay which closed the burette tap which had hitherto remained open.

Solenoid-operated burette taps permitting both a fast and slow rate of titrant delivery, the latter being introduced near the end-point, have been detailed by Audran and Dighton[53]. That shown in Fig. 2.25 comprises two different sized jets connected by thin-walled plastic tubing to a branched glass tube forming the base of the burette. Operation of a solenoid interrupts the titrant flow by pressing the roller against the plastic tubing. Two solenoid clamps are used, one for each jet. Switching from fast

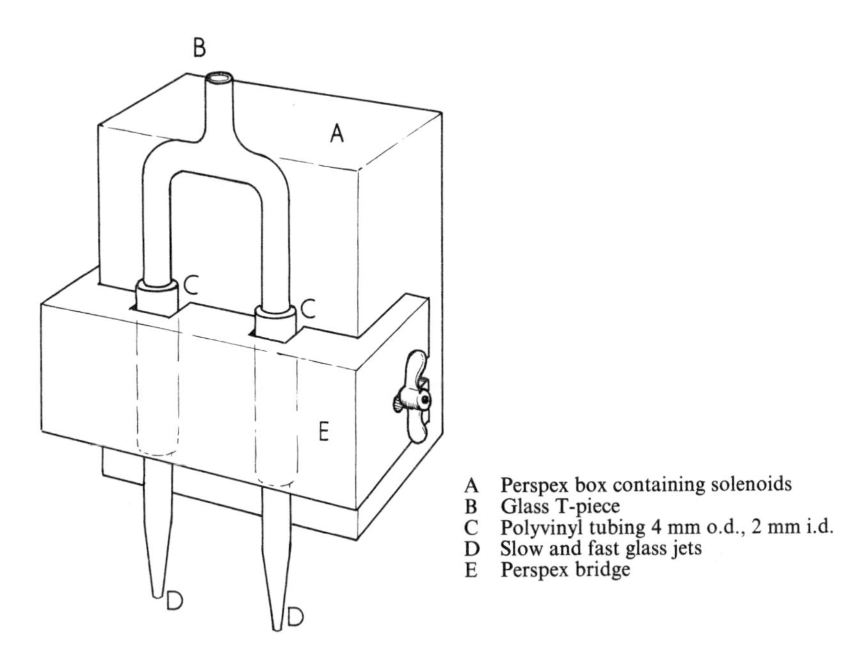

Fig. 2.25 Solenoid-operated burette tap using two jets for fast and slow titrant delivery. *Reproduced with permission from Audran and Dighton*[53] *and The Institute of Physics.*

delivery of titrant to slow (1 drop every 5 sec) is achieved by a relay which comes into operation when the measured electrode potential is 100 mV from the predetermined end-point potential. At the end-point potential a second relay terminates the titration. Fig. 2.26 shows a single-jet burette which permits two rates of flow. The position of the soft iron armature in the centre glass chamber is controlled by two coils capable of being energized by relays. In the upper position a reagent flow-rate of 0·25 ml/sec results, in the centre position the flow is restricted to one drop every few seconds and in the lower position the burette is closed by bringing the ground surface of the glass chamber into contact with the burette jet.

An automatic potentiometric titrimeter of almost universal applicability is that due to Lingane[54]. It is effectively applicable to all types of reactions and has no limitations as to the type of electrode system used. Either the complete titration curve may be plotted or the titration may be terminated at the end-point. Because the cell design provides a measure of end-point anticipation the instrument is applicable to slow reactions.

The essential features of the titrimeter are shown in Fig. 2.27. Titrant is delivered from a 50-ml hypodermic syringe driven through a screw by a constant-speed synchronous motor. The gear train between the motor

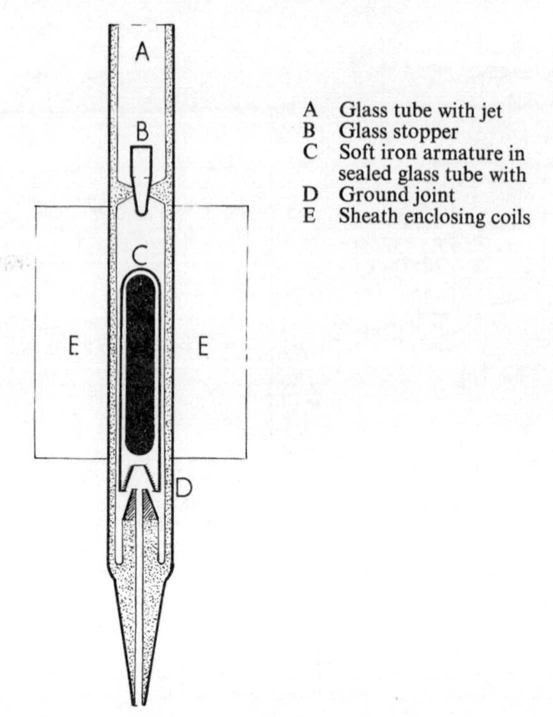

Fig. 2.26 Solenoid-operated burette tap allowing two titrant flow rates using a single jet. *Reproduced with permission from Audran and Dighton[53] and The Institute of Physics.*

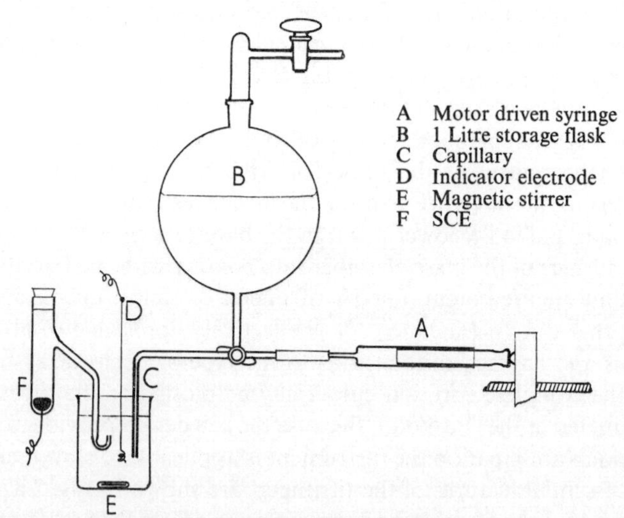

Fig. 2.27 Automatic potentiometric titrimeter layout.
Reproduced with permission from Lingane[54] and American Chemical Society.

and the syringe provides slow speeds for titrant delivery and fast speeds for refilling the syringe from the reservoir connected to it through a three-way tap. Contact between the screw drive and the end of the syringe plunger is provided by a brass nut held vertical by two spring arms sliding on parallel brass rods. To ensure uniform contact between the nut and the plunger the point-to-plane method is used, a hemispherical peg on the nut bearing on a flat glass plate cemented to the end of the plunger and arranged to be at exactly 90° to the screw shaft. A spring retaining mechanism prevents the plunger from creeping ahead of the nut. Because the syringe is driven at constant speed, the volume delivered, after initial calibration by weighing water dispensed over measured time intervals or revolutions of a counter attached to the screw shaft, can be calculated by time of delivery or from the revolution counter. Delivery rates were in the range 1·3–4·1 ml/min.

Where a full titration curve is required the potential across the indicating electrodes is applied to the Brown Electronik recorder of a Sargent Model XX recording polarograph. To stop the titration at the equivalence point, an enclosed-tip type of mercury switch, actuated by the recording potentiometer, is incorporated in the syringe circuit. The switch is attached by an adjustable clamp to the shaft of the drum which drives the slide-wire recorder contact. The angular position of the switch is set to correspond to the end-point potential and the delivery of titrant is stopped when the circuit is broken. The potential setting is accurate to $\pm 0{\cdot}005$ V.

Avoidance of over-titration is important in automatic batch titrations. It is achieved in the Lingane titrimeter by attention to the cell design. By immersion of the capillary tip of the syringe burette in the solution being titrated and mounting of the indicator electrode (metal wire) immediately adjacent to it, the recorded potential is slightly in advance of the true state of titration even in a well-stirred solution. Consequently a measure of end-point anticipation is provided and the titration will stop temporarily before true equivalence is reached. Stirring then brings the solution composition to equilibrium, causing the potential to drift away from the end-point value, whereupon the mercury switch on the recorder is closed and further titrant is delivered. This sequence repeats until a stable end-point is set up.

Typical recorder responses for the full titration curve and end-point titration termination are shown in Fig. 2.28, for the titration of iodide with silver. For titrations at concentrations of $6 \times 10^{-3} M$ and above, the automatic titrimeter yielded accuracy and precision equal to those obtained by equivalent manual titrations. The demonstrational titrations were of ferrous and vanadyl ions with ceric ion, and of chloride, iodide and a mixture of the two with silver. In the latter instance the two end-

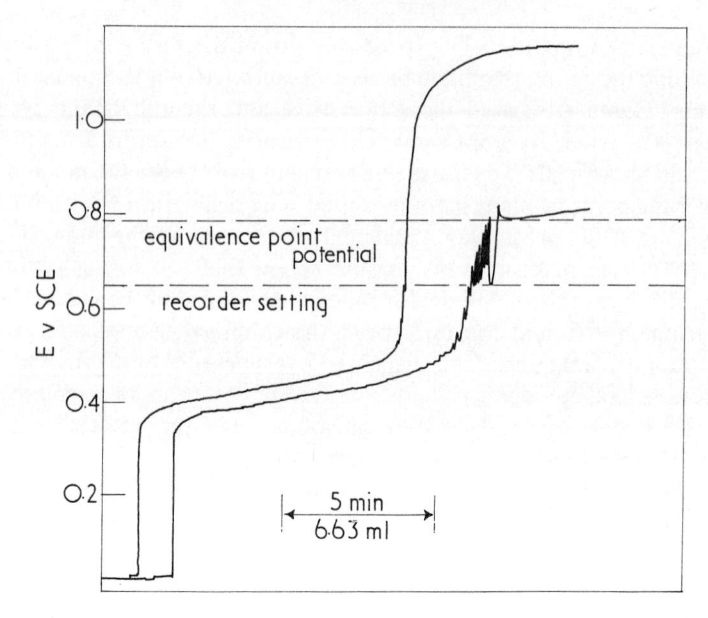

Fig. 2.28 Typical auto-titration responses showing both the full curve and end-point termination.
Reproduced with permission from Lingane[54] *and American Chemical Society.*

points are satisfactorily resolved. Because the titrant is protected from aerial oxidation the apparatus can be used with advantage for performing titrations involving air-sensitive titrants, as Lingane[55] showed for titrations of ferric ion and titanic ion with chromous ion.

The automatic titrimeters described above either yield a trace of the full titration curve from which the end-point must be calculated, or the titration is stopped at a predetermined end-point potential. If the control circuitry is arranged to produce a derivative of the titration curve no prior knowledge of the end-point potential is necessary and consequently no instrument settings are required before the titration is begun. Derivative titrimetry locates the point of inflection of the titration curve and not the true chemical equivalence point. Nevertheless the two are usually very close together and in practice calibration of the titrimeter with standard solutions enables the true results to be obtained from the inflection point. A titrimeter which yields the second derivative of the potentiometric titration curve has been designed by Malmstadt and Fett[56]; it is inexpensive and compact and can be used with a variety of reference electrodes without knowledge of their potentials. The control unit consists of an amplifier/differentiator circuit which produces an output voltage proportional to the second derivative of the electrode potential fed to it and a relay system which utilizes the second derivative voltage to terminate the titration at the inflection point. The amplifier/differentiator circuit is basically a twin triode tube (6SL7) and two *RC* differentiators; it is shown in full in Fig. 2.29 together with typical voltage curves pertaining to each stage of the circuit. The electrode voltage is fed to the control grid of one half of the twin triode and the resulting amplified voltage is applied across R_1C_1 to give the first stage of differentiation. The process is then repeated across

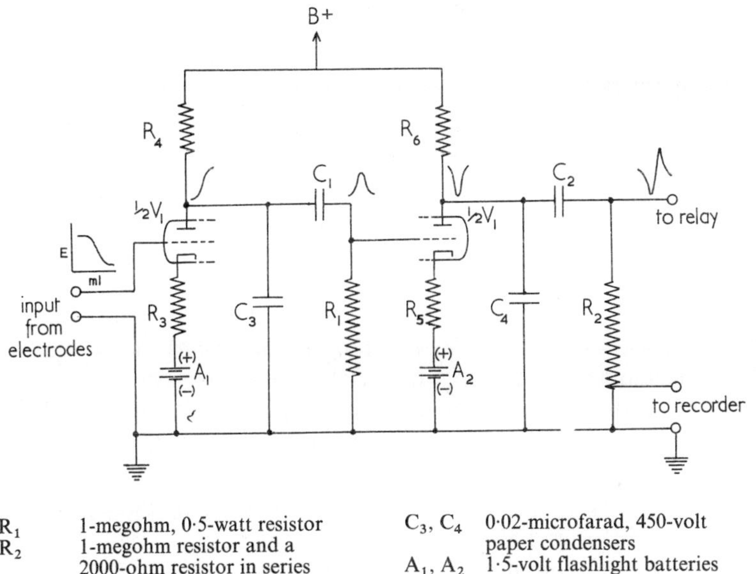

R_1	1-megohm, 0·5-watt resistor	C_3, C_4	0·02-microfarad, 450-volt paper condensers
R_2	1-megohm resistor and a 2000-ohm resistor in series	A_1, A_2	1·5-volt flashlight batteries
R_3, R_4	2000-ohm, 0·5-watt resistors	B^+	270-volt B batteries or electronic-regulated supply
R_5, R_6	1-megohm, wire-wound, 0·5-watt resistors	V_1	6SL7 twin-triode tube
C_1, C_2	0·5-microfarad, 300 WVDC, polystyrene condensers		

Fig. 2.29 Circuit for second-derivative potentiometric titrimeter.
Reproduced with permission from Malmstadt and Fett[56] and American Chemical Society.

the second half of the twin triode and the second differentiator R_2C_2. For relatively sharp end-points the resultant second differential voltage is of the order of 15 V, which can be used directly to operate the relay circuit controlling the titrant addition. The relay circuit is arranged so that an a.c. voltage applied to the burette motor maintains the burette open until the inflection, represented by discharge of a condenser, cancels the a.c. voltage and terminates the titration. Burettes giving constant or uniform flow-rate are suitable for use with this titrimeter. Titrations were performed in a 150-ml beaker, the solution being stirred by a motor-driven glass propeller. The disposition of the burette tip was such that titrant was directed at the propeller shaft above the level of the blades. The indicating electrodes were placed at the opposite side of the beaker close to the bottom and side-wall.

The performance of the titrimeter was established for titrations of dichromate and ceric ion with ferrous ion over the 0·01–0·1*M* concentration range. A platinum indicator electrode with either a calomel or platinum/10% rhodium reference electrode proved satisfactory. For further evaluation of the titrimeter, precipitation, redox, complex-formation and aqueous and non-aqueous acid–base titrations were used[57], with particular reference to the suitability of various indicator–reference electrode systems. Several electrode pairs which, through shift of their absolute potential from one titration to another, are unsuitable for manual titrations, are nevertheless satisfactory for derivative use which does not depend on a knowledge of absolute potential. These include electrodes of platinum/rhodium and pure rhodium. Graphite may be used as a reference electrode. For non-aqueous acid–base titrations in solvents such as acetic acid and benzene–methanol a platinum/10% rhodium indicator electrode in conjunction with a graphite reference electrode is satisfactory. The response characteristics of the electrode pairs determine the maximum titrant flow-rate at which they can be used; all of the pairs tested were satisfactory at rates up to 3 ml/min and those based on platinum/10% rhodium, which exhibit a rapid response, could be used with flow-rates up to 10 ml/min.

The second-derivative end-point detection technique has been utilized by Hieftje and Mandarano[58] to demonstrate the performance of a directly digital automatic titrimeter which can, in fact, be used with any means of end-point detection. The direct digital read-out is achieved through the design of a novel titrant delivery technique in which the titrant is supplied, not as a continuous metered stream, but as individual uniform small droplets of sub-microlitre volume. Hence the measurement by volume is replaced by counting the number of drops delivered up to the end-point. Since the drops are generated via discrete voltage pulses the summation of these is digital in form. In consequence it is possible to simplify the titrimeter design to the extent that analogue-to-digital conversion circuitry is not required.

The complete apparatus comprises three units, the titrant delivery system, the end-point detection unit and digital logic circuitry for control of titrant delivery and provision of digital read-out.

The basic requirement of the titrant delivery system is to disperse the titrant into uniformly-sized and spaced droplets of which a selected fraction can be directed into the titration cell and their number counted.

The titrant, from a reservoir, is forced through a hypodermic needle at constant flow-rate. The uniform droplets are produced by applying an intense, periodic disturbance to the jet by vibrating the needle in a direction normal to the jet[59, 60]. This is achieved piezoelectrically by using a bimorph strip operated by a variable-frequency power oscillator. The droplets then pass through a circular charging electrode to which a periodic voltage pulse is applied. In this manner charges are repelled back into the jet and the droplets breaking off during the voltage application are charge-deficient and can be deflected by passing the entire stream of droplets through a high voltage d.c. field. Control of the proportion of droplets extracted from the complete stream can therefore be accomplished through the voltage application to the jet. A pulsed voltage enables single droplets to be extracted whereas a continuous voltage selects a large number of drops. The deflected droplets pass through a titrant port into the titration vessel while uncharged droplets fall to the bottom of the titrimeter from which they can be drained to waste or recycled to the reservoir according to choice. Fig. 2.30 illustrates the titrant delivery

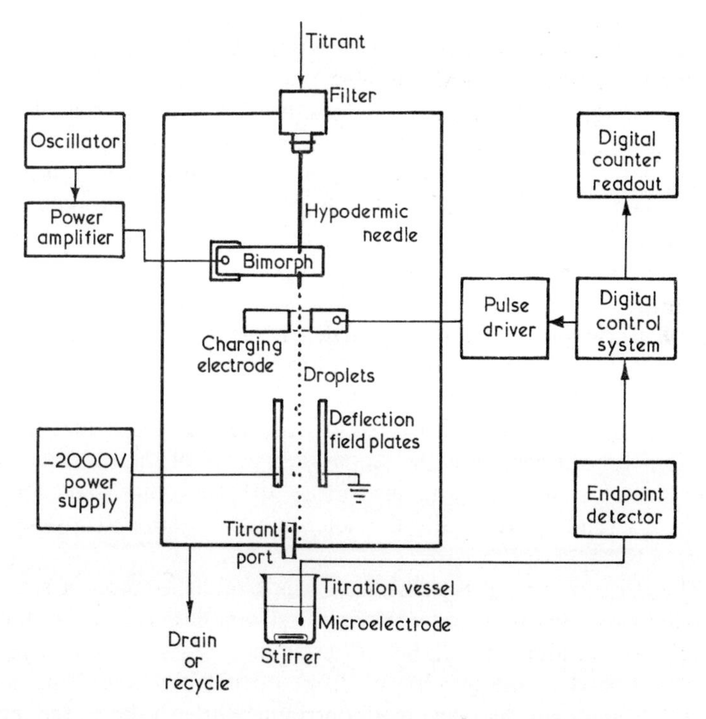

Fig. 2.30 Block diagram of equipment for generation of droplets of titrant.
Reproduced with permission from Hieftje and Mandarano[58] and American Chemical Society.

system in schematic form. The necessary measurement data are obtained from the pulsed voltage; because the droplets are of uniform size the volume of titrant delivered is in direct proportion to the total number of droplets deflected by the electric field, which is equal to the number of pulses applied, which can be accumulated in a digital counter incorporated in the digital control system and the rate of delivery of titrant obtained by digital measurement of the pulsing frequency. An undoubted design advantage of this approach to titrant delivery is that it involves the use of only one moving part, the bimorph, and this has a virtually infinite lifetime. The authors employed a simple pressurized reservoir to deliver titrant to the jet at constant rate. The performance of the titrimeter depends critically on the maintenance of the drop size uniformity. A 'Millipore' filter is inserted in the titrant line immediately before the jet to maintain the latter in a clean condition and free from partial blockage. The size of the droplets depends upon the rate of titrant flow to the jet and upon the frequency at which they are formed. Initial selection of droplet size is best achieved by adjustment of titrant flow-rate and by choice of jet dimensions at constant frequency. Attempts to achieve this control through frequency variation revealed instability of droplet size at certain frequencies. Droplet size variation was checked in a series of experiments in which a known number of drops ($\sim 1{\cdot}5 \times 10^4$) of distilled water was weighed. The relative standard deviation was less than 0·1%.

Logic circuitry is used to control the voltage pulse application and to provide end-point anticipation to improve the precision of titrations. A 300-V pulse is applied to the charging electrode and the pulse delay and pulse width are variable from 0·1 to 1·0 msec. End-point anticipation is achieved by decreasing the rate of droplet addition by reducing the pulsing rate. The frequency of the square-wave input is divisible by any power of 10 by means of a decade scaler. If the value of this power is n then every $(10^n)^{th}$ droplet passes to the titration vessel. It was found to be adequate in practice that a single decade reduction provides adequate protection against end-point overshoot. The circuitry provides for this reduction to be triggered as the second derivative of the titration curve passes through zero immediately before the end-point. The second-derivative circuit is essentially that of Malmstad and Fett[56] described above.

The performance of the titrimeter was established from a series of strong acid – strong base and weak acid – strong base titrations. In each case relative standard deviations of better than 1% were obtained; the values improved as the number of drops of titrant required increased. Evidently the titrant delivery error contributes little to the overall error.

Olsen and Adamo[61] described a simple and inexpensive apparatus for automatic potentiometric titration of metal ions with diethylenetrinirilo-penta-acetic acid (DTPA) as a chelometric titrant. It involves the use of a silver–DTPA indicating electrode; this is a pM electrode and it yields almost symmetrical titration curves for a number of metal ions. Titrations of several alkaline earth metal ions at concentrations of 5×10^{-4}–$10^{-3}M$ were conducted in a beaker, with a silver billet indicating electrode and a fibre-tipped reference SCE, the electrode potential being plotted on a recorder electrometer (Heath Model EUW 301). The burette consisted of an 18-in. length of thick-walled polythene tubing fitted with a fine capillary tip delivering 2–3 drops/min. The individual drop volume was calibrated daily by weighing fractions comprising 10 drops of 0·01*M* DTPA titrant. A simple drop-counter made up of two parallel platinum wires $\frac{1}{8}$ in. apart mounted horizontally and parallel to each other was employed. Each drop of titrant momentarily made contact with the two wires, producing a voltage pulse at the electrometer. A pulse appears on the recorder each time a drop is detected and the volume of titrant delivered at the end-point is calculated as the product of drop weight and number of drops recorded. Despite the simplicity of the apparatus the results it produces are of good accuracy and precision. Thus for concentrations of Mg^{2+}, Ca^{2+} and Ba^{2+} of 0·01–0·6m*M*, relative standard derivations are better than 0·5% except at the lowest concentration level examined.

2.3.2 Continuous Potentiometric Titrimetry

Continuous titrimetry is most conveniently performed by arranging that a constant flow of sample is mixed, before measurement, with a stream of titrant, the flow-rate of which can be varied. The signal from the sensing element is used to increase or decrease the rate of flow of titrant until equivalence is obtained. The desired sample concentration is then a linear function of the rate at which titrant is being delivered at the equivalence point. After calibration of the apparatus with a single standard of sample material, unknown sample concentrations may be determined by referring titrant delivery rate measurements to the calibration line. This approach has several advantages; in particular, no accurate volume measurements are required and a number of readings of the delivery rate may be taken, thereby improving the precision of measurement, and calibration requirements are minimal.

A continuous automated buretteless titrimeter suitable for this type of study was developed by Blaedel and Laessig[62]; it is shown in block diagram form in Fig. 2.31. Its performance has been evaluated in some detail for several applications. An essential requirement for continuous titrimetry is an electrode system which responds rapidly to changes in

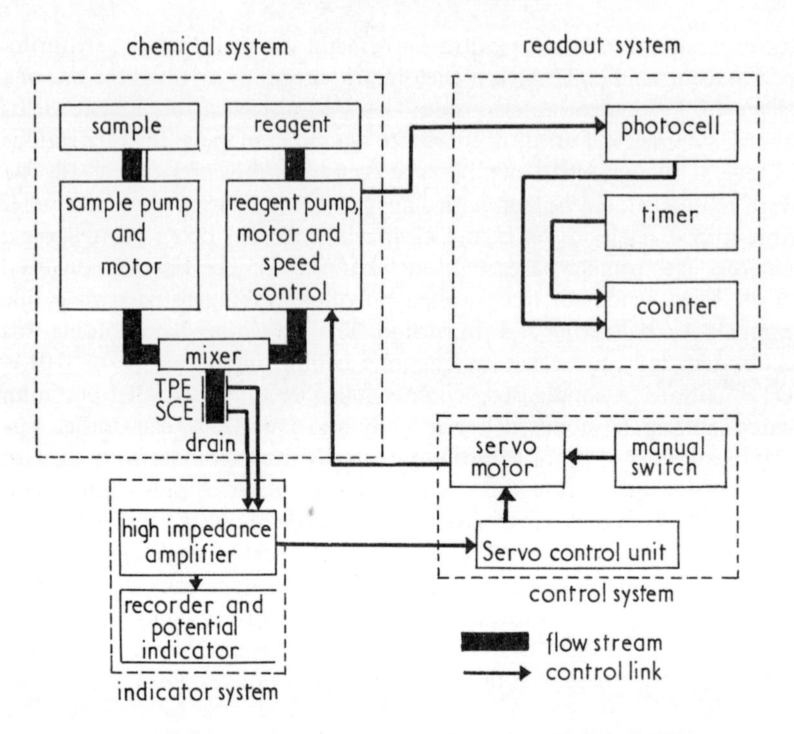

Fig. 2.31 Block diagram of automatic potentiometric continuous titrimeter. *Reproduced with permission from Blaedel and Laessig*[62] *and American Chemical Society.*

chemical composition of the flowing stream. The authors utilized the tubular platinum electrode, which is both highly sensitive and rapid in response, in a measuring cell of minimal hold-up volume.

The sample stream is pumped peristaltically with a fixed-speed pump operating at 56 rpm (Model PA-56, New Brunswick Scientific Co.). 'Tygon' tubing is used and by appropriate choice of tubing diameter flow-rates in the range 0·5–15 ml/min can be achieved. The variable-speed reagent pump is also of the peristaltic type (Model 500–1200, Harvard Apparatus Company). It is fitted with a feedback control circuit to maintain the pump at any fixed speed. The reagent pumping speed is determined from the rotational speed of the pump shaft. In this application equipment was designed to enable the pump speed, and hence the sample concentration, to be read digitally.

The sample and reagent streams are fed to a mixing vessel and then to the electrode assembly. To avoid electrode-response fluctuations at the equivalence point it is essential that the mixer should perform efficiently. It comprises a low-volume (0·7 ml) cavity sealed in a 'Teflon' chamber and it contains a 'Teflon'-enclosed magnetic stirrer driven at 1700–2400 rpm from below. Entrance to, and exit from, the mixer is by 'Tygon' tubing fitted, by moistening with cyclohexanone, into slightly undersized holes drilled in the 'Teflon' casing.

The mixed streams pass directly into the electrode system which is a TPE/SCE pair. The TPE indicator electrode is a platinum tube of 0·04 in. bore and 0·15 in. long, sealed at each end into soft glass capillary tubing and then into the 'Tygon' tubing of the flow system. The reference SCE is situated immediately downstream of the TPE. The stream flows through a length of 7-mm o.d. porous glass tube around which is the saturated KCl solution, calomel paste and mercury pool, contained in a 'Teflon' vessel. Contact with the mercury pool is made by a length of platinum wire enclosed in 'Tygon' and sealed into the 'Teflon'. Loss of solution from the electrode by evaporation is minimized by fitting a 'Perspex' cover, and filling is necessary only at monthly intervals. The response of the electrode system is fed to a potentiometer recorder through a high impedance operational amplifier to prevent excessive current drain at the electrodes when the recorder is unbalanced during the early stages of a titration.

In operating the titrimeter the end-point potential is selected from a manual titration curve. The difference between this voltage and that detected by the indicator electrode is amplified by a servo-controller through which it is used to control the speed of the reagent pump. Two control bands are set up – an outer band, well away from the end-point voltage, in which coarse continuous control is provided, and an inner band which controls intermittently. The voltage span of the inner control band is selected to be narrow enough to prevent the intermittent control from becoming operative too early in the titration, yet wide enough to prevent overshooting of the end-point zone, which is not subject to control. Fig. 2.32 shows the control band settings optimized for the titration of $3{\cdot}4 \times 10^{-3}M$ Fe(II) with $7{\cdot}5 \times 10^{-4}M$ Ce(IV).

The measurement of rate of rotation of the reagent pump shaft and the equivalence point can be carried out conveniently by using a meter or a recorder, but for more precise results a digital method is preferred. Blaedel and Laessig constructed a unit based on counting the interruptions of a light-beam falling on a photocell, by attaching to the shaft an aluminium disc having four holes drilled 90° apart near the circumference. Two pulse counters were employed (Model 5424 Berkeley Division, Beckman Instruments Co.), one controlling the timing interval over which pulses were counted and the second performing the actual pulse counting. The provision of a variable time interval enabled the system to be adjusted to give direct read-out of sample concentration.

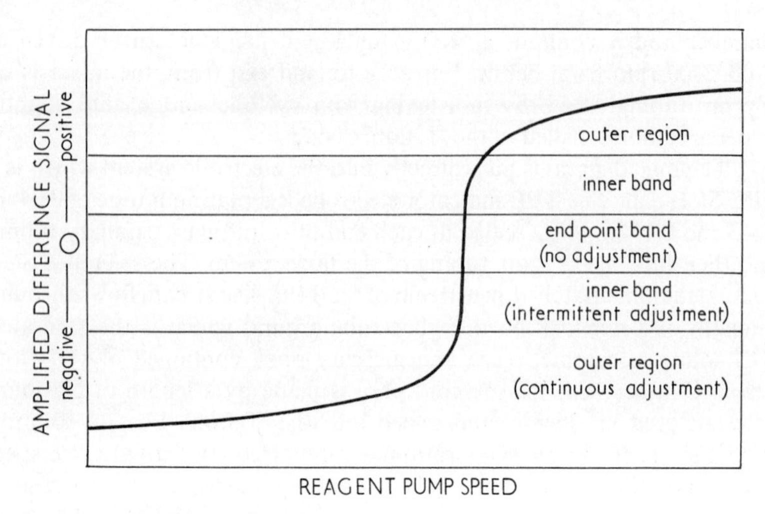

Fig. 2.32 Control-band settings.
Reproduced with permission from Blaedel and Laessig[62] *and American Chemical Society.*

In evaluating the overall performance of the titrimeter, the constancy of the pumping motor and of the volume delivered per revolution of the pump shaft were tested. Over the complete range of pumping speeds (79–216 rpm) 20 consecutive readings of about 2 min duration were taken and the maximum variation was less than 0·5% in all cases and less than 0·2% at the higher pumping rates. At constant pumping speed the relative standard deviation of volume delivered was about 0·05%. A small, irreproducible variation of delivered volume as a function of pump speed was noted; taking this into account the overall standard deviation was 0·17%. No significant error arose from the read-out equipment since the timing-interval pulse-counter is accurate to $\frac{1}{60}$ sec, i.e. less than 0·03% for a 1-min interval of pulse counting.

Titrations of Fe(II) with Ce(IV) served to evaluate the overall performance of the titrimeter. Over the concentration range $7 \times 10^{-5}M$–$6 \times 10^{-2}M$ the relative standard deviation on a single result ranged from 0·5 to 0·2%. The upper concentration limit is governed by the tendency to overshoot the equivalence region and concentrations below 0·01*M* are recommended. A volume of 15–30 ml of Fe(II) solution can be titrated within 6 min. A typical recorder trace of indicator electrode response is reproduced in Fig. 2.33, showing the approach to equivalence from point *A* and occurrences of overshoot at points *B*, *C* and *D*.

Titrimeters based on the principle above are potentially applicable to a wide range of redox titrations. However, a facet that may require

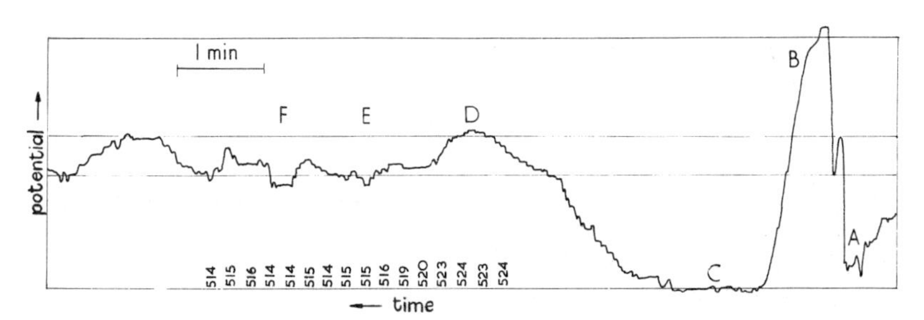

Fig. 2.33 Automatic Continuous Titrimeter: indicator electrode response showing approach to equivalence point and subsequent end-point control.
Reproduced with permission from Blaedel and Laessig[62] and American Chemical Society.

further evaluation is the time lag in the servo system, i.e. the time required for a change in reagent pump speed to be recorded as a change in electrode potential. For fast reactions such as Fe(II) with Ce(IV) the lag is almost solely due to hold-up of solution in the pump lines and mixer; it is independent of concentration, and of minimal significance. Concentration-dependent lag might be introduced if slow reactions are studied or if side-effects due to chemical reactions or adsorption occur.

Modification and extension of the titrimeter design has permitted its application to chelometric titration of metal ions with reagents such as EDTA. For this purpose a pM-indicating electrode is required, and Blaedel and Laessig[63] have utilized the mercury–EDTA electrode, the theory and experimental performance of which have been described in detail by Reilley and Schmid[64, 65]. To establish the mercury–EDTA electrode it is necessary to have a small concentration of Hg(II) in the flowing solution and this is incorporated in the EDTA titrant stream at $10^{-6}M$. The indicating electrode is mercury coated tubular platinum (MTPE)[30] used in conjunction with a reference SCE. In addition to the sample and reagent pumps an additional constant-speed pump is required to carry buffer solution to maintain constancy of pH and ionic strength. Buffering is provided to pH 9·7 [$NH_3/(NH_4)_2SO_4$]. In all other respects the titrimeter is that described above: Cd^{2+}, Ba^{2+}, Zn^{2+}, Ca^{2+}, Cu^{2+}, Mg^{2+} and Ni^{2+} may be satisfactorily titrated over the concentration range 0·003–0·04M with a relative standard deviation of 0·5%.

The MTPE must be used at pH > 5, preferably at pH 10. At pH 5 irregular responses result owing to aging of the MTPE. To extend the pH range of continuous EDTA titrations to pH 5 Blaedel and Laessig[66] replaced the MTPE by a flow-through DME[18]. Thirteen different metal ions were satisfactorily determined in the 0·003–0·006M range. At lower metal ion concentrations the flow-through DME yields erroneously high results. The MTPE, however, can be used satisfactorily down to

$10^{-5}M$ concentrations of metal ion subject to the pH limitation mentioned earlier. In this context it has been used to determine total hardness in water[67]. Certain multivalent ions, notably Cr(III), Al(III) and Zr(IV) react too slowly with EDTA to be determined by the direct continuous titration method. However, the equipment can be modified to operate according to the back-titration method[68] by which such ions are normally determined manually. Essentially the procedure is a two-stage one; three constant-speed peristaltic pumps feed the sample, EDTA (in excess of the metal ion concentration in the sample), and buffer, to a mixer as described above and thence to a heated delay coil to enable the chelating reaction to proceed virtually to completion. The mixed solution is then merged with two further streams, one containing a metal ion which reacts rapidly with EDTA (Cu^{2+} or Zn^{2+}) at a concentration in excess of the residual EDTA, and the second containing EDTA titrant. After agitation in a second mixer the solution passes through the Hg–EDTA/SCE electrode system and the servo-control adjusts the titrant pump speed so that the equivalence point is attained. From the speed of the titrant pump at the end-point the concentration of residual EDTA from the initial reaction may be calculated and hence also the concentration of the multivalent ion. With the pH buffered to 9·7, Ce(III), La(III) and Pr(III) at concentrations up to 0·035*M* can be accurately titrated by providing a 10-sec delay at room temperature. Similarly Th(IV) and In(III) can be titrated stoichiometrically at pH 4·8, but complete reaction of Zr(IV) requires a delay of 1 min with the mixer immersed in a heating bath. Cr(III) and Al(III) react even more slowly with EDTA than Zr(IV), the reactions being 93% and 91% complete respectively under these conditions.

Differential potentiometric titration with two similar indicating electrodes was first investigated nearly 50 years ago and apparatus involving twin[69] or single cells[70–73] has been described for titration of discrete samples. This approach has been further developed by Nicholson[74] for the analysis of flowing streams. The cell unit is depicted in Fig. 2.34; both sample and titrant flow are divided and enter the two identical titration cells through glass capillaries. These are so constructed that slightly different sample fractions are titrated in each cell. The method involves maintaining a constant preselected applied voltage ΔE between the two indicator electrodes by regulating the flow of titrant through stainless-steel needle valve V. This is achieved by a detector-controller unit D which consists of a microvoltmeter amplifier and a proportional controller which operates valve V. M is a flowmeter which indicates the titrant addition rate. Evaluation of the titrimeter by titrating dichromate with ferrous ion revealed that care must be exercised in the selection of control conditions. Thus stable indicator electrode responses were

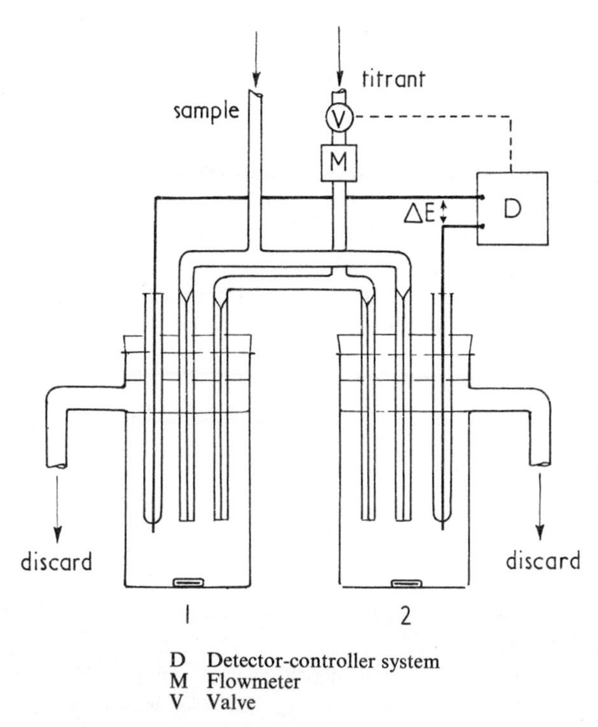

D Detector-controller system
M Flowmeter
V Valve

Fig. 2.34 Continuous differential potentiometric titrimeter.
Reproduced with permission from Nicholson[74] *and American Chemical Society.*

obtained on the ferrous-rich side of equivalence but not on the dichromate-rich side where the potential-determining process is irreversible. It may be inferred, therefore, that the method would be of limited applicability in monitoring slow reactions. To minimize overshoot of the end-point, control is best established on a shoulder of the titration curve. Given that these limitations do not apply or can be overcome, the technique is attractive for monitoring sample streams in which the solute concentration can change in a continuous manner.

2.3.3 Automatic Potentiometric Reaction Rate Measurement

Reaction rate analysis by the variable-time method involves measurement of the time taken for a predetermined extent of reaction to occur. Provided that solution conditions are chosen such that reaction of the species sought is the rate-determining process, that the ionic strength of the solution is sufficient to eliminate the need for corrections due to activity coefficient variations, and that the measurement conditions are such that the consumption of sample material is insignificant, then the concentration of the sample material is directly proportional to the reciprocal of the time taken for the chosen extent of reaction to occur. Unknown samples may be analysed by reference to a calibration curve or directly by first experimentally determining the constant of proportionality. The extent of reaction is selected by suitable choice of initial and final values of an analytical parameter. Malmstadt and Pardue[75] utilized a fixed change of potential between the sample and a reference solution and established that glucose in the concentration range 5–500 ppm could be determined enzymatically with a relative error of $\pm 1\%$. The reaction sequence used was

$$\text{Glucose} + O_2 \xrightarrow[\text{oxidase}]{\text{glucose}} \text{gluconic acid} + H_2O_2 \qquad (1)$$

$$H_2O_2 + 2H^+ + 2I^- \xrightarrow{\text{molybdate}} I_2 + 2H_2O \qquad (2)$$

Since reaction (2) is much faster than (1) the concentration of glucose is related to the rate of production of iodine. Therefore the analysis consists of measuring the time for a predetermined change in potential, due to the build-up of iodine, to occur.

Since the measurement is made over a small fraction of the total reaction the potential change for analysis over a tenfold concentration range is small (6·00 mV was selected and consequent measurement times were between 10 and 100 sec). It is obviously essential that the measuring equipment be capable of sensitive and reproducible potential measurements. A reproducibility of $\pm 0{\cdot}02$ mV was achieved with a Sargent Model Q concentration comparator which was augmented by auxiliary relays to provide the automatic measurement and control, the latter including temperature of the sample and reference solutions which must be controlled to $\pm 0{\cdot}1\%$ if precise measurements are to be made. The reaction time was read from a timer capable of discrimination to 0·02 sec and which was started and stopped by the relay circuits responding to the potential change.

The cell unit is shown in Fig. 2.35. A 100-ml tall-form beaker formed an outer cell containing reference solution (buffer, glucose, oxidase and iodide/iodine) and a test-tube constituted an inner cell in which the reaction occurred. Electrical contact between the two parts was provided by a fibre sealed in the base of the test-tube; solution diffusion across it did not occur. Composite reagent (1·00 ml) of buffer, glucose oxidase and iodide was first placed in the test-tube, followed by addition of 1·00 ml of the sample solution. The immediate decrease in iodide concentration triggered the relay system and the timer operated until the 6-mV potential

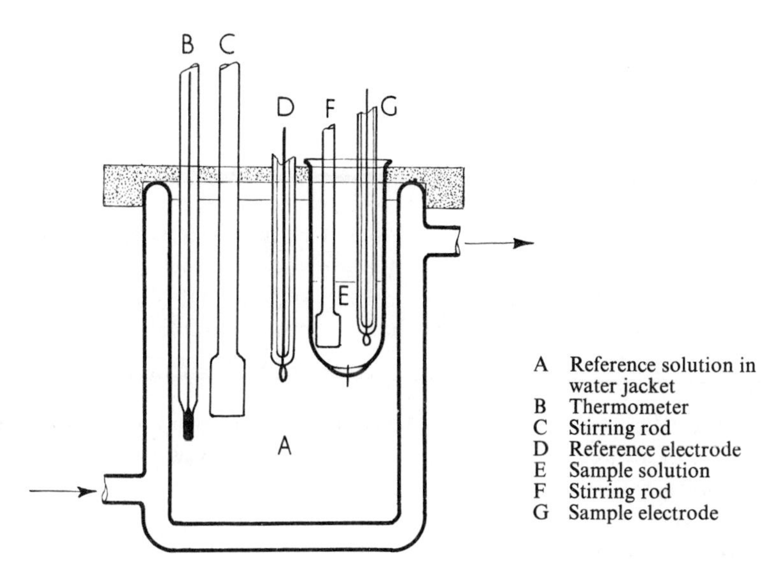

Fig. 2.35 Cell for automatic potentiometric analysis by the reaction rate method. *Reproduced with permission from Malmstadt and Pardue*[75] *and American Chemical Society.*

change had occurred. Although manual addition of reagent and sample was employed the control system had facilities for controlling an automatic reagent- and sample-dispensing device.

The applicability of the technique to the determination of glucose in blood serum and plasma has been demonstrated[76]. Glucose concentrations in the range 54–186 mg/100 ml were successfully determined; recoveries were between 94 and 103% and the maximum relative standard deviation was 3%.

The reaction sequence of equations (1) and (2) above can also be used for the automatic determination of glucose oxidase activity, as demonstrated by Malmstadt and Pardue[77] using essentially the equipment described above.

2.4 Coulometry

A wide range of both oxidizing and reducing species can be generated coulometrically under conditions of effectively 100% efficiency, and in consequence a coulometric generator may be used instead of a volumetric dispenser in automatic analysis. Coulometry has been used successfully in titrimetry with potentiometric, amperometric and photometric end-point sensors. Indeed coulometry offers certain advantages over volumetric reagent addition. In both constant-current and constant-potential coulometry the quantity of reagent added is calculated from two variables, current and time: in constant-current coulometry the number of coulombs involved is simply the product of current and time, in constant-potential coulometry it is the integral of the current–time relationship. Both current and time are capable of being measured precisely and with high accuracy. Therefore unless problems related to slow reactions or slow end-point indicator response are encountered, coulometric titrant generation should result in accurate and precise results. In automatic coulometric titrations equipment for titrant dispensing and transport and associated mechanical control mechanisms are eliminated, but electronic control circuitry is required between the indicating electrodes and the coulometric generator electrode.

Within the field of automatic analysis, coulometric titrant generation has been utilized in a variety of ways, both for discrete and continuous sample analysis, with internal or external titrant generation with both fixed and variable generation rates, and with constant-current and constant-potential generating circuits.

2.4.1 Continuous Coulometric Titration

The first automatic continuous titrimeter was developed during the 1939–45 war for the determination of mustard gas in air[78]; it employed internally-generated bromine as titrant. The titration cell is depicted in Fig. 2.36; essentially it comprises two compartments, an inner one in which absorption of the air sample and the coulometric titration occur, and an outer one which serves as a receiver of absorbent, which is cleaned by passing through a layer of charcoal at the base of the cell before entering the inner compartment. The two compartments are connected by holes in the base of the inner one and an overflow at the top of the inner compartment into the outer one. As air enters the base of the inner cell the small holes create a stream of bubbles which, as they rise, cause vigorous mixing and also carry reacted solution across the overflow into the outer compartment. The solution used was $3N$ H_2SO_4 and $0{\cdot}05N$ KBr and served both as absorbent and reduced form of titrant.

The indicator electrodes were of platinum, one being situated in each compartment. The generator electrodes, also platinum, were also situated one in each compartment, the bromine-generating anode being in the inner one. The chemical system was arranged so that, with clean air passing through the cell, a slight excess of bromine was present and permanently detected by the indicating electrodes. When air containing mustard gas entered the cell the bromine concentration decreased and was brought to its original concentration by generating an electrolysis current across

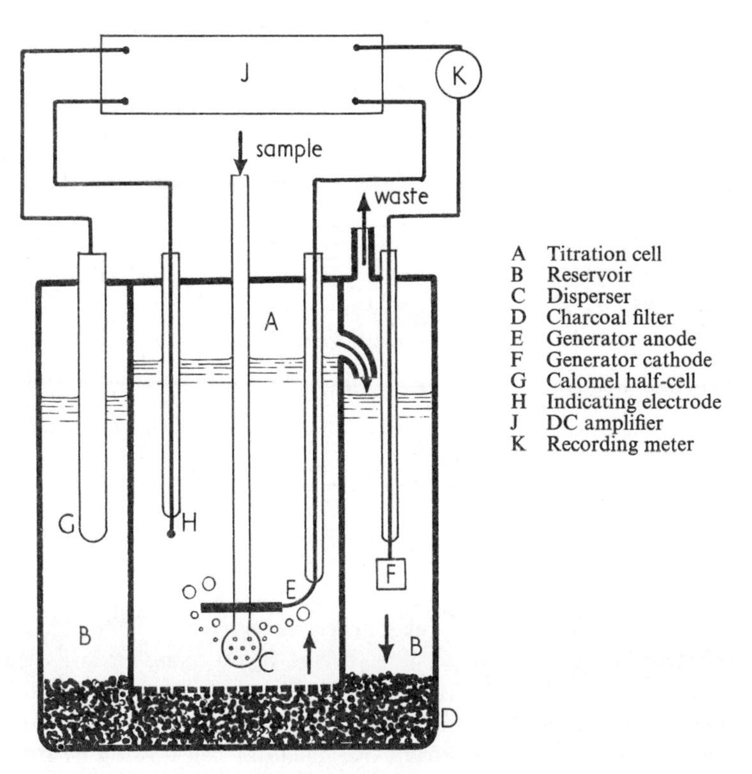

Fig. 2.36 Apparatus for the continuous coulometric analysis of air for mustard gas. *Reproduced with permission from Shaffer, Briglio and Brockman*[78] *and American Chemical Society.*

the generator electrodes. This was achieved by connecting the indicator electrodes to the input terminals of a power amplifier and the generator pair to the output terminals. The relative phases of the input and output signals were such that when a decrease in bromine concentration was detected more current was forced to flow through the output circuit to form more bromine. The concentration of mustard gas in the air sample is proportional to the rate of electrolysis of bromide, which is indicated by a suitable current-recording or metering device in series with the generator electrodes.

Calibration was carried out by using standards based on known air flows and the saturation vapour presence of mustard gas and prepared in a gas-saturation line. The operational range of the instrument is 0·1 ppm v/v mustard gas and upwards. It can be applied to the analysis of other oxidizable vapours such as H_2S, SO_2 and acrolein.

An example of a continuous coulometric titrimeter utilizing external

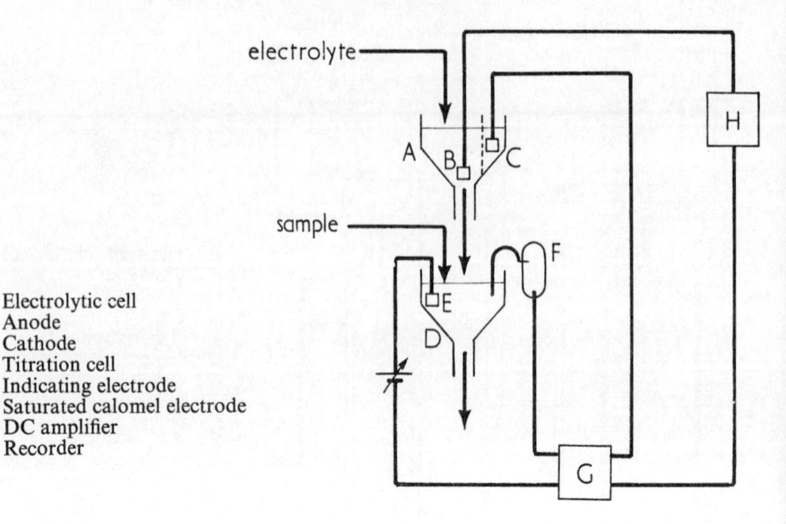

Fig. 2.37 System for automatic continuous coulometry using external generation of titrant.

reagent generation is that of Takahashi and co-workers[79, 80]. This was designed for titration of oxidizable species, again using electrolytically generated bromine. Fig. 2.37 is a schematic diagram of the apparatus. The reagent generation and titration sequence is performed in two separate vessels, the bromine solution from the generator cell flowing by gravity into the titration cell. The design of the two cells is shown in Figs. 2.38 and 2.39. As with the titrimeter for mustard gas described above the response of a pair of platinum indicator electrodes is used, after rectification and amplification, to create a corresponding current across the platinum generator electrodes.

A novel continuous coulometric titrimeter has been described in detail by Eckfeldt[81]. It involves maintaining the working electrode at a constant predetermined potential throughout the analysis. Whereas in batch coulometric titrimetry at constant potential it is necessary to integrate the current–time relationship, in analysis of a flowing stream conditions may be arranged such that the instantaneous current developed is a measure of the concentration of the reacting species. The essential condition is that the species being determined should be completely reacted during passage through the electrode system. If the sample flow-rate R is known, then the normality N of the sample is related to the electrolysis current I by the equation

$$N = \frac{I}{96{,}500R}$$

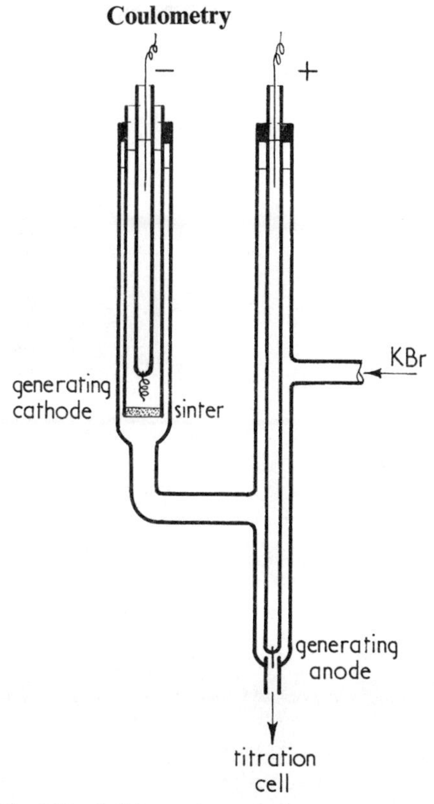

Fig. 2.38 Cell for coulometric titrant generation.

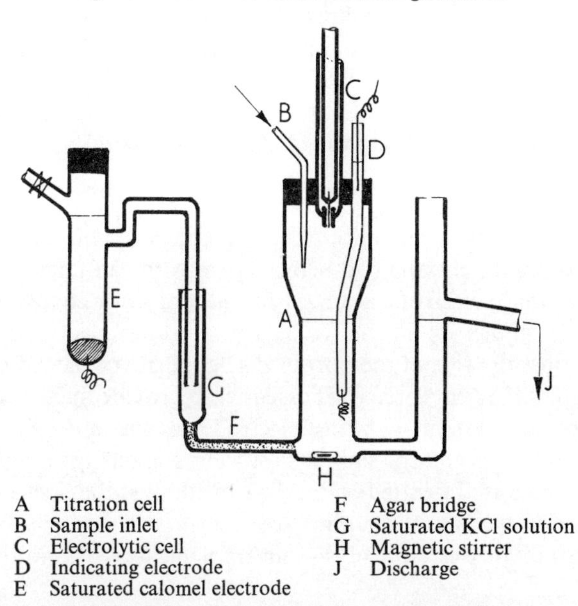

A Titration cell
B Sample inlet
C Electrolytic cell
D Indicating electrode
E Saturated calomel electrode
F Agar bridge
G Saturated KCl solution
H Magnetic stirrer
J Discharge

Figure 2.39 Detail of titration cell of Fig. 2.37.

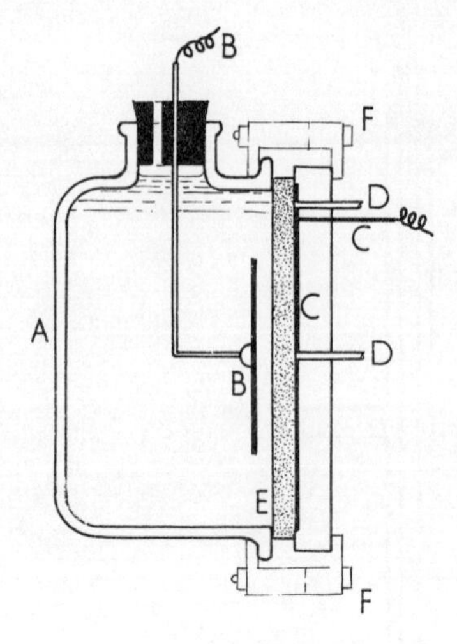

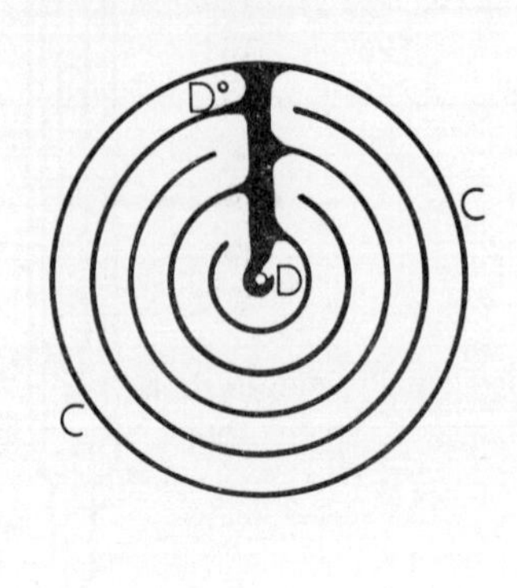

A Reference electrode chamber containing reference solution
B Reference electrode
C Working electrode
D Solution inlet and outlet
E Porous ceramic disc
F Clamps

Fig. 2.40 Cell design for constant potential coulometric analysis of flowing streams, showing detail of generating electrode.
Reproduced with permission from Eckfeldt[81] *and American Chemical Society.*

This concept offers two advantages; it obviates the need for complex electrical control circuitry, the only necessity being the maintenance of a constant potential at the working electrode, and by appropriate choice of potential it is possible to eliminate interference from ions which are reduced only at higher potentials. The fundamental problem of such a titrimeter lies in the design of the electrodes to ensure that essentially complete reduction or oxidation of the sample species occurs during transit across them.

Fig. 2.40 shows the design of the cell and electrode system. The cell comprises two parts, a reference electrode compartment and a working electrode compartment, the two being separated by an unglazed ceramic disc. The reference electrode is a 3×1 in. sheet of silver coated with silver chloride and immersed in $2N$ KCl. The working electrode, which is abutted against the side of the porous disc opposite to the reference side, is of pure gold of 38 cm^2 area. It is essential to provide maximum contact between the sample and working electrode to maximize the extent of reaction, and to this end the sample flows across a labyrinth milled in the electrode surface as illustrated in Fig. 2.40. In addition the electrode surface is roughened to increase the surface area.

The instantaneous electrolysis current is measured from the voltage drop across a calibrated resistor, by using a recorder. To permit ready change-over between anodic and cathodic current measurements, a reversing switch is incorporated in the circuit. The constant applied potential within the working range from $-1{\cdot}5$ to $+3$ V is set manually, or automatically by driving the appropriate rheostat shaft by a reversible clock-motor. The potential is chosen to be on the plateau of preplotted current–voltage curves for each couple studied.

Sample flow is maintained constant by using a Marriotte-bottle system and the solution is stirred by applying a pulsating air pressure to the solution as it enters the cell. Demonstrational studies on the determination of iodine, iodide and oxygen over the concentration range $0{\cdot}5 \times 10^{-4}$–$10^{-3}M$ showed that under optimized conditions of flow-rate (1–5 ml/min) and with adequate stirring, cell efficiencies approaching 100% could be achieved. Subsequently a slightly modified cell design was developed specifically for determining oxygen[82].

An apparatus for automatic continuous coulometric titration has been described by de Kainlis, Merigot and Pourcel[83] and evaluated for the determination of Fe^{2+} by reaction with Mn^{3+}. The cell has anodic and cathodic compartments separated by a porous membrane and is fitted with an overflow. The sample stream and base electrolyte (acidified $MnSO_4$) are introduced separately into the anodic compartment by means of Milton Roy 'Miliroyal D' pumps. The generator electrodes, situated one in each compartment, are 2 cm^2 sheets of platinum, and the indicator pair, both situated in the anodic compartment are platinum/calomel. These provide potentiometric indication of the equivalence point; the electrolysis current required to maintain the equivalence is directly proportional to the instantaneous Fe^{2+} concentration in the sample.

2.4.2 Automatic Coulometric Titration of Discrete Samples

Two approaches to automatic coulometric titrations are possible. Either the generating current may be maintained at a known and constant value, in which case a measurement of titration time is all that is required to yield the total number of coulombs concerned, or a variable current can be employed and a means of current integration built into the electrolysis circuit. Automatic coulometric titrimeters based on both principles have been reported.

Lingane[84] has demonstrated that the Beckman automatic titrimeter, which utilizes potentiometric end-point detection and a solenoid-operated burette to deliver titrant, is readily adaptable to coulometric generation of titrant. The 110-V a.c. leads to the solenoid valve were connected instead to a double-pole–double-throw relay of 700 ohms resistance. One side of this relay acted as a single-throw switch to control the electric stopclock which indicated the titration time, the other side functioned as a single-pole double-throw switch which enabled current to flow either through the cell or through a dummy resistance (1000 ohms) the purpose of which was to maintain a constant drain on the power supply while the cell was switched out of the circuit. The constant-current power supply, capable of generating up to 250 mA at 350 V, was derived from a standard d.c. power supply with full-wave rectification, fed from a 110-V a.c. line through a constant-voltage transformer. The supply was stabilized by the use of a large capacitance-input filter in the output line and also a 4000-ohm high-voltage resistor connected across the output.

The titrimeter functions by titrating continuously to a preset potential just in anticipation of the end-point and thereafter the end-point is approached by incremental additions of titrant. A small negative error was shown to be inherent in the incremental procedure and a Veeder Root counter was provided to count the number of increments to enable a correction to be made if desired. Provided that the number of increments was less than 50 no correction was required.

In addition to acid–alkali and redox titrations the equipment was specifically evaluated for the determination of milligram amounts of

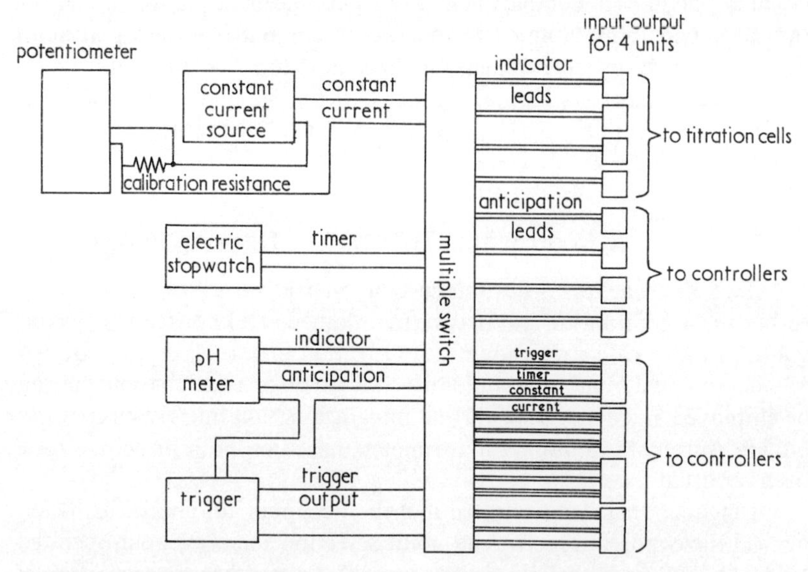

Fig. 2.41 Block diagram of equipment for multiple automatic coulometric titration. *Reproduced with permission from Carson*[85] *and American Chemical Society.*

halide ions by precipitation with electrolytically generated silver. Determinations were carried out in a 75-ml weighing bottle, using a generator anode of pure silver or silver-coated platinum with a shielded platinum-helix cathode. The indicator electrode was a silver or platinum wire used in conjunction with a calomel electrode, the latter being connected through a 0·1*M* $NaNO_3$–agar gel salt bridge to minimize chloride contamination. For quantities of chloride, bromide and iodide ranging from 0·2 to 10 mg the average error was ±0·005 mg.

A constant-current automatic coulometric titrimeter has been described by Carson[85]. It was designed primarily for microtitrations using generating currents in the range 1–20 mA, i.e. 10^{-9}–2×10^{-7} equiv/sec, and it incorporates a considerable degree of automatic control. It comprises a constant-current power supply, a control unit and a titration assembly.

The power supply is the most expensive part of the equipment, but with the aid of a multiple switch it can be used to drive up to four titration units one at a time. The arrangement is shown in block diagram form in Fig. 2.41. The delivered-current value is obtained from the *IR* drop across a precision resistor in the output of the constant-current source. The end-point is detected by using an a.c. pH-meter modified so that its impedance is matched to the triggering circuit which terminates the titration. Potentiometric end-point detection is used and the predetermined end-point potential is set on an adjustable potentiometer in the trigger circuit (Fig. 2.42). The voltage developed by the pH-meter during the course of the titration is compared with it. During the titration the potential difference between the potentiometer and the pH-meter signal decreases until the end-point is reached; it then reverses in polarity and increases. These changes are followed by a galvanometer switch which

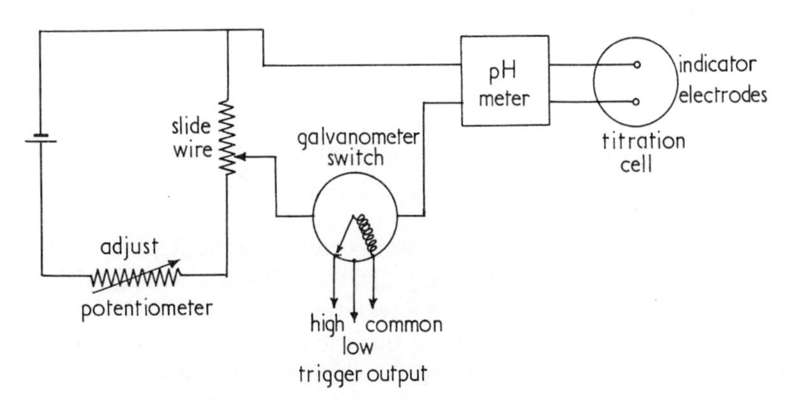

Fig. 2.42 Trigger circuit for end-point detection.
Reproduced with permission from Carson[85] and American Chemical Society.

at the end-point opens one contact and closes another, so providing the trigger output to stop the current generation. A chopper-type amplifier (Brown Electronik) amplifies the trigger output and the polarity change is detected by a thyratron. The titration time is read from a standard electric stop-watch of appropriate impedance. Some improvements in and modifications to the trigger circuit have been described[86].

The entire titration process is controlled and operated by the controller. It controls both the electrolysis and stop-watch currents and it also incorporates a delay time to eliminate spurious results arising from transient overshoot of the end-point. In such circumstances the end-point trigger operates but if, during the delay period, the potential drifts away from the end-point value, the titration is automatically restarted by a relay unit. Only when the end-point persists for the duration of the delay time is the titration complete, and then the controller flashes a warning light to this effect. In the titration of uranium, utilizing the oxidation of uranium(IV) to uranium(VI) with electrolytically generated ferric ion[87] the solution must be heated to $95 \pm 1°$ C and the controller incorporates facilities for heating and temperature control and also for control of stirrer speed. It is designed to prevent the titration from starting until the operating temperature is attained.

The titration assembly design is not critical. The generator and indicator electrode systems are chosen to suit the needs of the reaction being performed. In general the electrodes are separated by a low-resistance (1000–2000 ohms) salt bridge; saturated KCl/agar is suitable for titrations at room temperature and H_2SO_4/silica gel is satisfactory for heated reactions. Motor-driven or magnetic stirrers are satisfactory.

An alternative means of end-point anticipation has been developed by Carson[88] which is of universal applicability to automatic titrimeters for discrete samples. It involves withdrawing a proportion (10–20%) of the sample from the titration vessel, titrating the remainder rapidly, re-introducing the withdrawn portion and completing the titration slowly. A solenoid-operated syringe is used to withdraw the appropriate volume of sample and its operation can be built into the control unit described above.

The use of a low-inertia integrating motor in the electrolysis circuit obviates the need for circuitry capable of controlling the generated current to within the close limits, typically $\pm 0{\cdot}01\%$ on currents of 1–100 mA, required for precise analysis. In an integrating motor the speed of rotation of the shaft is a linear function of the applied voltage, the latter being obtained by passing the electrolysis current across a standard resistor. The counter reading on the motor is then proportional to the total quantity of electricity that has passed through the cell. In order that the motor should respond rapidly to changes in applied voltage, a low resistance (about 1 ohm) is used for the standard.

Automatic coulometers utilizing integrating motors have been described by Bett, Nock and Morris[89] and by Smythe[90]. The former authors were concerned essentially with macro-titrations, typically samples of concentration 0·01–0·1N. Two levels of current generation were provided, 1·3 A for the majority of the titration and 250 mA for the approach to the end-point. The potential across the indicating electrodes was applied to a direct-reading pH-meter. This voltage was compared with two preset voltages on the slide-wires of two potentiometers, these values representing the end-point voltage and a voltage in anticipation of it. At the latter voltage a relay caused the generating current to be switched to the lower current value and at the former the titration was terminated by a second relay. A simple titration cell was used, the titrant being generated externally by the technique of DeFord, Pitts and Johns[91]. Titrations of 25 ml of 0·1N acid could be performed with a relative standard deviation of 0·2%, on a routine basis.

Smythe's titrimeter[90] has many design similarities to the one described above but the emphasis was placed on titrations at lower concentration levels. A versatile electrolysis-current generation-circuit allowed a choice of six current values in the range 5–100 mA and also 1 mA for the approach to the end-point or for titrating microgram quantities of the sample constituent. Potentiometric end-point detection was used, the electrode pair being either platinum/SCE or platinum/tungsten. In the titration of dichromate ion with electrolytically generated ferrous ion the relative mean error varied from −0·2% for 17 mg of chromium to +5·0% for 17 μg of chromium.

Coulometric titrimetry is advantageous when the titrant is unstable, because it is generated *in situ* and undergoes reaction immediately. In this context iron(II) ethylenediaminetetra-acetate is of interest. Schmid and Reilly[92] demonstrated that it can be produced electrolytically from iron(III) ethylenediaminetetra-acetate by using a platinum generator cathode. Its redox potential is pH-dependent but above pH 2 its reducing power is similar to that of titanous ion; indeed the two are complementary in that they function over different pH ranges; both are unstable in air but coulometric generation overcomes this disadvantage. Schmid and Reilley showed that the reagent could be used to determine iron(III) at concentrations of 0·05–0·1M with a resulting error of ±1·0% in automatic titrations.

Karl Fischer determination of water in phosphate fertilizers has been reported[93]. Iodine is generated coulometrically in the presence of pyridine and methanol. The analytical precision over a wide range of water contents is claimed to be marginally better than in the azeotropic distillation method.

Concentrations of nitrogen dioxide in air ranging from 0 to 5000 ppm can be determined continuously by coulometry based upon the reaction

$$NO_2 + 2H^+ + 2Br^- \rightarrow NO + H_2O + Br_2$$

This forms the basis of the Mast 724–11 nitrogen dioxide analyser[94]. The electrode system consists of a pillar wound with many turns of a platinum wire cathode and a single turn of platinum wire anode. Air at 140 ml/min is pumped into the cell by a constant-volume piston pump, where it meets the bromide solution pumped at 2·5 ml/hr by a $\frac{1}{60}$ rpm motor-driven pump. The solution passes through a narrow annulus in intimate contact with the electrodes. Application of a potential across the electrodes causes a current to flow which is measured by the microcoulometer.

2.5 Ion-selective Electrodes

In recent years a range of electrodes capable of responding to one particular cation or anion has been developed[95, 96]. Although by no means wholly specific for the ion in question, they nevertheless show a high degree of discrimination against other ions and selectivity factors can readily be determined experimentally. For analytical use in aqueous systems the ion of interest must be capable of incorporation in a water-insoluble electrode. This has been achieved in four ways: (*a*) by using a single crystal or disc containing the responsive ion (solid-state electrodes), (*b*) by dispersing the appropriate compound in an inert membrane such as silicone rubber (heterogeneous membrane electrodes), (*c*) as glass electrodes, these being the earliest form of cation-selective electrode, and (*d*) with the ion bound to a high molecular weight organic molecule (liquid ion-exchangers). The list of ion-selective electrodes commercially available is large and growing; it includes H^+, Na^+, K^+, NH_4^+, Ag^+, Ca^{2+}, Cu^{2+}, Pb^{2+}, Cd^{2+}, F^-, Cl^-, Br^-, I^-, ClO_4^-, NO_3^-, CN^-, BF_4^- and S^{2-}.

When the electrode and reference electrode are immersed in a test solution containing the ion of interest, a potential response is obtained which is related by the Nernst equation to the activity of that ion. To permit measurements of ionic concentration the analysis is usually performed in such a way that both sample and standards are contained in identical buffer solutions of high ionic strength. Because the measuring circuitry is extremely simple the use of ion-selective electrodes in automated analytical systems is attractive and several typical applications are described below.

Krijgsman, Mansveld and Griepink[97] have developed a simple automatic titrimeter for the determination of halide ions at the 40–1500 neq level, using a silver sulphide electrode (Orion Ltd.). The sample solution is titrated continuously with 0·002*M* silver nitrate, with an LKB pump as the automatic burette. To achieve precise results at the low halide ion concentrations of interest, titrations were performed in 75% acetic acid solution. The response of the indicator–reference electrode pair is amplified and fed to a millivolt recorder. The end-point potential band is preselected on the recorder which then terminates the titration when this potential is attained. A relative standard deviation of the order of 1% is reported. The only significant interference is caused by sulphide ion, and can be eliminated by adding some hydrogen peroxide to the titration medium.

The ion-selective electrode is a valuable addition to the range of analytical detection systems which can be used to monitor treated sample

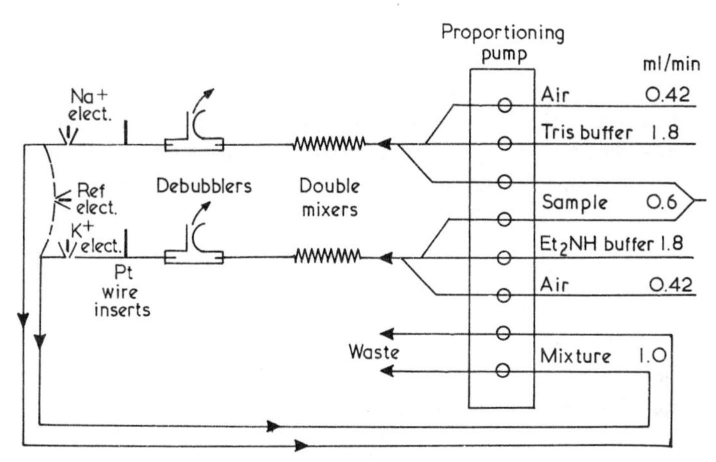

Fig. 2.43 Manifold for utilization of ion selective electrodes for the determination of Na and K in dog urine.
Reproduced with permission from Jacobson[98] and New York Academy of Sciences.

streams in the Technicon AutoAnalyzer. Fig. 2.43 depicts the manifold employed by Jacobson[98] to determine sodium and potassium in samples of dog urine. The sample stream is divided in two and mixed with an appropriate buffer solution before passing the indicator electrodes. For sodium the Beckman sodium-electrode was used and for potassium the Beckman cation-selective electrode. A common reference electrode was used for the two streams. For sodium-ion measurement a buffer at pH 8·15 of 2-amino-2-(hydroxymethyl)-1,3-propanediol (tris) and acetic acid was employed and for potassium a mixture of acetic acid and diethylamine at pH 11·55. The cation-selective electrode exhibits sufficient response to sodium to require a correction factor to be applied. This was obtained by determining the selectivity coefficient $k_{K^+Na^+}$ by using standards and calculating the potassium concentration from the equation

$$E = E^{\circ} + \frac{RT}{nF} \ln a_{K^+} + k_{K^+Na^+} a_{Na^+}$$

The performance of the equipment was assessed from determinations of sodium saccharinate in a proprietary product, using only that part of the manifold necessary for sodium measurement. For standard solutions prepared from either sodium chloride or sodium saccharinate recoveries of 100 ± 2% were recorded.

The platinum wire inserts shown following the debubbler in Fig. 2.43 were earthed to the recording pH-meter and served to eliminate electrical noise due to the proportioning pump.

The use of ion-selective electrodes for the automatic determination of inorganic constituents in blood has been evaluated by Dahms[99]. Equipment was developed which provided automatic sampling, calibration and data acquisition and processing for Na^+, K^+, pH and Cl^-. The alternative method for Na^+ and K^+ is automatic flame photometry which is capable of greater sample throughput than the potentiometric method using ion-selective electrodes, but the equipment designed by Dahms has several important advantages. Not only does it provide simultaneous measurements of pH and Cl^- in addition to Na^+ and K^+, but it yields results in terms of activities rather than concentrations, which is important physiologically. Further, the potentiometric method is non-destructive and is capable of handling samples of whole blood as well as serum. With careful equipment design highly precise results can be achieved, thus for typical blood concentrations relative standard deviations of ±0·2% and ±1·1% were obtained for Na^+ and K^+ respectively. For pH and Cl^- the corresponding relative standard deviations were ±0·4% and ±0·3%.

The equipment provides for continuous sequential pumping of sample and two calibration standards through the electrodes mounted in series. The response of each electrode can be selectively monitored for subsequent computation. Fig. 2.44 shows a block diagram of the complete apparatus, that part enclosed within the broken line being thermostatically controlled at 38° C. The flow-path of the solutions is depicted in Fig. 2.45. The valve which sequentially selects sample and calibration standards is of the rotary type, driven by a slow-speed synchronous motor. The valve itself is constructed of 'Delrin' and connections to it are made of 'Teflon'. A standard

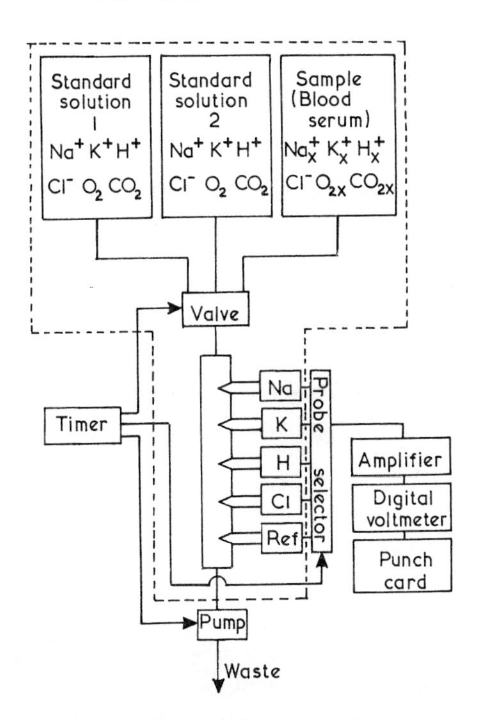

Fig. 2.44 Block diagram for the automatic determination of Na^+, K^+, Cl^- and pH, using ion selective electrodes.
Reproduced with permission from Dahms[99] and American Association of Clinical Chemists.

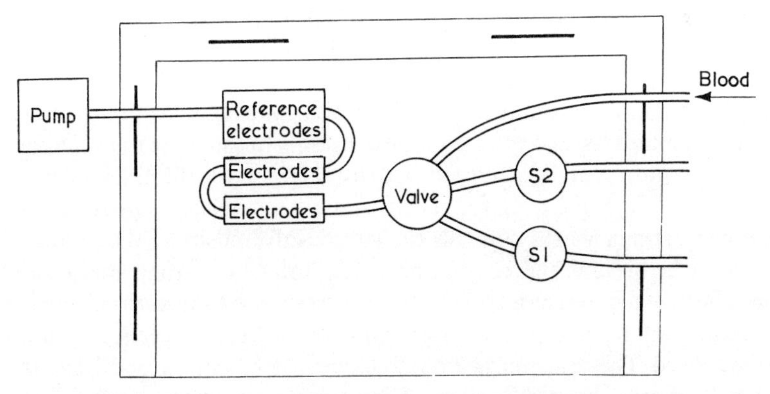

Fig. 2.45 Solution flow path for equipment of Fig. 2.44. S1 and S2 are standard solutions.
Reproduced with permission from Dahms[99] and American Association of Clinical Chemists.

peristaltic pump, suitably geared down, provides the sample transport. The electrode design for Na^+, K^+ and pH measurement is illustrated in Fig. 2.46, which shows the Na^+ and K^+ electrodes in a common 'Teflon' housing, glass capillaries of the appropriate composition (NAS_{11-18} for Na^+, NAS_{27-4} for K^+ and 015 for pH, all from Corning Glass Works) together with an AgCl-coated silver wire being sealed through rubber into glass tubing containing 0·1*N* HCl. Design of the reference electrode is

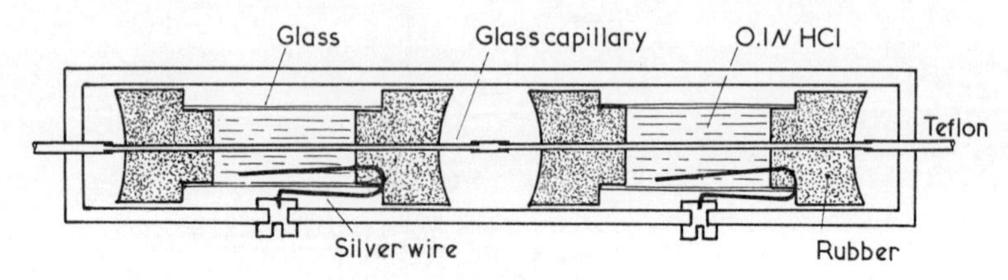

Fig. 2.46 Mounting for Na^+ and K^+ electrodes.
Reproduced with permission from Dahms[99] and American Association of Clinical Chemists.

critical in relation to the overall precision of results because liquid-junction potentials contribute to the reproducibility of ±1–2 mV, which represents ±3% on the activity measurement. This could be reduced to better than ±0·1 mV by using a membrane of wide-pore cellophane to form a sharp boundary between the two solutions which allows free passage of inorganic ions and thereby limits the development of a Donnan potential.

The voltage response from the electrodes is fed to a probe selector comprising either an automatic switch or a relay bank, which monitors each electrode in turn. The signal is then amplified, converted into a digital output with a 4-digit digital voltmeter, and finally fed to a card-punch. The calculation of activities requires two stages, evaluation of the Nernst equation constants for the electrodes by using responses from two standard solutions, and the use of the constants to calculate ionic activities for the samples; it is accomplished by using a FORTRAN program. The automatic sequencing for each phase of the analysis is provided by a pneumatic timer which controls the rotary valve position, the pump, the probe selector and the card-punch. The total cycle time for a single electrode measurement is 65 sec, the finite response time of the electrodes, particularly for Na^+, precludes reduction of this time if precision is not to be sacrificed. The drift characteristics of the instrument are good, less than 0·1 mV/hr, and enable batches of 10 samples to be analysed between calibration checks.

Dahms's approach to automatic analysis by using ion-selective electrodes has subsequently been updated[100] to an instrument termed ABA (Automatic Blood Analyser). The chloride electrode is omitted and pO_2 and pCO_2 electrodes are incorporated. Standard electrodes (E 5036 and E 5046) from Radiometer A/S are used for the latter measurements. The flow system differs little from that of Dahms but the rotary valve is replaced by solenoid-operated pinch valves and electronic timing is used instead of pneumatic. To avoid formation of deposits from the blood samples within the flow lines an automatic flushing is provided after each sample analysis, whereby the lines are cleansed with a standard solution. A time-shared IBM 1800 digital computer is used for all data processing and report presentation.

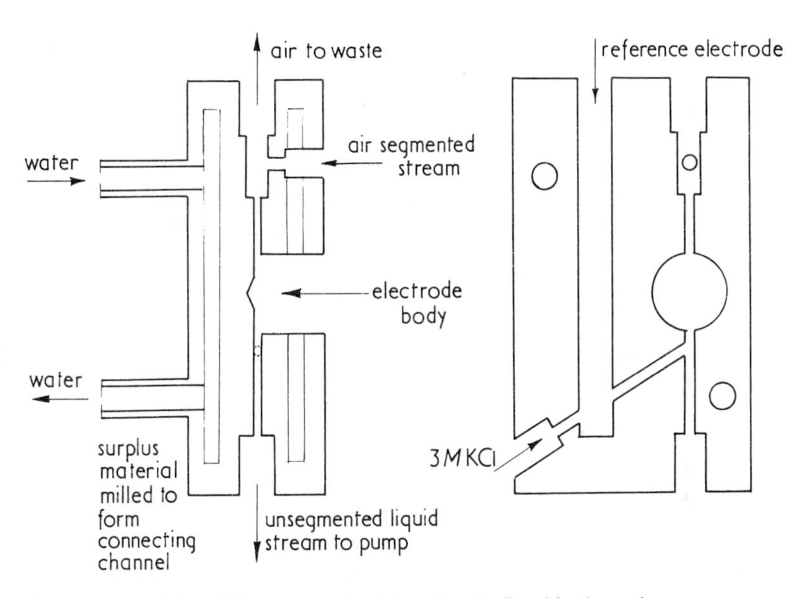

Fig. 2.47 Flow cell for continuous analysis using the fluoride electrode.
Reproduced with permission from Sawyer and Foreman[102] and United Trade Press Ltd.

The fluoride-selective electrode[101], which exhibits Nernstian behaviour at concentrations down to $10^{-6}M$ and which suffers interference only from hydroxyl ion, has already proved of considerable analytical value in simplifying the determination of traces of fluoride. Its adaptability to automatic analysis has been demonstrated by Sawyer and Foreman[102] in an adaptation to use with the 'AutoAnalyzer'. The design of flow cell employed is shown in Fig. 2.47, (*a*) indicating the sample electrolyte pathway in which air-segmented samples are deaerated at a T-junction. The released air, together with some of the sample solution, flows to waste from one exit, and the sample stream flows from the other past the face of an Orion fluoride-electrode. It then flows through a narrow-bore channel past a calomel reference electrode and is finally pumped to waste. The disposition of the fluoride-electrode with respect to the reference electrode, to which 1*M* KCl is pumped continuously from the manifold, is shown in (*b*). The holder for the electrodes is fabricated from 'Perspex' sheet. The whole unit is encapsulated in an outer 'Perspex' jacket and maintained at constant temperature by circulation of water from a constant-temperature bath. The complete assembly is 1 × 1·5 × 3 in.

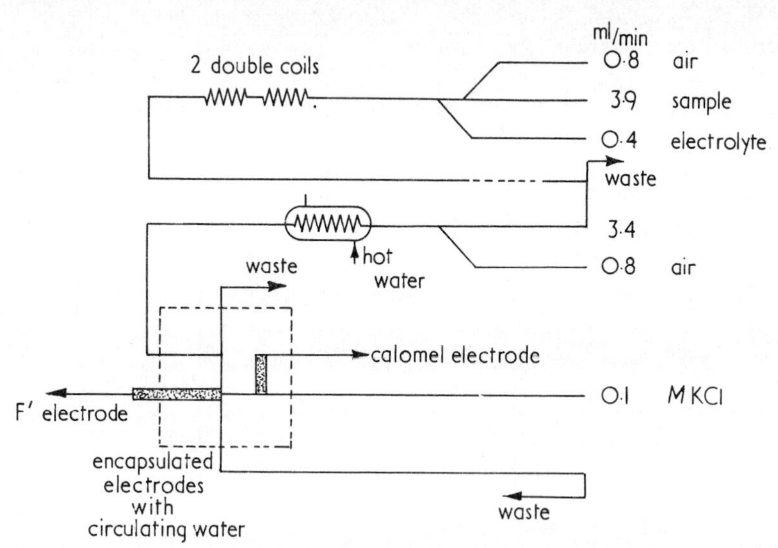

Fig. 2.48 Manifold for continuous determination of fluoride using an ion selective electrode.
Reproduced with permission from Sawyer and Foreman[102] *and United Trade Press Ltd.*

The system was evaluated for the determination of fluoride over the concentration range 0·2–2 mg/l in aqueous leach solutions from dental silicate cements. The manifold design is illustrated in Fig. 2.48. The leach solutions contained aluminium, and an electrolyte functioning both as a buffer and a decomplexing agent was required. Fluoride recoveries in the presence of up to 5 mg of aluminium per litre were satisfactory when an electrolyte comprising either 0·25*M* KH_2PO_4, 0·25*M* Na_2HPO_4, 0·01*M* trisodium citrate, 0·15*M* disodium ethylenediaminetetra-acetate at pH 6·4 or glycine/NaCl/NaOH at pH 9·2 was used. Operation of the apparatus

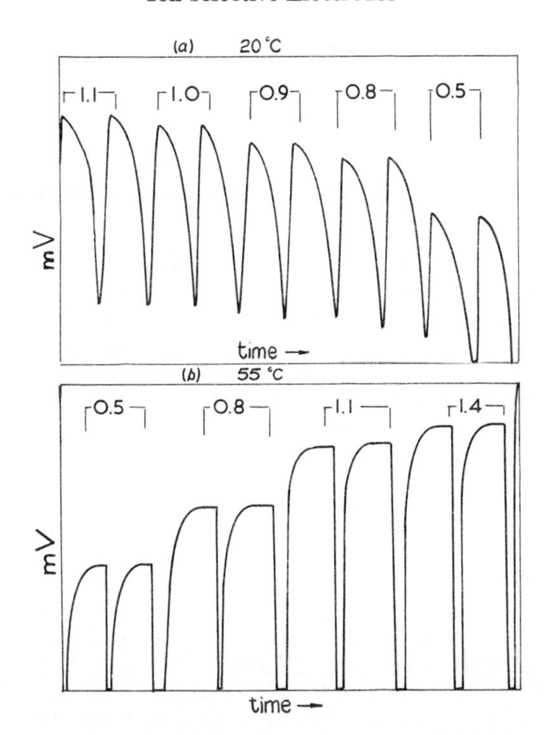

Fig. 2.49 Effect of temperature on performance of fluoride electrode in continuous analysis over the fluoride concentration range 0·5 to 1·4 mg/l.
Reproduced with permission from Sawyer and Foreman[102] and United Trade Press Ltd.

at a sample throughput of 20 per hour at 20° C yielded recorder responses typified by Fig. 2.49 (*a*), from which it is evident that the electrode response had not reached equilibrium. Fig. 2.49 (*b*) demonstrates that operation at 55° C overcomes this problem.

References

1. Manning, W. M. *Ecology*, 1940, **21**, 509.
2. Ingols, R. S. *Ind. Eng. Chem., Anal. Ed.*, 1942, **14**, 256.
3. Ingols, R. S. *Sewage Works J.*, 1941, **13**, 1097.
4. Spoor, W. A. *Science*, 1948, **108**, 421.
5. Beckman, P. *Chem. Ind., London*, 1948, 791.
6. Laitinen, H. A., Higuchi, I. and Czuha, M. *J. Am. Chem. Soc.*, 1948, **70**, 561.
7. Wise, W. S. *Chem. Ind., London*, 1948, 37.
8. Wilson, L. D. and Smith, R. J. *Anal. Chem.*, 1953, **25**, 218.
9. Müller, O. H. *J. Am. Chem. Soc.*, 1947, **69**, 2992.

10. Mann, C. K. *Anal. Chem.*, 1957, **29**, 1385.
11. Drake, B. *Acta Chem. Scand.*, 1950, **4**, 554.
12. Kemula, W. *Roczniki Chem.*, 1955, **29**, 1153.
13. Blaedel, W. J. and Todd, J. W. *Anal. Chem.*, 1958, **30**, 1821.
14. Blaedel, W. J. and Todd, J. W. *Anal. Chem.*, 1961, **33**, 205.
15. Rebertus, R. L., Cappell, R. J. and Bond, G. W. *Anal. Chem.*, 1958, **30**, 1825.
16. Kraus, K. A. and Moore, G. E. *J. Am. Chem. Soc.*, 1953, **75**, 1460.
17. Blaedel, W. J. and Strohl, J. H. *Anal. Chem.*, 1961, **33**, 1631.
18. Blaedel, W. J. and Strohl, J. H. *Anal. Chem.*, 1964, **36**, 445.
19. Jura, W. H. *Anal. Chem.*, 1954, **26**, 1121.
20. Lewis, J. A. and Overton, K. C. *Analyst*, 1954, **79**, 293.
21. Airey, L. and Smales, A. A. *Analyst*, 1950, **75**, 287.
22. Bertram, H. W., Lerner, M. W., Petretic, G. J., Roszcowski, E. S. and Rodden, C. J. *Anal. Chem.*, 1958, **30**, 354.
23. Kolthoff, I. M. and Lingane, J. J. *Polarography*, 2nd ed., Vol. I, p. 331. Interscience, New York, 1952.
24. Blaedel, W. J., Olson, C. L. and Sharma, L. R. *Anal. Chem.*, 1963, **35**, 2100.
25. Levich, V. G. *Physicochemical Hydrodynamics*, Prentice-Hall, Englewood Cliffs, N.J., 1962.
26. Kolthoff, I. M. and Jordon, J. *J. Am. Chem. Soc.*, 1954, **76**, 3843.
27. Jordan, J., Javick, R. A. and Ranz, W. E. *J. Am. Chem. Soc.*, 1958, **80**, 3846.
28. Marple, T. L. and Rogers, L. B. *Anal. Chem.*, 1953, **25**, 1351.
29. Cooke, W. D. *Anal. Chem.*, 1953, **25**, 215.
30. Oesterling, T. C. and Olson, C. L. *Anal. Chem.*, 1967, **39**, 1543.
31. Ramaly, L., Brubaker, R. L. and Enke, C. G. *Anal. Chem.*, 1963, **35**, 1088.
32. Booth, M. D., Fleet, B., Soe, Win and West, T. S. *Anal. Chim. Acta*, 1969, **48**, 329.
33. Fano, V. and Scalvini, M. *Microchem. J.*, 1970, **15**, 97.
34. Fleet, B., Ho, A. Y. W. and Tenygl, J. *Talanta*, 1972, **19**, 317.
35. Fleet, B., Ho, A. Y. W. and Tenygl, J. *Analyst*, 1972, **97**, 321.
36. Juliard, A. L. *Anal. Chem.*, 1958, **30**, 137.
37. Murayama, K. *U.S. Patent* 2,834, 564, Feb. 1st, 1954.
38. Gonzales Barredo, R. *J. Electrochem. Soc.*, 1947, **92**, 303.
39. Myers, S. A. and Swann, W. B. *Talanta*, 1965, **12**, 133.
40. Kubota, H. and Surak, J. G. *Anal. Chem.*, 1963, **35**, 1715.
41. Olson, E. C. and Elving, P. J. *Anal. Chem.*, 1954, **26**, 1747.
42. Rechnitz, G. A. and Mark, H. B. *Kinetics in Analytical Chemistry*, Interscience, New York, 1968.
43. Pardue, H. L. *Anal. Chem.*, 1963, **35**, 1240.
44. Pardue, H. L. and Simon, R. K. *Anal. Biochem.*, 1964, **9**, 204.
45. Pardue, H. L. and Frings, C. S. *J. Electroanal. Chem.*, 1964, **7**, 398.
46. Blaedel, W. J. and Olson, C. *Anal. Chem.*, 1964, **36**, 343.
47. Easty, D. B., Blaedel, W. J. and Anderson, L. *Anal. Chem.*, 1971, **43**, 509.
48. Blaedel, W. J. and Boyer, S. L. *Anal. Chem.*, 1971, **43**, 1538.
49. Bomstein, J., Shepp, J. M., Dawson, S. T. and Blaedel, W. J. *J. Pharm. Sci.*, 1966, **55**, 94.
50. Robinson, H. A. *Trans Electrochem. Soc.*, 1947, **92**, 445.
51. Kelley, M. T., Horton, J. M., Tallackson, J. R. and Miller, F. J. *Proc. Instrum. Soc. Amer.*, 1952, **7**, 63.

52. Shain, I. and Huber, C. O. *Anal. Chem.*, 1958, **30**, 1286.
53. Audran, R. and Dighton, D. T. R. *J. Sci. Instr.*, 1956, **33**, 92.
54. Lingane, J. J. *Anal. Chem.*, 1948, **20**, 285.
55. Lingane, J. J. *Anal. Chem.*, 1948, **20**, 797.
56. Malmstadt, H. V. and Fett, E. R. *Anal. Chem.*, 1954, **26**, 1348.
57. Malmstadt, H. V. and Fett, E. R. *Anal. Chem.*, 1955, **27**, 1757.
58. Hieftje, G. M. and Mandarano, B. M. *Anal. Chem.*, 1972, **44**, 1616.
59. Lord Rayleigh. *Proc. London. Math. Soc.*, 1879, **10**, 4.
60. Lindblad, N. R. and Schneider, J. M. *Rev. Sci. Instr.*, 1967, **38**, 325.
61. Olsen, E. D. and Adamo, F. S., *Anal. Chem.*, 1967, **39**, 31.
62. Blaedel, W. J. and Laessig, R. H. *Anal. Chem.*, 1964, **36**, 1617.
63. Blaedel, W. J. and Laessig, R. H. *Anal. Chem.*, 1965, **37**, 333.
64. Reilley, C. N. and Schmid, R. W. *Anal. Chem.*, 1958, **30**, 953.
65. Reilley, C. N. and Schmid, R. W. and Lamson, D. W. *Anal. Chem.*, 1958, **30**, 953.
66. Blaedel, W. J. and Laessig, R. H. *Anal. Chem.*, 1965, **37**, 1255.
67. Blaedel, W. J. and Laessig, R. H. *Anal. Chem.*, 1966, **38**, 187.
68. Blaedel, W. J. and Laessig, R. H. *Anal. Chem.*, 1965, **37**, 1651.
69. Cox, D. C. *J. Am. Chem. Soc.*, 1925, **47**, 2138.
70. MacInnes, D. A. *Z. Physik. Chem.*, 1927, **130A**, 217.
71. MacInnes, D. A. and Cowperthwaite, I. A. *J. Am. Chem. Soc.*, 1931, **53**, 555.
72. MacInnes, D. A. and Dole, M. *J. Am. Chem. Soc.*, 1929, **51**, 1119.
73. MacInnes, D. A. and Jones, J. T. *J. Am. Chem. Soc.*, 1926, **48**, 2831.
74. Nicholson, M. M. *Anal. Chem.*, 1961, **33**, 1328.
75. Malmstadt, H. V. and Pardue, M. L. *Anal. Chem.*, 1961, **33**, 1040
76. Pardue, H. L., Simon, R. K. and Malmstadt, H. V. *Anal. Chem.*, 1964, **36**, 735.
77. Malmstadt, H. V. and Pardue, H. L. *Clin. Chem.*, 1962, **8**, 606.
78. Shaffer, P. A., Jr., Briglio, A., Jr. and Brockman, J. A. *Anal. Chem.*, 1948, **20**, 1008.
79. Takahashi, T., Miki, E., and Sahurai, H. *J. Electroanal. Chem.*, 1962, **3**, 371.
80. Takahashi, T. and Niki, E. *J. Electroanal. Chem.*, 1962, **3**, 381.
81. Eckfeldt, E. L. *Anal. Chem.*, 1959, **31**, 1453.
82. Eckfeldt, E. L. and Shaffer, E. W., Jr. *Anal. Chem.*, 1964, **36**, 2008.
83. de Kainlis, G., Merigot, D. and Pourcel, C. *Chim. Anal., Paris*, 1971, **53**, 696.
84. Lingane, J. J. *Anal. Chem.*, 1954, **26**, 622.
85. Carson, W. N., Jr. *Anal. Chem.*, 1953, **25**, 226.
86. Carson, W. N., Jr. *Anal. Chem.*, 1954, **26**, 1673.
87. Carson, W. N., Jr. *Anal. Chem.*, 1953, **25**, 467.
88. Carson, W. N., Jr. *Anal. Chem.*, 1953, **25**, 1733.
89. Bett, N., Nock, W., and Morris, G. *Analyst*, 1954, **79**, 607.
90. Smythe, L. E. *Analyst*, 1956, **82**, 228.
91. De Ford, D. D., Pitts, J. N. and Johns, C. F. *Anal. Chem.*, 1951, **23**, 938.
92. Schmid, R. W. and Reilly, C. N. *Anal. Chem.*, 1956, **28**, 520.
93. Jordan, D. E. and Hoyt, J. L. *J. Assoc. Off. Anal. Chem.*, 1969, **52**, 569.
94. Rostenbach, R. E. and King, R. G. *J. Assoc. Air Pollution Cont.*, 1962, **12**, 459.
95. Covington, A. K. *Chem. Brit.*, 1969, **5**, 388.
96. Florence, T. M. *Proc. Roy. Aust. Chem. Inst.*, 1970, **37**, 261.
97. Krijgsman, W., Mansveld, J. F. and Griepink, B. F. A. *Z. Anal. Chem.*, 1970, **249**, 368.

98. Jacobson, H. *Ann. N.Y. Acad. Sci.*, 1968, **153**, 486.
99. Dahms, H. *Clin. Chem.*, 1967, **13**, 437.
100. Neff, G. W., Radke, W. A., Sambucetti, C. F. and Widdowson, G. M. *Clin. Chem.*, 1970, **16**, 566.
101. Frant, M. S. and Ross, J. W. *Science*, 1966, **154**, 1553.
102. Sawyer, R. and Foreman, J. K. *Lab. Practice*, 1969, **18**, 35.

Chapter 3

Colorimetric Methods

Colorimetry is one of the most extensively used analytical techniques for both organic and inorganic materials. The concentration of a substance may be determined directly if it is coloured or by reaction with a chromogenic reagent if it is colourless. By suitable choice of reagent, concentrations ranging from per cent to parts per million and below can be determined.

Largely because of the impetus from the field of clinical analysis the automation of colorimetry has been intensively studied, perhaps more so than any other technique. In consequence there is a wide range of commercial instruments available for conducting automated colorimetry both by the continuous and discrete methods. They include instruments which provide automation of the colour measurement only and those which automatically pretreat the sample before analysis. The latter may be fully automatic or may involve manual intervention between treatment stages.

3.1 **Automated Colorimetry**

3.1.1 Discrete Sample Analysis

When a number of samples are to be submitted sequentially for measurement of radiation absorption two general approaches are possible; either each successive sample (or blank or standard) is conveyed to the absorptiometer cell and removed from it after the measurement has been made, or the sample vessel constitutes the cell and each one is mechanically moved into the light-beam in sequence. The latter technique eliminates completely any cross-contamination between samples but it requires a large number of vessels of highly reproducible dimensions if accurate results are to be achieved. While this is entirely feasible for measurements at wavelengths in the visible range where plastic or glass vessels can be utilized, it is uneconomic for ultraviolet measurements where quartz vessels are essential. Both principles are illustrated below by reference to several typical commercial instruments.

The Unicam SP 3000 automatic ultraviolet spectrophotometer is fitted with an automatic sample-changer capable of holding up to 50 glass sample-tubes and the contents of each tube are transferred sequentially to the measuring cell within the spectrophotometer. Two sample-transfer mechanisms are available, the choice between them depending on the nature of the analysis and the volume of sample available. The SP 3002 AU Autocell is based upon an external suction pump which transfers each sample to the measuring cell and flushes it to waste when the measurement is complete. The SP 3002P system, which is best suited to aqueous solution where the available volume is limited, uses an integral metering pump which delivers the sample to the cell and returns it, after measurement, to the sample tube. For a 10-mm path length the volume of sample needed for absorptiometric measurement is 0·6 ml. The manufacturers claim that for the SP 3002 AU system cross-contamination between successive samples is equivalent to an absorbance difference of 0·2% between successive samples. The overall cycle time is approximately 40 sec. The SP 3000 is a single-beam instrument in which the reference cell, containing standard or blank solution, and the sample cell are successively introduced into the light beam. Optical balancing is achieved with the aid of an auxiliary tungsten lamp. The radiation intensity transmitted by the reference cell is attenuated until it is equal to that of the modulated auxiliary standard source; the transmission of the sample cell is then determined as a function of the auxiliary source intensity and this function is converted into absorbance or transmittance. Use of the auxiliary source eliminates errors due to drift in the response of the multiplier phototube, factors which might become significant in a single-beam instrument when a large number of samples are being processed.

Digital display of each result is achieved by an automatic ratio measurement using a servo-driven potentiometer to obtain a null balance. A coded disc is attached to the potentiometer spindle and this converts the angular displacement of the spindle into a four-digit result. Digital print-out, which is essential when many samples are to be analysed, is provided by an electric typewriter and serializer unit. The print-out takes the form of an absorbance (or transmittance) reading preceded by an index number which is generated sequentially by the automatic sample-changer, thereby enabling each result to be unequivocally associated with the sample producing it.

The SP 3000 can also be used to follow automatically the course of kinetic studies. A programming unit generates measurements at 12-sec intervals and also at every $2\frac{1}{2}$, 5 or 10 min. The unit can be programmed to omit intermediate measurements in the case of slow reactions.

A simple, yet versatile, automatic colorimeter is marketed by Evans Electroselenium Limited as the EEL Model 171. The photometer covers the visual range 400–700 nm, wavelength selection being achieved either by a variable interference filter of 30-nm half-bandwidth (400–700 nm) or by interchangeable interference filters of 10-nm half bandwidth (340–650 nm). The useful absorbance range is 0–0·5; for solutions of absorbance greater than 0·5 prior dilution is necessary. Digital read-out (3 digits) or print-out directly as sample concentration is achieved by a log–linear converter which produces an integrated d.c. output proportional to sample concentration. This potential is compared by a balancing amplifier with a reference voltage developed across a servo motor-driven potentiometer. The rotation of the servo-motor is digitized through its connection to the shafts of a digital indicator.

The means of transferring each sample to the cell, and the cell system itself, are shown in Fig. 3.1. The cell is a glass vessel into which sample is drawn by a motor-driven syringe; the lower section *C*, of nominal 5-mm path length, comprises the measuring cell, and the upper section *D* acts as a reservoir. As the sample is drawn in through the inlet *A* any residue from the previous sample is swept out of the light-path into the upper reservoir. Carry-over from one sample to the next is claimed to be equivalent to 1% of the previous sample reading. After drawing sample for $3\frac{1}{2}$ sec

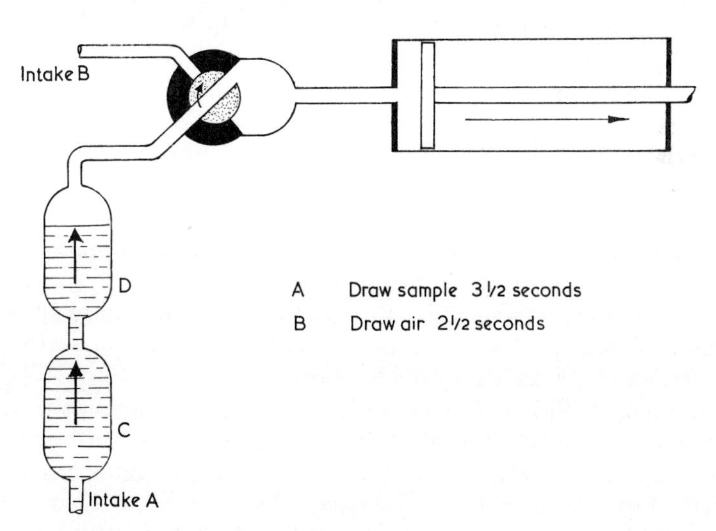

Fig. 3.1 Cell-filling method for the EEL 171 Automatic Colorimeter. *Reproduced by permission of Corning-Eel.*

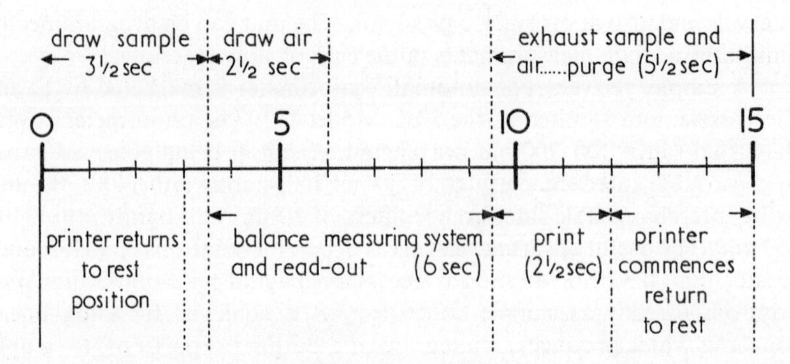

Fig. 3.2 Time-sequencing of sample flow and read-out for the EEL 171 Automatic Colorimeter.
Reproduced by permission of Corning-Eel.

the syringe draws air through inlet *B* for $2\frac{1}{2}$ sec, the air being necessary to eject the sample from the cell after the optical measurement has been completed. Fig. 3.2 depicts the sequencing of events both for filling and exhausting the cell and the concurrent operations of optical measurement, print-out and return of the printer to its rest position in readiness for the next sample. A cam-timer controls the sequencing.

An automatic sampler, the EEL 178, is marketed for use in conjunction with the EEL 171 automatic colorimeter. It has a capacity of 48 sample-tubes loaded in four quadrant racks on a rotating turntable. The turntable rotates until the alignment of a sample-tube with the sampling probe is detected photoelectrically. The signal so produced is used to stop the turntable, actuate the descent of the sampling probe and to start the cam-timer of the colorimeter. The sample-transfer stages occupy 3 sec and the measurement cycle (Fig. 3.2) 15 sec, or 30 sec if an alternative cell of long path length (5 cm) is employed. Identifying tags for sample, standard blank and final tube in a sequence are provided. By being attached at specified different heights on each tube each tag is identified by an appropriately sited photocell on reaching the sampling probe.

The colorimeter is provided with facilities for automatic calibration and zero correction and with manual over-ride switches for performing analyses out of the programmed sequence.

Although the EEL 171 is not intended to offer the photometric range or accuracy of the Unicam SP 3000 described above it nevertheless provides a rugged and relatively inexpensive approach to automatic colorimetry and its reproducibility is adequate for very many analytical needs.

Both of the automatic photometers described above utilize one cell for all sample analyses. The alternative approach, in which the sample vessel constitutes the photometer cell, is embodied in the LKB 7400 Calculating Absorptiometer. The 10-mm glass or disposable polystyrene cells used are fed to the light-beam in linear racks holding 9, 10 or 11 cells; the racks are slotted to allow passage of light through the sample. The 9-cell rack allows measurements to be made on 6 samples together with reference cells, the rack containing 10 cells is for examining 5 samples each with its own reference solution, and the 11-cell rack is designed to allow 10 samples to be measured against one reference solution. The volume examined can range from 0·7 to 3·6 ml. Cross-contamination between samples is eliminated if care is taken to clean and dry the cells before use.

The photometer comprises a highly stabilized prefocused tungsten lamp and a vacuum phototube detector; it covers the wavelength range 400–700 nm, wavelength selection being achieved by using one of ten interference filters of standard bandwidth 15 mm and mounted on a rotating disc. For each selected wavelength, calibration must be performed before analysis of samples is commenced so that the required calibration factors can be held in the data handling unit. The theoretical processing capacity of the absorptiometer is 1200 per hour, at which level automatic data processing is essential if full economic use of the high throughput is to be achieved. Of the several alternatives provided, the simplest is an ADDO printer which prints the measured value in 4 digits together with a two-digit sample number and two-digit analysis code. The printer can be used in parallel with either a programmed or unprogrammed paper-tape punch. When the programmed punch is used an input–output printer is used and additional information can be added manually through the keyboard. Necessary interfacing is supplied to enable results to be processed by an on-line computer; both analogue and digital methods of data transfer can be used.

Shapiro and Masson[1] constructed a sample changer having a capacity of 72 100-ml beakers which they used in conjunction with a Beckman DB-G spectrophotometer for performing automatic colorimetric measurements on samples of geological interest.

3.1.2 Continuous Sample Analysis

A simple flow-cell is all that is required when measurements are to be made on a flowing stream and most commercial spectrophotometer manufacturers offer a flow-cell as an accessory. Results can be presented as a recorder trace or peak areas can be digitally integrated.

The Labotron UDC1 universal flow-through colorimeter is a modular instrument capable of performing up to six flow-through colorimetric measurements simultaneously. It comprises a centrally mounted lamp with facilities for mounting up to six colorimeters circumferentially. The flow-through cells are inserted between the lamp and colorimeter. Wavelength selection in the colorimeters is achieved by interference filters and two detectors cover the wavelength ranges 340–600 and 420–1100 nm. Results can conveniently be presented on a multipoint recorder. Alternatively the channels may be monitored consecutively by means of a multichannel switching unit. Three ranges of flow-through cell are provided, (*a*) micro cells of optical path length 20, 10, 5, 4 or 1 mm, constructed of PVC or stainless steel and fitted with optical glass windows, (*b*) thermoregulated cells of 10, 5, 2 and 1 mm optical path length, fabricated from stainless steel and fitted with optical glass windows and (*c*) flow-through cells for higher flow-rates and having 5, 2 and 0·8 mm optical paths.

3.2 Colorimetric Analysers with Automatic Sample Pretreatment

The instruments described in the preceding section are concerned solely with automation of the colour measurement process. In this section a number of analysers are described in which the complete analysis is partially or wholly automated. The partially automated systems are more correctly described as mechanized analysers; individual stages such as sampling, dilution, addition of reagents and colorimetric measurement are performed mechanically but intervention by the operator is required to transfer the partially treated samples from one stage to the next. Mechanized analysers have economic advantages over the more costly fully automated type if sample loads are not excessive, because the cost of occasional operator attention is cheaper than that of automated sample-transfer equipment. Several commercial mechanized analysers are available for discrete sample processing. The mechanization approach is not compatible with the continuous flow method in that interruption of the flow sequence is contrary to the basic principle.

3.2.1 The BTL 'Analmatic'

The Baird and Tatlock Ltd 'Analmatic' is designed to provide automatic analysis of clinical samples by the discrete method in batch sizes of up to 100. Provided that the analysis comprises sampling, dilution, addition of up to three reagents and colour measurement the procedure can be performed fully automatically. If more than three reagents are needed the dispensing syringes must be changed manually, and if deproteinization is necessary the samples must be manually transferred to a specially designed centrifuge. The instrument is modular in construction and consists of a sample preparation unit, a double-beam colorimeter (or dual-channel flame photometer), a printer and a centrifuge.

Sampling and addition of reagents are performed in the preparation unit. Up to 100 samples in a 10×10 configuration can be processed. They are held in a water-bath, the temperature of which can be maintained at a selected value between 25 and 75° C by a circulatory system. A warm-up time of 30 min is required for control at 37° C and 60 min for 57° C. Samples are contained in plastic cups and the analytical operations are performed in glass or PTFE tubes. Each reaction tube is adjacent to the sample tube. Sample volumes between 10 μl and 1·5 ml can be conveniently handled, the upper limit being set by the need to wash each sample with five times its volume of diluent when it is transferred to the reaction tube. Sample transfer is performed through a dipping probe by a valve-operated syringe, the volume of which is pre-set. Diluent and reagents are also added by valve-operated syringes mounted on a syringe plate. Syringes of volumes ranging from 0·1 to 5 ml are provided and are selected to meet the chemical requirements of the analysis. Thus the 'Analmatic' is basically a single-channel discrete-sample analyser, but the change from one analysis to another can be rapidly made by replacing one syringe plate with another pre-prepared one. The sampling and reagent-addition assembly traverses the sample tubes successively until each sample has been treated. The cycle time for sampling, diluting and reagent dispensing is 12 sec so that a full batch of 100 samples can be processed in 20 min. The sample volume repeatability specification for the preparation unit is 3, 2 and $< 1\%$ for volumes of 10, 20 and > 50 μl respectively. For reagent addition it is $< 0{\cdot}15$, $< 0{\cdot}1$ and $< 0{\cdot}05\%$ for delivered volumes of 0·5, 1·0 and 5·0 ml.

The colorimetric measurement is performed in a double-beam photometer operating over the wavelength range 400–700 nm. The desired wavelength is selected by means of a graded interference filter, the bandwidth being 5% of the wavelength setting. The one-piece double-beam flow-through cuvette is filled by aspiration from the reaction tubes contained in the preparation unit. The sample-transfer probe is carried on the same head as the sampling and dispensing probes so that processing and measurement can proceed simultaneously; to this end the measuring cycle is matched to the 12-sec preparation cycle. The cuvette volume is 0·35 ml with a 6-mm optical path length. A 3·5-ml volume is drawn through the cuvette, providing a wash-to-measurement volume ratio of 8 : 1 to minimize errors due to carry-over between successive samples. The double-beam optics enable each sample to be measured against a blank. Where one blank will serve for a range of samples it can be isolated in one arm of the cuvette and used throughout the run. Where each sample must be read against its own blank, two matched sample syringes are used in the preparation unit; these are so mounted that two identical samples are drawn from one sample cup and delivered to two reaction tubes, one of which is processed as the sample and the other as the blank. Two colorimeter probes then transfer the sample and blank simultaneously to the cuvette. The photometer reading is integrated for a nominal 1 sec and the results recorded on the print unit which provides a three-digit reading including an adjustable decimal point together with batch number and sample number (both 01–99) and an analysis code. Where calibrations are linear a two-point calibration is used and the instrument is set to print out sample concentrations direct. The photometer is provided with a linearizer to correct non-linearity in curves concave relative to the concentration axis. Deviations from linearity of up to 15% of the maximum extinction value can be corrected.

To provide for deproteinization a 4000-rpm centrifuge capable of treating 100 samples in 3 min is provided. To avoid the need for balancing, the reaction-tube rack of the preparation unit is made up from a frame and four tube holders, two holding 20 tubes and two holding 30 tubes each. Tubes can therefore be mounted in the centrifuge in two sets of 50. The centrifuge samples are then returned to the preparation unit for completion of the analysis. To avoid confusion of sample identity during manual use of the centrifuge a locating system is provided to ensure a unique rack-mounting geometry.

The 'Analmatic' can be used to process small and large sample batches, the upper limit being 100. A wide range of physiological fluids can be treated, the only limitation being that sample should not be so viscous as to prejudice operation of the transfer probes.

3.2.2 The 'Mecolab'

Joyce, Loebl and Co. Ltd. introduced the 'Mecolab'[2] in 1968. It is designed to perform rapid clinical analyses on sample batches of limited size. It is capable of meeting the needs of small and medium sized hospitals and clinics where the sample load is insufficient to utilize the full capacity of a larger fully automatic analyser and therefore does not justify the cost of the latter. The 'Mecolab' processes samples in batches of 15; each individual stage is performed automatically but transfer of sample between the various units is done manually.

The samples are processed in a turntable which is transferred from one analytical module to the next. It is indexed in such a way that it can only fit one way into the various modules, thus ensuring that the samples cannot be treated in the wrong order. The 'Mecolab' comprises five modules, a sampler, reagent-addition unit, a centrifuge, a supernatant control unit and an autocolorimeter. The turntable holds 15 sample-tubes in an outer ring and a concentric inner ring contains 15 tubes into which a fixed volume of sample is transferred for the analysis. The inner ring is detachable and fits into the centrifuge.

The sampling unit performs two functions, pipetting a small volume of sample into the reaction tube and adding up to three reagents, one of which washes the sample from the pipette. The volume of sample taken is very small, as little as 15 μl, which permits several different analyses to be performed on an initially small sample of blood or serum and deproteinization can often be avoided at the high dilutions possible (2000:1). The use of such small samples places critical emphasis upon the reproducibility of sampling. Part of the sampling action therefore consists of wiping the outside of the sampling probe free of adhering liquid and 240 samples can be pipetted in an hour with an error of less than 1%. The sampling cycle occupies 15 sec, at the end of which the turntable rotates to the next position. The normal programming cycle causes reagents to be added simultaneously, but for special purposes, such as kinetic studies, the programme can be modified to operate the dispensers differentially. In its standard format the 'Mecolab' is a four-channel instrument so that four successive analyses can be performed on each of the 15 samples without manual adjustment of the dispensing system; 0·12 ml of sample is adequate for four analyses.

The reagent-addition unit, also four-channel, is necessary only when more than two reagents are required or reagents need to be added following centrifuging. Its operational principle is the same as that of the sampling unit.

The supernatant control unit fulfils the function of transferring an appropriate volume of treated sample (3–5 ml) to the photometer sample-carrier. A single pipette is used and rinsing facilities, controlled by a push-button, are provided.

Colour measurement is by double-beam autocolorimeter; the optics and flow-through cuvette are made of quartz and wavelengths down to 250 nm can be used. Sample is transferred to the cuvette by vacuum from the sample table, four-fifths of the sample being used to wash out the previous one. On completion of the washing stage a valve closes and the remainder of the sample is used to fill the cuvette. It is retained there for 5 sec while the photometric measurement is made. The cuvette is then automatically emptied ready to receive the next sample. A sequencing switch on the sample table advances it to the next sample position. The total cycle time is 12 sec. Results are presented as a recorder trace and a retransmitting slidewire transmits a signal to an analogue-to-digital converter. Use of a balancing wedge in the colorimeter yields a signal proportional to absorbance, which is linearly related to concentration when Beer's law is obeyed. Therefore, provided that scaling factors are fed to the printer, a direct concentration read-out is obtained. To avoid unnecessary pen fluctuations on the recorder trace the recorder drive is switched off during these operations. The measurement is fed to the analogue-to-digital converter when a steady-state reading is obtained, usually after $2\frac{1}{2}$ sec. The printed value is an average taken over a 1-sec period. An advantage of this read-out technique is that minimal damping is required. The result is printed as three digits without a decimal point. Also printed are the batch number (one digit) and sample number (two digits).

3.2.3 THE 'SACAS'

The 'SACAS' semi-automated analyser is marketed by Beckman Instruments specifically for clinical analysis[3]. It has some formal similarities to the 'Mecolab' instrument but it possesses greater versatility in that it can provide up to 10 analytical channels operating simultaneously, though ultimately all samples pass through a single measuring unit which can be either a spectrophotometer or a flame photometer. Up to 300 measurements per hour can be performed. In order to maintain sample identification throughout the procedure each sample is accompanied by a patient-record card which serves also as the sample label. On completion of the analysis the results are printed, with appropriate test and patient identification, on to self-adhesive labels which can be attached to the patient-record card. Samples are transported through the analyser in tubes contained in racks which are both colour-coded and pin-coded, the former for test identification and the latter to ensure correct sequencing of sample tubes throughout the analytical stages. In addition the individual racks can be loaded into a specially designed centrifuge head. Since the centrifuge is a separate module, samples may be centrifuged before analysis (to produce serum or plasma) or during the analysis (as a physical separation stage). The racks are moved from one analytical stage to the next in transport modules which can be assembled in a series of parallel configurations depending on the analytical requirement.

The mode of operation of the 'SACAS' analyser is as follows. A rack of centrifuged samples is loaded into the dispensing unit, which removes sample aliquots of preselected volume. The dispenser operates on 10 samples simultaneously and can take up to 10 separate aliquots from a given sample. The operational accuracy of the dispenser is stated to be better than $\pm 1\%$. All racks of one colour are collected on a pin-coded process tray and transported to the process unit. This comprises a series of modules which automatically perform the functions of reagent addition, transfer and mixing. Each analytical method requires a separate process unit. The reagent addition module plugs on to the channel of the process module which provides the power to operate it. It consists of an electromechanical syringe which can perform three functions, addition of reagent, sample or supernatant transfer, and transfer of sample to the measuring unit. The volume range for reagent addition is 25–5000 μl and a repeatability of better than 0·5% is claimed for the syringe performance. Up to 6 reagent units can be plugged into a transport module and this therefore sets the limit on the number of reagents that can be added at one time. Also attached to the transport module is a high-speed impeller for mixing the contents of each tube. Where incubation or centrifugation is required the racks are transported to a constant-temperature water-bath or to the centrifuge. Finally the samples are transferred, by the electromechanical syringe, to a micro flow-cell in a Beckman DB Spectrophotometer. Both single- and dual-flow cells are provided, the latter for simultaneous blank correction. Alternatively the treated sample is fed to a Beckman 105 flame photometer for the determination of sodium and potassium. Results are presented in concentration units on a digital printer.

A control unit, based on self-programming logic units, performs the sequencing operations in accordance with the requirements for each individual analysis. It also provides visual and audible warning of completion of run and of instrument malfunction.

3.3 Fully Automatic Discrete Sample Analysers

Development of fully automatic discrete sample analysers has advanced to a stage where a number of proven commercial products are now available. They vary considerably in their design, mode of operation, sample capacity and versatility, but they have all been developed in response to the increasing demand for clinical analysis. For the purpose of this presentation the analysers are considered in two broad groups, those which offer chemical or physical separation facilities in addition to sampling, reagent addition and colour measurement, and those which do not. The incorporation of automatic separation facilities into discrete sample analysers poses formidable design problems and inevitably adds additional mechanical complexity. It is fortunate that, in the clinical field, most of the commonly used analyses do not require chemical pretreatment; in consequence the lack of reliable separation facilities has not proved a major drawback in the use of discrete sample analysers, and in general their ability to process large sample numbers outweighs their limited analytical applicability. Nevertheless more widespread acceptance of automatic discrete sample analysers is dependent upon the development of techniques for performing more complex analyses. Some progress, reported below, has already been made and further progress can be expected. Hitherto the design approach to discrete sample analysers has been to seek mechanical means of performing the analytical stages exactly as they are done manually. Evidently a more novel approach is called for and recent developments suggest considerable future activity in this direction.

3.3.1 Analysers Without Separation Facilities

Several well-established systems exist and their general principles are set out in this section.

3.3.1.1 *The 'AC 60'*

Pye Unicam have developed the AC 60 sample processing system to be compatible with their SP range of spectrophotometers, the SP600 visible range spectrophotometer (340–1000 nm), the SP 500 ultraviolet and visible instrument (186–1000 nm) and the double-beam SP 800.

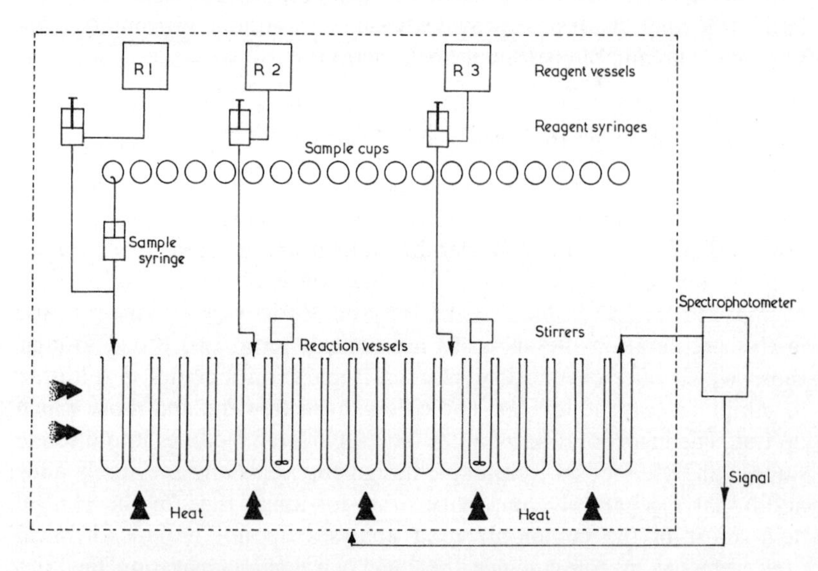

Fig. 3.3 Schematic configuration of the AC60 sample processing system. *Reproduced by permission of Pye Unicam Ltd.*

The functions performed automatically by the AC 60 are shown diagrammatically in Fig. 3.3, namely sampling, reagent addition, heating, stirring, transfer to the flow-through spectrophotometer cell and return to the sample vessel. The AC 60 is a single-channel instrument but change-over to another analysis can be made in a few minutes by modifying the sampling and reagent addition facilities. The full capacity of the unit is complete processing of 120 samples per hour; replacement of analysed samples in the rack by new ones need not await completion of the run.

The 120 sample-cups and reaction-tubes are mounted in parallel on a continuous rectangular thermostatted track which mechanically moves each sample to the sampling and reagent-addition stations. The rate of movement is maintained constant and the timing sequence for reagent addition is determined by appropriate positioning of the sampling and reagent-addition stations round the track. The selection of sample and reagent volumes is predetermined and is achieved by means of a preformed metal template. Four stations are provided, one for sampling, two for reagent addition and one for transfer of treated sample to the spectrophotometer. Stations not required can be switched out. Each is pneumatically operated by a compressor and vacuum pump. The sampler withdraws a preset volume (normally 10–300 μl) from the sample-cup and transfers it to the adjacent reaction-tube. The reagent-addition stations dispense an appropriate volume of reagent and mix the solution with a mechanical stirrer. The first reagent addition serves to wash the sample into the reaction-vessel. A dipping probe transfers the required volume (4·0–4·5 ml) to a flow-through spectrophotometer cell of 5-mm optical path. It has a low residual volume to permit efficient washing of the cell by the incoming sample (half the sample volume is used in cell-washing to minimize carry-over between samples). Carry-over is said by the manufacturers to be less than 1·5% of the difference in absorbance between successive samples. After measurement of the sample absorbance the sample is ejected into the reaction-tube by the transfer module. Blank solutions are processed at the beginning of the run and the spectrophotometer zero-adjustment performed while the blank is in the flow-cell; thereafter each sample passes through the spectrophotometer without further operator-attention. Results are shown on a recorder trace with a digital display or as a print-out. A typical recorder response for inorganic phosphorus in blood serum, determined by using the AC 60 in conjunction with the SP 600 spectrophotometer, is shown in Fig. 3.4.

Automatic enzyme analyses can be performed by using a slightly modified AC 60 system in conjunction with an SP 1800 ultraviolet spectrophotometer; this combination is designated the AC 1800. Samples enter the AC 60 processing unit in plastic cups, and a measured volume is

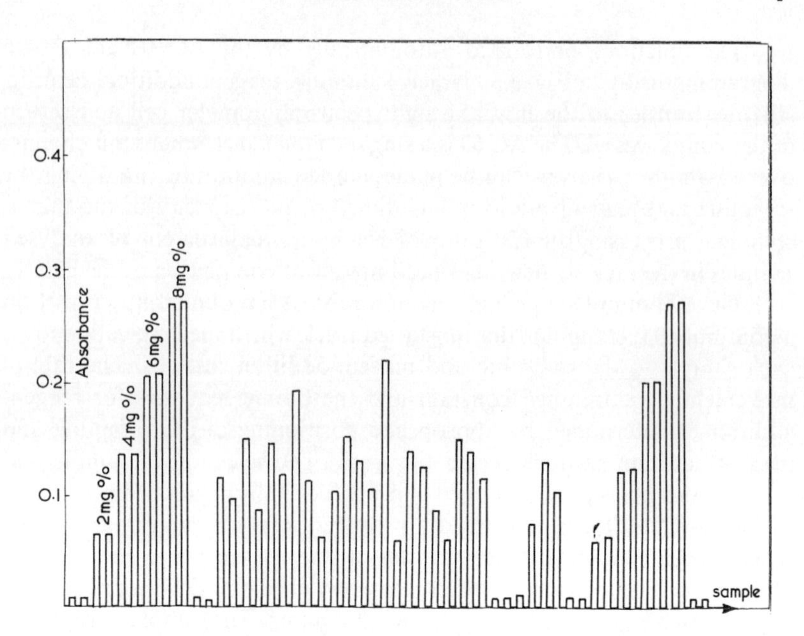

Fig. 3.4 Typical response of the AC 60: determination of inorganic phosphorus in blood. *Reproduced by permission of Pye Unicam Ltd.*

withdrawn at the first station and washed into a reaction vessel alongside the sample cup. Thereafter buffer, reagents, substrate and co-enzyme are added at subsequent stations. The reaction is initiated upon addition of substrate, and to allow for the possibility that the initial phase of the reaction is non-linear the time delay before addition of co-enzyme can be varied so that a linear rate may be established. This is achieved by movement of the addition station along a timer-bar. The solution is stirred mechanically after each reagent addition. A dipping probe transfers the reacting solution to a thermostatted cell in the SP 1800 spectrophotometer, which can operate over absorbance ranges of 0·05–0·10 or 0·2–0·5; automatic switching between them is provided should full scale on the more sensitive range be exceeded. The absorbance change is plotted on a flat-bed recorder. The length of time for which the reaction is studied can be varied from 5 sec to 5 min by an adjustable timer. This is incorporated in a programmer, AC 62, which also provides facilities for automatically resetting the zero between sample measurements, automatic absorbance range-switching, and adjusting the full-scale deflection on the recorder during the initial setting-up procedure.

The versatility of the AC 60 automatic processing unit has been increased by the provision of two interfacing kits to enable it to be used in conjunction with a range of Pye Unicam spectrophotometers. One provides facilities for linking it with the SP 500 and SP 600 spectrophotometers and the other performs a similar function in respect of the SP 1800 and SP 8000 models. Applications of several of these combinations have been described. In the pharmaceutical field methods have been developed for the assay of trypsin and chymotrypsin activity, for the assay of tetracycline[4] and for the assay of penicillin[5]. In the latter case coefficients of variation of the order of $\pm 1{\cdot}5\%$ were obtained. In food analysis the automatic determination of glycolytic intermediates has been successfully accomplished[6]. Methods are also available for the determination of aluminium, manganese and silicon in mild and carbon steels[7]; coefficients of variation are similar to those achieved by the corresponding manual procedures.

3.3.1.2 *The 'Autolab'*

The 'Autolab' was developed in Sweden and is marketed by AB Lars Ljunberg and Company. It is a single channel instrument capable of processing 240 samples per hour. A motor-driven chain is used to transport samples to the sampling and reagent addition points. The chain links consist of interconnecting plastic tubes which hold test tubes of 4, 7 or 10 ml capacity. Two chain systems are provided, one carrying the sample and an initially parallel one containing the corresponding tube in which the analysis is performed. The chain geometry for the complete instrument, which is modular in construction, is shown in Fig. 3.5. The chain is driven at constant speed and flexibility of time sequencing between analytical stages is achieved by the spacing of samples between the sampling and analysis stations.

When the sample reaches the sampling station an aliquot of sample is transferred to the adjacent reaction tube. Also the first reagent is added.

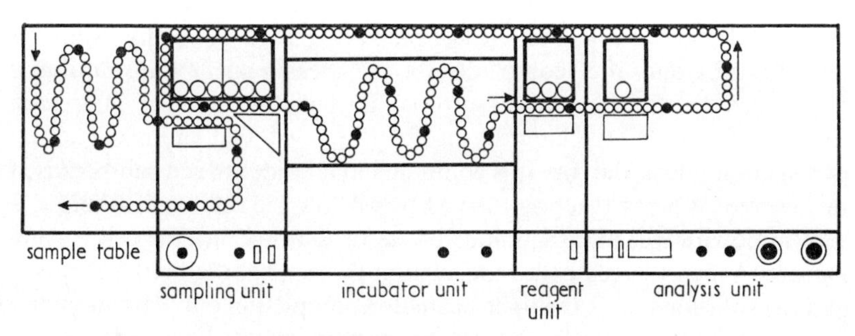

Fig. 3.5 Schematic configuration of the 'Autolab' automatic colorimeter. *Reproduced by permission of AB Lars Cjundberg & Co.*

The sampling is performed by a pipette attached to a vertically mounted precision syringe. The reproducibility of the syringe is 2 μl for a 2 ml sample. The reagent is added through a second syringe which is fitted with mechanically operated valves at its inlet and outlet; the inlet is connected to the reagent supply bottle and the outlet to the sample pipette. By opening and closing of the two valves in turn the reagent is drawn into the syringe on the down-stroke and ejected through the pipette into the reaction-tube. By arranging that the reagent volume is at least double the sample volume, efficient washing of the pipette is obtained. In order to remove droplets of liquid from the outside of the pipette it is drawn, at the end of the sampling operation, through a Teflon ring fitted with a groove in its inner circumference. The groove is connected to a suction pump at one point and to a water reservoir at another. In this way a stream of water is kept flowing through the groove and this cleans the outside of the pipette. For samples of blood serum the carry-over between successive samples is claimed to be less than 1% of the concentration difference between them. On completion of the sampling and reagent-addition operations the chain is advanced to the next position.

The chain then moves to the incubator unit which is a thermostatically controlled water-bath. The plastic tubes carrying the reaction-tubes are bottomless so that on reaching the edge of the bath the reaction-tubes drop into the water. Incubation times between 2 and 30 min can be selected according to the number of chain sections between the sampling and analysis unit. Where incubation is not required the incubator unit can be omitted. A second reagent-addition station is provided after the incubator, and up to three reagents can be added by valve-operated syringes. At each reagent-addition point mechanical stirring can be performed by automatically lowering stirrers into the reaction-tubes.

The photometric measurement is carried out in a single-beam filter photometer; narrow-band interference filters are used to select an appropriate waveband in the visible spectrum. The measuring cell is cylindrical and has a 10-mm optical path; it is filled by suction from the reaction-tube and after completion of the measurement the solution is discharged to the same reaction-tube. To facilitate transfer of treated sample into the cell the latter has a tip at its lower end and is connected to a suction line at the top. It is contained in a holder which can be raised or lowered. When a reaction-tube arrives below the cell the cell-holder is moved downwards and suction of 1·5 ml of solution into the cell is commenced. When the cell is full the suction is stopped, the holder is raised and the solution enters the light-beam. On completion of the photometric measurement the holder is lowered, the solution is ejected into the reaction-tube and finally the cell-holder is raised again. Setting-up and calibration of the photometer involves the following stages before the sample run is started. With the cell in the upper position and empty an 'air blank' is taken, and the signal is stored in a memory. Then a blank solution is inserted and the resulting signal is balanced manually against that in the memory, thereby producing a zero for the print-out unit. A standard solution is then introduced into the cell and the signal adjusted to yield a digital value corresponding to the concentration of the standard. Samples are then fed in by the automatic operation of the instrument and the digitized signal corresponds to the sample concentration. Both the sample number and analytical results are printed. The digitized results can also be fed to a paper-tape or card punch or direct to an on-line computer.

3.3.1.3 *The 'Clinomak'*

The Clinomak range of discrete automatic analysers[8] is of Italian design (Polimak s.r.l.). The essential components are a sampling system and a continuous interference-filter colorimeter. A digital print-out unit, the Data-mak, is also available.

The sampling unit is a circular plate driven by a motor and has a capacity of 90 samples. Samples are contained in an inner ring of cups and the sampling mechanism transfers a preset aliquot of the sample to a corresponding glass cuvette in which the remainder of the analysis, including the colorimetric measurement, is carried out. These cuvettes are mounted in an outer concentric ring. Sampling is performed by aspiration through a needle held in the sampling arm. It is operated by a high-precision piston pump having adjustable settings which enable preset sample volumes of 10, 20, 50, 75 and 100 μl to be taken. After the sample has been delivered the first reagent is added through the sample needle to wash the sample through and minimize carry-over from one sample to the next. To complete the cycle, 20 μl of air are aspirated through the needle to ensure separation of samples. Other reagents can be added at subsequent stations around the periphery of the sampling plate. Each reagent-dispenser is operated by electromagnetic valves. Flow of reagent is started by activation of the valve and the volume of reagent added is controlled by the time for which the valve remains open. The electric timers can be preset from the instrument controls. It is necessary to calibrate the control settings externally to determine the volume of reagent delivered. Although the 'Clinomak' is a single-channel instrument, the change-over to another method is rapid, involving only the disconnection of the reagent lines from the valves and connection of the new reagent reservoirs.

Treated samples pass individually through the light-path of the colorimeter, the time interval from sampling to colorimetric measurement being 7 min. It is a single-channel unit capable of operation over the wavelength range 400–700 nm, wavelength selection being achieved by using a continuous interference filter. For enzymatic analyses a UV filter attachment is available which enables a wavelength of 366 nm to be utilized. The colorimeter cycle time is 12 sec, the same as the sampling and reagent-addition time, so that the maximum output of the instrument is 300 measurements per hour. This does not correspond to 300 analyses per hour; because the colorimeter is a single-channel device certain of the sample positions must be reserved for blank and reference solutions. However, the 'Clinomak' colorimeter has two advantages; first each sample is measured in separate cuvettes and interaction between samples is eliminated, and secondly the measured sample is not discharged to waste, so samples may be remeasured if necessary. Results are presented on a recorder or printed at 12-sec intervals on the Data-Mak printer, which can be adjusted to present results as sample concentrations.

The automatic operation of the analyser is synchronized and timed by an electric programmer. Two programming devices are inserted into the outer ring of the sampling plate, one before the first sample and one after the final sample. These control the number of samples examined, the reaction times and the automatic operation of the individual analytical units, including recording of results, and ultimate shut-down of the instrument.

3.3.1.4 *The 'Robot Chemist'*

The Warner-Chilcott 'Robot Chemist' was introduced in the U.S.A. market in 1966 as a general-purpose automatic analyser. It is a single-channel instrument capable of analysing 120 samples per hour.

Samples to be processed are contained in test-tubes held in racks; the rack-handling mechanism advances the tubes sequentially to the dispensing point and transfers depleted sample-tubes to a storage area before their removal from the instrument. The sample-dispenser is mounted on a swinging arm which is first positioned over the sample-tube and the appropriate volume of sample (from a few μl to 12.5 ml) is withdrawn by means of a piston pump with micrometer control settings. The dispensing arm then moves across to a turntable and discharges the sample aliquot into one of a series of tubes. The quoted error of sample pick-up is $\pm 1\%$ of the volume setting or ± 1 μl, whichever is the greater, with a reproducibility of $\pm 0{\cdot}2\%$ or ± 1 μl whichever is greater. Diluent, or the first reagent, is added through the same pipette to ensure that the latter is cleansed of sample. A diluent-to-sample ratio of at least 5:1 is recommended. The reaction-tubes held in the turntable are immersed in a water-bath and incubation can be controlled at temperatures from ambient to 95° C. Rotation of the turntable brings each tube in turn to the reagent-addition units, of which up to 7 can be incorporated in the standard instrument. Preselected volumes of reagents are dispensed by piston pump, the available volume range being up to 12·5 ml. Reagent delivery to an accuracy of ±0·5% or ±2 μl, whichever is greater, is claimed together with a reproducibility of ±0·2% or ±2 μl whichever is greater. High-speed stirring facilities (4000 rpm) are provided on the turntable unit; the stirrer blades are of non-wetting Teflon and in consequence carry-over between samples is minimized.

Spectrophotometric measurements are made in a grating spectrometer having an optical range of 340–1000 nm. The optical path contains a beam splitter which enables the sample absorbance to be compared with that of a reference solution. The cuvettes are glass tubes fitted with plungers attached to pistons and withdrawal of the plunger draws sample into the tube. The tubes are washed free from previous sample by discharging the first few volumes. Absorbance readings are digitized and printed-out, three digits being allowed for sample identification and four for the absorbance reading. All sequencing operations are controlled by a matrix-board programmer.

3.3.1.5 *The C4 Automatic Analyser*

Designed specifically for clinical analysis, the Perkin-Elmer C4 is a 4-channel instrument capable of performing 120 analyses per hour when used as a single-channel instrument. Since all treated samples pass through the same spectrophotometer, the capability in terms of completed samples per hour is reduced two-, three- and four-fold if the analyser is operated as a two-, three- or four-channel instrument respectively.

The analytical reactions are carried out in polypropylene tubes of approximately 3-ml capacity mounted in a concentric 4 × 60 configuration on a turntable. The tubes can be incubated in a water-bath at temperatures from 15 to 60° C. Samples are presented to the analyser in tubes accompanied by a punched card for patient identification, previously prepared in the hospital ward. The cards are fed to the analyser with the samples and they are interpreted simultaneously with the photometric measurement; the analytical results in concentration units, patient identification and test identification are presented as a digital print-out. Alternatively results may be processed by digital computer in either on-line or off-line mode.

Sample aliquots, up to four per sample, are delivered to the reaction-tubes by means of a valveless piston pump. The number of sample aliquots is selected by a programmed control unit. The sampling pump has a stepped-motor drive and before each sample is delivered it is flushed with a volume of sample. The size of aliquot can be independently programmed for each of the four channels. Volumes from 0 to 50 μl can be selected in steps of 5 μl, from 60 to 95 μl in steps of 15 μl, from 100 to 500 μl in steps of 50 μl and from 600 to 950 μl in steps of 150 μl. The accuracy of sampling is $\pm 2\%$ for the range 10–40 μl range and $\pm 1\%$ for the range 45–950 μl. The precision of delivery is 1% for the 10–40 μl range and 0·5% for the 45–950 μl range. Reagents are added by means of valve piston pumps having a mechanically adjustable stroke. Up to 16 such pumps can be mounted on the turntable, four for each channel. They are normally activated automatically from the programmer but manual operation is also possible by over-riding the automatic program. The delivered volume is continuously adjustable between 30 and 2500 μl. Over the volume range 30–95 μl the accuracy of delivery is $\pm 2\%$ and the precision is 1%; for the 100–2500 μl range the figures are $\pm 1\%$ and 0·5% respectively.

Absorbance measurements are made with a Perkin-Elmer–Hitachi grating spectrophotometer which can be used over the wavelength range 340–600 nm in the automatic mode or 200–950 nm in manual operation. The spectrophotometer is fitted with an 8-position cell carriage which permits samples on all four channels to be measured against individual reference solutions.

3.3.1.6 *The 'Multichannel* 300'

Vickers Ltd. have designed and developed the 'Multichannel 300' analyser primarily to meet the need for detailed analysis of blood samples. It is a multi-channel instrument, the standard model being provided with 6, 8 or 12 individual channels, though alternative configurations are possible because once a sample is received each channel functions as a separate unit. The instrument has been conceived to discharge high workloads such as would be met in large hospitals or in institutions concerned with health-screening of the general population, as opposed to hospital patients only. In theory there are few ultimate limitations to the number of channels which could be incorporated. Apart from local considerations such as space, complexity and cost, the more general limitations are likely to be the number of subsamples which can be taken from the initial sample and the number of methods which have been fully proven for automatic operation. With the exception of a central sample-distribution line, there are no other common facilities in the chemical stages of the analyser, each channel having its own individual reagent-addition and photometric-measurement units. Since each channel is capable of processing 300 samples per hour, a 12-channel instrument is theoretically able to yield 3600 results in every hour. In an analyser of this type the reliability and reproducibility of individual components is of the utmost importance in order to minimize potentially high maintenance costs. Nevertheless the approach is advantageous in that a breakdown of one channel is confined to that channel and the results from others are unaffected.

Before individual analyses are commenced, the blood samples, contained in labelled closed rectangular vials, are centrifuged and then loaded into magazines. Unequivocal coding of samples is of paramount importance when processing the number of samples of which the Multi-channel '300' is capable; a 12-digit system is used, six for the patient's name and six for local departmental identification. The sample-loading mechanism holds several magazines. The vials are delivered from the magazine to the next stage by a system of pawls which open a gate and allow the vial, held against the gate by a constant-tension spring, to be pushed out. As soon as a magazine is emptied the next one is automatically indexed into position. An optical sensing technique rejects vials which are underfilled or which contain haemolysed or lipaemic samples. Rejected samples are discharged to separate vial holders. The rejection system can be manually over-ridden if a sample that has been designated as unsatisfactory may nevertheless be suitable for a limited range of analyses. Samples accepted by the checking procedure move to the primary sampling and dilution stage. The sampling-probe unit transfers a volume of sample, together with diluent, into an open-topped vessel. The vertical positioning of the sampling head is controlled by a cam; when it is lowered the probe breaks the vial seal and draws in sample. The sampling head is in two parts in the form of a rotary valve. The lower part, which remains stationary, is connected to the supply of diluent. The upper part carries two capillary probes and rotates in half-turns. Thus one probe dips into a vial to withdraw sample while the second is delivering sample, followed by diluent, taken from the previous vial. The holders for diluted sample are contained in a distribution assembly which serves the individual analysis channels. When the diluted sample reaches an analysis channel, termed a reaction console, a secondary sampler and diluter removes an aliquot into a cavity in a thermostatted reaction rotor. All further treatment is performed in the cavity. Each rotor contains 60 such cavities, or 120 if blank facilities are required. The inner surface of the cavities is non-wetting, and chemically inert. The time-sequencing is such that a complete rotation of the rotor takes 12 min. During rotation each cavity passes reagent dispensers situated round the periphery; the geometric positioning of the dispensers depends on the analysis being performed and the incubation times required. At the final sample position is situated a suction probe which transfers the sample to the colorimeter cuvette (or to the atomizer of a flame photometer). The cavity is subsequently washed free from any adhering residue and dried in readiness for the next sample.

The colorimeter provides double-beam operation so that simultaneous blank corrections can be carried out. This eliminates errors due to variations in the light-source emission. The double beam is obtained from a single lamp by using a beam-splitting prism. Results are presented on a teletype together with a punched paper tape which can be used for further data processing. Results are presented accompanied by the vial code which is obtained by moving the sample vial to a reader which detects the digits impressed on the label.

3.3.1.7 *The GeMSAEC Fast Analyser*

The automatic analysers so far described in this chapter process a batch of samples sequentially. The GeMSAEC system (General Medical Sciences – Atomic Energy Commission) developed at Oak Ridge National Laboratory is designed for very fast rates of processing of clinical samples and treats a batch of samples in parallel. It utilizes centrifugal force to mix solutions and transfer them to cuvette rotors which intercept the light-beam of a photometer as they rotate[9].

Fig. 3.6 shows the basic principle of GeMSAEC. The sample and reagent holders are fabricated from Teflon, as is the cuvette rotor in which the photometric measurement is made. The design of the indentations for holding sample and reagents is such that no mixing occurs when the apparatus is stationary but on spinning the rotor at 500–1500 rpm complete mixing occurs within 10–15 sec, and the mixture is transferred to the cuvette. The equipment is designed to handle small volumes of sample and reagents, usually of 0·2–0·3 ml. In early studies a 15-cuvette rotor was used but subsequently a 42-cuvette rotor has been designed[10]. At the high speeds of rotation each sample is photometrically measured over a very short period of time, about 0·05 sec. Therefore if one cuvette contains a reagent blank solution a simultaneous double-beam analysis of 41 samples is obtained. Allowing for complete mixing of sample and reagents to occur, the analytical results on a sample batch can be obtained some 30 sec after the start of the experiment. Drainage and wash-out of the rotor take a further 90 sec. The rotor unit can then be removed and replaced by another one ready loaded with sample and reagents.

The GeMSAEC analyser can be operated either with premeasured

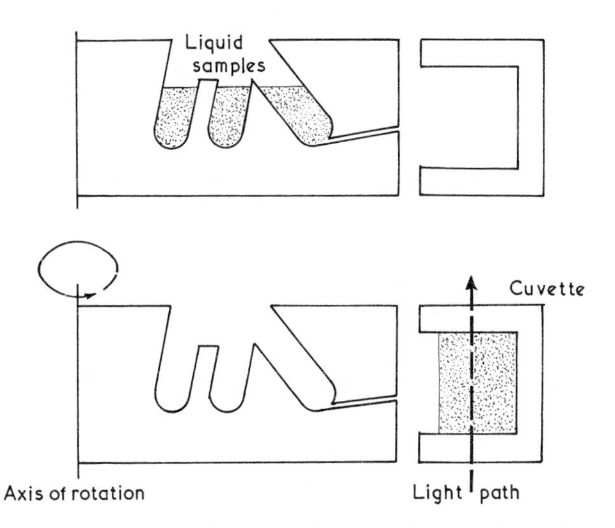

Fig. 3.6 The GeMSAEC automatic colorimeter: method of centrifugal mixing of sample and reagents, and transfer to cuvette.
Reproduced by permission from Anderson[9].

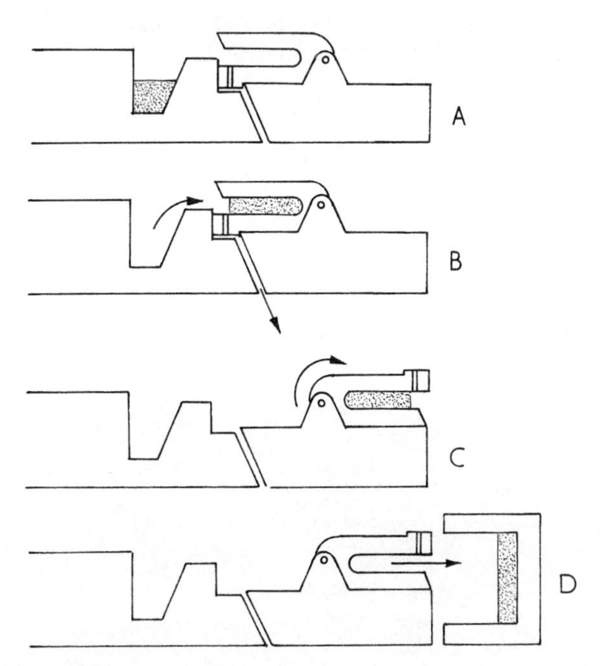

Fig. 3.7 The GeMSAEC automatic colorimeter: centrifugal method of fixed volume sampling and transfer.
Reproduced by permission from Anderson[9].

volumes of sample and reagent, or with unmeasured but sufficient volumes. In the first case standard automatic micropipettes can be used to load the rotors. In the second case the equipment must be modified to enable a fixed final volume to be measured and transferred to the cuvette. One means of achieving this is shown in Fig. 3.7. On spinning the rotor the solution is forced into a transfer tube and excess liquid is drained away. The measuring tube is then mechanically turned through 180° and the measured volume delivered to the cuvette. Alternatively a series of siphons can be used to measure and transfer solutions as shown diagrammatically in Fig. 3.8.

Since all the cuvettes intercept the light-beam over a period of about 0·05 sec the results can conveniently be displayed on an oscilloscope and then photographed to provide a permanent record. For the workloads for which GeMSAEC is designed, computer processing of results is clearly justified and interfacing of the system with a PDP8/I computer has been described[11].

The ability of GeMSAEC to produce readings within seconds of the initiation of a reaction makes it ideally suited for the study of kinetic systems. With the aid of suitable software[12], plots of extent of reaction as a function of time can be generated and the approach has been successfully applied to enzyme kinetic studies[13].

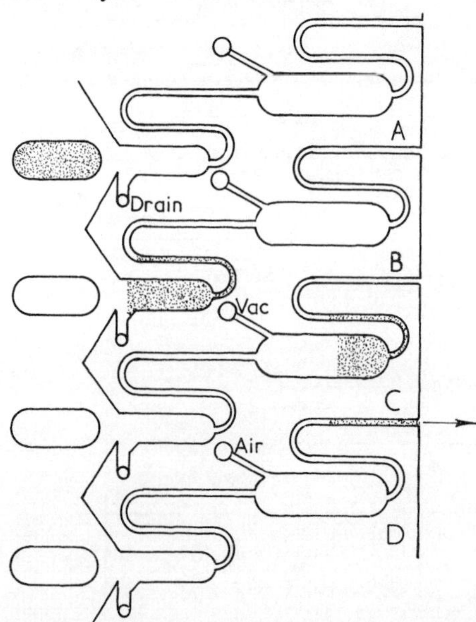

Fig. 3.8 The GeMSAEC automatic colorimeter: siphon method of fixed volume sampling and transfer.
Reproduced by permission from Anderson[9].

3.3.2 Fully Automatic Analysers with Separation Facilities

The useful range of an automatic analyser can be considerably enhanced if provision can be incorporated for performing physical or chemical separation facilities. As mentioned in the introduction to section 3.3 this has so far proved mechanically difficult in discrete-sample analysers. Nevertheless some progress has been made and three commercially available automatic analysers offering some degree of separation capability are described in this section.

3.3.2.1 *The 'DSA 560'*

Beckman Instruments introduced the 'DSA 560' in 1968. It is a discrete-sample analyser primarily intended for analysis of blood and having particular advantages when the available sample volume is small, as is frequently the case with geriatric and pediatric patients. In addition to automatic sampling, sample transfer, reagent addition and colour measurement, it provides an automatic filtration facility. It can be used as a single- or dual-channel instrument; in the single-channel mode the sample-processing rate is 120 per hour and when two channels are used the rate is 80 per hour. The lower rate for dual-channel operation is imposed by the time taken in operation and subsequent electronic restabilization of the automatic wavelength changer.

Samples are presented to the analyser in a 40-position rotating tray. Rotation of the tray brings each sample in turn to the dispensing station at which sample aliquots of between 5 and 50 μl are withdrawn by a pump and transferred to a plastic sample-capsule in which the subsequent analytical reactions are performed. Coefficients of variation for the reproducibility of sampling are $\pm 0{\cdot}5\%$ for volumes in the range 5–20 μl and $\pm 0{\cdot}2\%$ for volumes greater than 20 μl. The sample-capsule contains 5 separate cups which may be used for sample or blank measurements. These are automatically moved across the analyser to the reagent-addition stations, the location of which can be varied to enable different incubation times to be used. Up to 7 reagents may be added at each channel. The plastic cups are kept in a water-bath, the temperature of which can be controlled over the range 35–50° C.

The filtration stage is carried out in one of the 5 cups which is larger in volume than the other four. It accommodates a filter 'hat' which is automatically lowered into it and suction is applied to the top of the 'hat'. The filtrate can then be transferred to another cup by a dipping probe. An air jet provides stirring after reagent addition. On completion of reagent addition the sample is pumped to a double-beam, two-channel colorimeter which permits absorbance measurements to be made against a reference solution over the wavelength range 340–700 nm. Results may be presented in analogue form on a chart recorder or digitally on a teletype printer which can simultaneously prepare a punched paper tape which can be processed by an off-line computer.

3.3.2.2 *The 'Chematic'*

The 'Chematic', a product of James A. Jobling Ltd., is a fully automatic single-channel discrete-sample analyser which provides for phase separation by means of a semi-continuous automatic centrifuge[14]. Known as the microcentrifuge, it is designed primarily for liquid–solid separation as in blood deproteinization but has some value as a liquid–liquid separator. It comprises a hollow disc which can be rotated at 3000 rpm, developing about 330 *g* in so doing. The upper liquid phase is removed to a collecting vessel by a hollow scoop which traverses the disc. The lower phase, solid or liquid, is taken direct to waste through a side-arm. The phase-separation stage can be performed satisfactorily regardless of the relative volumes of the two phases because the removal of the supernatant liquid is terminated by the opening of an adjustable valve which is preset to coincide with the phase boundary and then diverts the heavier phase to waste. The centrifugation time can be varied to provide optimum separation for the analysis being performed, but for a run of similar samples it is fixed beforehand at the appropriate value. Centrifuge components are fabricated from nylon, borosilicate glass and PTFE to provide resistance to solvents, acids and alkalis. The centrifuge is further described in Chapter 10.4.

The 'Chematic' has two turntables mounted one above the other; both have a capacity of 32 groups of two tubes. Dispensing, diluting and reagent addition are carried out by three sets of dipping probes mounted beside the turntables on detachable plates; this configuration permits ready interchange of units when a different analysis is required. The microcentrifuge is situated between the two turntables, the lower of which is provided with a water-bath to provide temperature control of samples from ambient temperature to 90° C. Control is stated to be better than ±0·3° C at 37° C and ±1° C at 80° C. Dispensing, diluting and addition of reagents are achieved by using pneumatically driven syringes, the settings of which can be readily varied. Mixing of solutions is provided by the force at which reagents are ejected from the syringe. To minimize sample carry-over and cross-contamination all probes are thoroughly rinsed with diluent between samples, also the centrifuge and scoop are washed with water and 80% of the final treated sample volume is used to wash the colorimeter cell. Five ranges of dispenser and diluter syringes cover the volume range 0·01–5 ml. Repeatability of their operation is stated to be ±0·001 ml at 0·1 ml and ±0·01 ml at 2 ml.

The operations performed at the upper turntable are dispensing, diluting and (if required) addition of deproteinizing reagents. Samples are held in the outer of the two rings of tubes and the dispenser transfers a selected volume into the corresponding inner ring tube. Reagents are added at the same station. A 180°-movement of the turntable brings the treated sample to the second probe which transfers the mixture, with diluent, to the microcentrifuge. The probe unit which serves the lower turntable transfers the supernatant liquid to the outer ring tube and dispenses one sample into the adjacent inner ring tube. Up to three reagents can be added to this stage. The colorimeter transfer-probe is situated at 180° to the dispensing probe. It transfers the contents of the tube to the colorimeter, which is usually either a Joyce Loebl or an EEL Model 171 automatic colorimeter, both of which are single-channel instruments. Alternatively an atomic-absorption spectrophotometer or flame photometer could be used. Absorbance measurements can be presented on a chart recorder or printed out in conjunction with the sample identification number. The maximum rate of sample processing is 96 per hour, but both small and large batches can be conveniently handled. If centrifuging is not required the automatic analysis can be performed with the lower turntable only.

3.3.2.3 *The Automatic Clinical Analyser* (*'aca'*)

The Du Pont 'aca' is a fully automated colorimetric analyser of novel design[15]. Although the sample-processing rate is slower than that of the analysers described above, about 100 or 50 per hour, depending on the analytical method, it is extremely versatile in that the standard format can handle 30 different methods and this range is capable of extension to a maximum of 62. A computer is used to provide the sequencing operations for the various methods. All reactions are carried out in a plastic pack which contains the appropriate reagents for each method. Separation of unwanted components can be achieved by ion-exchange or gel filtration in a column contained in the top of the pack. The lower part of the pack forms the optical cell for photometric measurements. Thus, once the sample has been added to the pack there is no further transfer of sample or reagents and in consequence sample interaction is eliminated. The packs are individually transported by a chain drive through the various stages of analyser operation which include reagent mixing, incubation and photometric measurement.

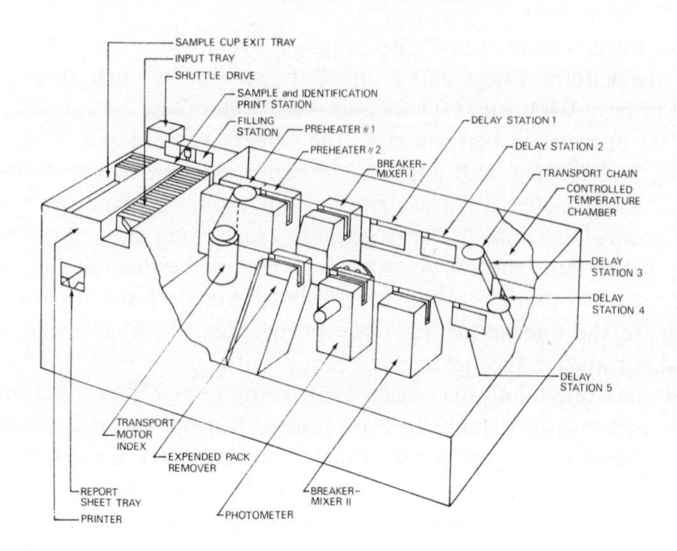

Fig. 3.9 Operational stages of the aca automatic colorimeter.

A schematic drawing of the 'aca' is shown in Fig. 3.9, and the design of reagent pack in Fig. 3.10. The packs are vertically supported on a header plate which has an access point for the sample and for diluent added by the analyser. In those packs in which a chromatographic step is carried out, the column is mounted horizontally across the header plate and two access ports are provided, one for sample and diluent immediately before the column and one for diluent immediately following it. The name of the test is written on top of the header plate for identification purposes, and embossed on it is a binary code for each analysis. The

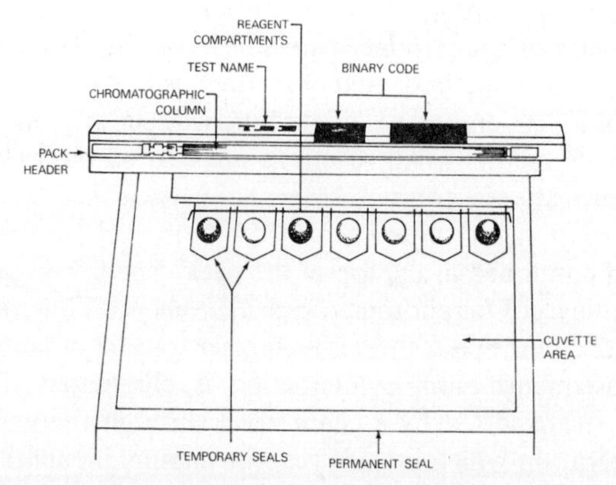

Fig. 3.10 Reagent pack for the aca automatic colorimeter.

analyser reads this code which selects the appropriate program from the computer. The program controls all subsequent analytical operations. Up to 7 reagents can be individually held in the pack by temporary seals which can be broken in programmed sequence by application of hydrostatic pressure.

Samples are introduced into the pack-filling station in cups, each cup being followed by the appropriate analytical pack. Attached to the sample cup is an identification card bearing the patient's name and any other information required. A spring-loaded follower feeds each sample in turn to the filling mechanism which transfers an aliquot of sample and diluent or buffer to the analytical pack. Here the analyser senses the binary code on the header plate to identify the test and to program the analyser with appropriate information including the volume of sample to be transferred, the selection of diluent and its volume, whether a chromatographic stage is involved and if so the choice and volume of eluent, and further details of the processing cycle. The filling system comprises a piston pump and eight precision microvalves having a total capacity of 5·0 ml. The piston is operated by a stepping-motor which withdraws or discharges 5·0-μl volumes at each stage. Its operation is controlled by the computer, the programming interval being 20 μl over the range from 20 μl to 3·0 ml. Of the eight valves one connects the pump to the pack, six are available for diluent supply and the other provides a water wash. Liquids are withdrawn or added through a Huber needle which can take up four positions, one for withdrawal of sample, one for addition of sample and eluent to the top of the chromatographic column, one for direct addition of sample and diluent to the pack and one to drain to permit washing of the needle. On completion of the filling cycle the spent sample-cups are removed to a tray and the prepared pack is moved into a preheating unit which brings the temperature of the pack liquid to $37 \pm 0{\cdot}1°$ C. Two heating units are used in series, the first electrically heated and the second air-heated; in each one heat-transfer to the pack is provided by two close-fitting metal plates. The pack then passes to the first chemical reaction stage which is performed by a breaker–mixer unit. This is designed to allow breakage of the seals containing reagents in four of the seven available positions, the remaining three being protected by a suitably designed bar. This impacts upon the pack and generates a hydrostatic pressure by forcing the liquid upwards, so breaking the seals and allowing the reagents, which may be solid or liquid, to mix with sample. Agitation is provided by oscillation of platens situated below the protector bar and in contact with the pack. The pack passes through five delay stations providing a preselected incubation time of 2 min 55 sec. Thereafter the pack passes through a second breaker–mixer to release reagents from the three remaining compartments. After an interval of 45·5 or 31·5 sec the pack is transferred to the photometer. Wavelength selection over the range 340–600 nm is provided by one of 12 narrow-band interference filters mounted on a wheel. The photometer unit houses a quartz–iodine lamp, a quartz beam-splitter, a regulator phototube and a measuring phototube. It also contains a cell-forming die which yields a reproducible cuvette between two quartz windows. It is pressure-operated, and in addition to forming the cell it also forms a pressure-relief cell away from the optical path. The photometer incorporates an analogue-to-digital conversion unit which is fed with calibration factors from the computer to yield results in concentration units. Results are reported on individual slips which contain, in addition to the analytical results and test identification, a photographic reproduction of the information card initially provided with the sample. This is obtained by an ultraviolet photographic unit which reads each identification card after the sample has been withdrawn at the filling station.

3.4 Continuous Automatic Analysers

Several examples of analysis of discrete samples by continuous methods have been detailed in Chapter 2 on automatic electrochemical methods. The principles apply equally to colorimetric methods, the treated sample stream being presented to a flow-through cell of a photometric device instead of to an electrode system. The continuous approach has the merit of mechanical simplicity in that both metering and liquid transport are provided by the pumping system. Nevertheless the accuracy and reproducibility of the pump are fundamentally important factors in determining method performance, as also is the design of the flow system to minimize surging and irregular dynamic flow conditions.

Within the field of automatic continuous colorimetry (and flame photometry and fluorimetry) the dominant influence has been the Technicon 'AutoAnalyzer', which has its origin in the work of Skeggs[16] on the automatic analysis of blood for urea nitrogen, glucose, calcium, chloride, alkaline phosphatase and acidity. His design of automatic analyser presented several novel features which have proved amenable to extensive further development and remain fundamental to the most widely used approach to automatic analysis by the continuous method. Detailed consideration of Skeggs's paper is omitted because the principles have been exploited and extended in the Technicon 'AutoAnalyzer' which is discussed more fully below; nevertheless the basic principles should first be recalled. Skeggs's studies predated the advent of the discrete-sample automatic analysers which have subsequently attained such prominence, particularly in biomedical analysis, and it is of interest that he considered the extensive automation of manual operations to be impracticable. Instead he designed and evaluated a continuous-flow system in which were performed the analytical operations and which also provided the means of sample transport without recourse to mechanical methods. The analyser comprised a series of modules each performing a specific function, e.g. sampling, sample transport, heating, dialysis and photometric measurement. The sample and each reagent were pumped along individual flexible plastic tubes by means of a multichannel peristaltic pump in which each tube, up to 8 in the original design, was aligned taut and parallel to the others beneath rollers chain-driven by a synchronous motor. Sample and reagents were merged in appropriate stages through T-connections and then pumped through the subsequent modules. Horizontally-mounted coils between the modules provided mixing by repetitive inversion of the liquid phases and the length of the coils governed the delay between stages. Heating and dialysis units, if required, were connected in series and the sample pumped through them. Finally the treated sample was pumped into a flow-through photometer cell and then to waste. Samples were placed in depressions around the circumference of a rotatable 'Plexiglas' plate and aspirated into the analyser by the action of the pump. The turntable advanced to the next position every two minutes, sample being aspirated throughout this period. The desired relative volumes of sample and reagents are obtained by selecting pump tubes of different internal diameters; consequently the analyser required no quantitative volumetric dispensing.

An operational feature of Skeggs's analyser which has proved of profound significance in the subsequent success of his concept is that, in addition to sample and reagents, air is drawn into the analyser through one of the pump tubes and produces segmentation of the liquid stream once it has been merged with it. This segmentation is retained through the succeeding stages of the analysis up to the photometric unit where the air is removed in a vented flow-through cell and a continuous solution phase is reformed. The introduction of air causes each individual sample to be divided into a number of small segments and this has several practical advantages. First, the air segments are responsible for maintaining a sharp concentration profile at the leading and following edges of each individual sample. In the absence of air the sample would 'tail' along the inner wall of the tubing thereby increasing the possibility of interaction between successive samples and increasing the time required for each sample to yield a steady-state reading at the photometer. As discussed below, both of these effects adversely influence the maximum rate at which samples can be processed. Secondly the presence of air bubbles promotes mixing of the phases. Each stream segment can invert efficiently as it rises and falls through each turn of the mixing coil. For maximum mixing efficiency the length of each liquid segment must be less than half the diameter of the coil. In addition the wiping action of the air along the tube wall discourages the build-up at the surface of residues from preceding liquid segments.

3.4.1 THE 'AUTOANALYZER'

The 'AutoAnalyzer' is the commercial manifestation by Technicon Corporation of the Skeggs automatic continuous analyser. It was introduced in 1957 and has found widespread use in almost every facet of analytical chemistry; its range and flexibility have been extended by the introduction of additional modules. The modular approach confers considerable flexibility but for certain requirements, notably biomedical analysis and process control analysis, more specific configurations have been developed. The number of literature references to 'AutoAnalyzer' applications runs into several thousands; over 1500 citations are listed in a bibliography[17] published by the manufacturers for the period 1957–1967. Method details can be conveniently expressed by a line diagram of the manifold construction and subsequent analytical stages.

The 'AutoAnalyzer' modules perform the following functions: sampling, pumping, separation of unwanted sample components, heating, detection and recording together with data presentation. In its initial form the 'AutoAnalyzer' consisted of one module for each purpose, but subsequently additional modules, which complement the original ones, incorporate improvements in technique and extend the applicability of the 'AutoAnalyzer' system, have been developed. For example the basic 'AutoAnalyzer' was limited to colorimetry in the visible-light range as the detecting method, but units for flame photometry, UV spectrophotometry and fluorimetry are now available. In addition to the advances emanating from the manufacturers, numerous workers have modified 'AutoAnalyzer' modules to meet specific requirements; several examples are quoted below. In principle the continuous-flow approach of the 'AutoAnalyzer' does not impose any limitations on the choice of detection technique other than those inherent in design compatibility. Consequently references can be found to the use, in conjunction with the 'AutoAnalyzer', of several detection techniques in addition to the commercially available ones. They include electrochemical detection techniques, discussed in Chapter 2, radiometry (Chapter 6) and flame-ionization detection (Chapter 7).

(*a*) *Sampling*

The first sampler module, Sampler I, comprises a plate with holes around the circumference to hold 40 plastic sample-tubes. It is fitted with a probe for withdrawing sample, which is done by lowering the probe into each tube in turn for a fixed time and then withdrawing it while the plate is advanced to position the next sample under it. Both the sampler-plate and the probe are operated, through appropriate cams and gears, by a motor which receives impulses from a cam-timer which controls the overall operation. The aspirated sample is drawn into the peristaltic proportioning pump whence it subsequently meets the reagents being pumped through the other tubes. While the sampler-plate is advancing to the next position the sample-probe aspirates air into the analyser. The sample-plate can be driven at three fixed speeds enabling 20, 40 or 60 samples per hour to be taken.

The Sampler I has several design shortcomings which, for some analyses, could be serious. By far the most significant is that no wash facilities are provided for the sampling-probe, the only cleaning between successive samples being provided by air drawn through it between sampling periods. In consequence interaction between successive samples can occur and incorrect results can be produced. The magnitude of this interaction is variable, being a function of the relative sample compositions in successive tubes. For a run of very similar samples the effect is small, but it can become unacceptably large when adjacent sample tubes contain widely differing concentrations of the species being determined.

Thiers and Oglesby[18] studied the interaction effect, using Sampler I, for determinations of carbon dioxide, chloride, potassium, sodium, urea, nitrogen and glucose. The interaction occurs between a sample and the one following it, the magnitude of the effect being proportional to the concentration of the preceding sample and independent of the concentration of the measured one. A correction can be applied by determining the extent of the effect as a percentage of concentration and applying this to each result. If alternate cups on the sampler are filled with water the interaction is effectively eliminated at the expense of halving the sample processing rate. The problem of interaction between successive samples is discussed more fully in section 3.6 below.

The ingress of air into the sample-line between samples is undesirable. Its presence distorts the regular pattern of sample-segments, the effect being particularly noticeable when large diameter sample-pump tubes are used. The surging so produced can reduce the level of discrimination between samples and thereby adversely affect results. This imposes a practical limitation of about 0·3 ml/min on the acceptable sampling rate. A further disadvantage which is smaller in magnitude and normally only significant when results of high accuracy are required, is related to the volume of sample placed in each cup. Because the sample-volume is not formally measured but is a function of aspiration-time there is, in principle, no need to fill all sample-tubes to the same volume; indeed the obviation of all quantitative volume measurement is claimed as an advantage of the 'AutoAnalyzer'. Thiers and Oglesby[18] demonstrated that the sample-volume withdrawn varies with the level of sample in the tube. The reason for this behaviour was not conclusively established but it may be conjectured that it is caused by the slow rate of insertion and withdrawal of the probe. The effect is eliminated by ensuring that all sample-tubes are filled to the same height.

A modification of Sampler I to enable it to feed two analytical lines simultaneously involves fitting a plastic holder to carry a second ring of sample-tubes and brazing a second sample-probe on to the existing one.

The Technicon Sampler II overcomes the limitations of Sampler I. The dipping probe mechanism has been redesigned to allow rapid insertion and removal from the sample-tube and the operational sequence modified to incorporate a washing stage between samples. A receptable is provided which contains water or other suitable 'blank' solution and the probe dips into it between successive samples. The time-ratio for aspirating sample and wash-liquid, the so-called sample-to-wash ratio, can be varied within the range from 6 : 1 to 1 : 6 by operation of a programmable cam. A second such cam controls the sampling rate, which, as with Sampler I, can be 20, 40 or 60 samples per hour. Whereas Sampler I accommodates only 2-ml sample-tubes, Sampler II can be used with 0·5, 2, 3 and 8-ml sizes. The capacity of Sampler II is 40 samples but a larger version is available holding 200 samples. This is of particular value where large sample throughputs are involved; indeed it was designed for use with the Technicon Automatic Blood Typing system.

An updated version of Sampler II, designated Sampler IV, incorporates several design advances which enhance its versatility. It has a capacity of 40 samples and can accommodate sample cups of 0·5, 2·0, 3·5, 4·0 and 5·0-ml capacity. The sampling program provides for the successive aspiration of sample, air and wash-liquid. The sampling rate can be varied from 30 to 90 per hour in increments of 10; for each rate separate interchangeable cams are provided which fit directly on to a 60-cycle synchronous motor. A 120 sample/hr cam is available for non-steady-state analyses. For each timing cam a choice of 11 sample-to-wash ratios is incorporated, spanning the range from 9 : 1 to 1 : 1. The wash-liquid receptable is fed from the proportioning-pump manifold to constantly replenish the solution and minimize interference from residues of previous samples. Sampler IV can be programmed to aspirate the same standard several times during the sampling sequence. This allows fuller utilization of the sampler capacity for processing samples. When the instrument-response is linear a single standard is adequate for setting up calibrations. An optional addition to Sampler IV is a rotary mixer to maintain homogeneity of serum samples before aspiration.

Where large numbers of samples are to be processed it is essential to eliminate the attendant risk of confusion between samples, especially in the case of clinical analyses. Positive sample-identification mechanisms are available[19, 20] for both Sampler II and Sampler IV. To each sample-tube is affixed a coded identification card; immediately following aspiration of the sample, the electronic reader identifies the code from the card and stores the information until the analytical result is ready for printing. The sample number is printed on the chart record for each sample and is also included in the digital print-out.

Technicon Samplers I, II and IV process samples in liquid or solution form, including absorbable gases (for which an absorption-column unit can be purchased). For solid samples the economic use of the 'AutoAnalyzer' is vitiated by the need to dissolve or suspend them manually before transfer to the sampler. To meet the requirement for automatic processing of solid samples, particularly for quality control analysis in the pharmaceutical industry, a solid sampler, termed the 'Solid-Prep'[21], was introduced in 1964. It overcomes some of the limitations of an earlier design[22] based on feeding individual tablets together with solvent into a Waring blender. The latter sampler provided automatic, timer-controlled mixing and automatic pipette sampling of the resulting solution or homogenate. However, the absence of intersample washing facilities renders the sampler susceptible to interaction between successive samples.

The Solid-Prep sampler is compatible with the 'AutoAnalyzer' system and its functions are briefly as follows. Preweighed samples are placed in plastic cups on a rotating turntable which is motor-driven at a selected speed which can be varied to suit the demands of the analysis. When a sample-cup is aligned with the homogenizer the cup is tilted and the sample descends through a hopper into the homogenizer. A preselected volume of solvent or diluent is ejected through nozzles mounted so that the cup and the hopper walls are thoroughly washed. The volume of solvent or diluent is adjustable in 1-ml increments over the range 50–200 ml by varying the stroke of a spring-loaded piston pump by using a dial on the control unit. Application of suction draws liquid into the pump and when the vacuum is released the spring ejects it. A chopper blade is rotated at 10,000 rpm to promote efficient mixing and the sample is either dissolved or brought into homogeneous suspension. The prepared sample is then aspirated from the homogenizer into the analytical train by the proportioning pump and is simultaneously segmented with air. During the aspiration stage the speed of the chopper blade is reduced to prevent solid material from being deposited. Finally further diluent is admitted to wash out the homogenizer and its contents are taken to drain through a solenoid-operated valve. A cam programmer operates the 12 switches which are required to actuate each stage of the sequence from turntable movement through to aspiration of the prepared sample and final washing. The program can be varied by adjustment of the appropriate cam. The total cycle time can be adjusted within the range 2–10 min.

A further extension of automatic solid sampling is possible if the samples can be weighed automatically rather than manually. This has proved possible with powders[21], which can be automatically sampled from a moving belt on to the pan of an automatic balance such as the Mettler type. When a predetermined weight of powder has accumulated on the pan the latter is tilted to discharge the contents into the homogenizer. The sampling-rate from the moving belt can be programmed to be compatible with the rate at which homogenization and analysis can be carried out. Clearly the principal field of application for this type of automatic sample preparation lies in 'on-line' production control analysis rather than laboratory analysis.

Jansen, Peters and Zelders[23] designed a sampling unit based on constant volume instead of constant time as in the Skeggs sampler. It

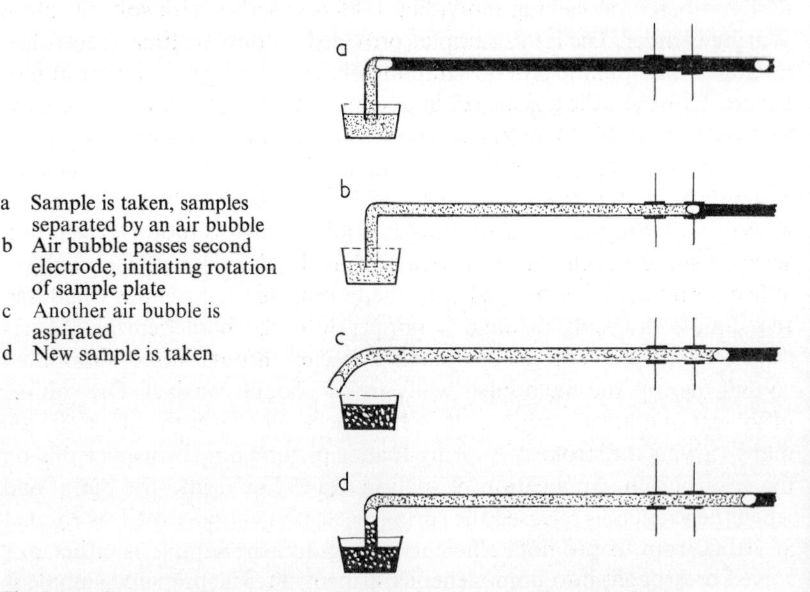

Fig. 3.11 Constant volume sampling techniques for continuous flow analysis. *Reproduced with permission from Jansen, Peters and Zelders[23] and Elsevier Publishing Company.*

eliminates errors due to variations of volume in the sample-tube. An almost twofold improvement in reproducibility, with respect to the constant-time method, is claimed. The volume of sample fed to the analyser is that enclosed between two constantly spaced air-bubbles. Two platinum-tube electrodes are placed in the sampling tube and these remain in electrical contact as long as conducting liquid flows between them but passage of an air-bubble breaks the contact. The sequence of events is shown in Fig. 3.11; the sample-tube is open to atmosphere during successive movements of the turntable holding the sample-tubes and an air-bubble is introduced, followed by sample drawn in from the next tube until the bubble reaches the electrodes and contact is broken. After the bubble passes through the electrode pair the current is restored and this actuates the next movement of the turntable whereupon the next air-bubble is drawn in. The plug of sample between the two air-bubbles then passes into the analyser. The actual volume taken need not be known and the amount of sample can be optimized with respect to the analytical method by using pump tubing of different internal diameters.

(b) Proportional Pumping

The 'AutoAnalyzer' is not dependent upon the metering of accurately measured volumes of sample and reagents; it is necessary only to introduce reagents and sample in the appropriate volume ratios. These ratios are preselected when assembling the pump-tube manifold for each analysis by using tubing of the correct diameter for each stream. The term 'manifold' refers to the complete layout of tubing for a given analysis, including connecting tubes and mixing coils. The pump-tubes are held taut and parallel on a spring-loaded platen fitted with slotted plastic end-pieces which maintain the tubes in fixed positions. A series of motor-driven rollers, carried on a chain, bear successively on the tubes and push the solutions in the tubes forward into the following stages of the analyser. The platen is spring-loaded to ensure that the pump tubes are pressed hard against the rollers. Fig. 3.12 shows a diagrammatic representation

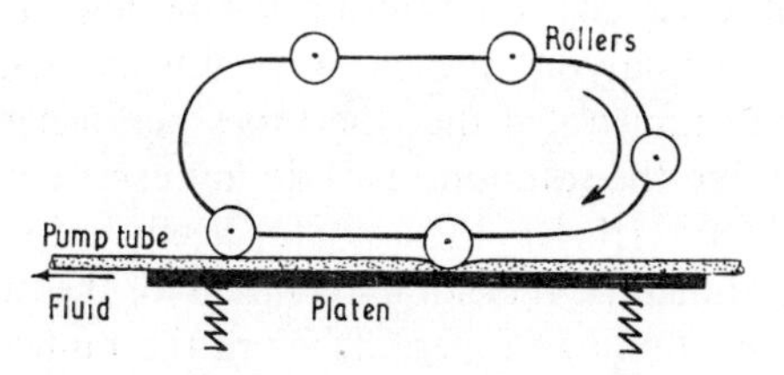

Fig. 3.12 Principle of operation of the 'AutoAnalyzer' peristaltic pump. *Reproduced by permission of Technicon Corporation.*

of the mode of operation of the proportioning pump. The spacing of the rollers is such that the pump-tubes are always compressed by at least one roller. The proportioning pump therefore serves two purposes, the introduction of both sample and reagents into the analyser and the transportation of solutions through the analytical system at a fixed speed. Because the pumping speed is fixed, the rate at which each solution is pumped is determined solely by the diameter of the pump-tube. For the standard pump-tubes 20 different sizes are provided. However, it is desirable in practice to design the manifold, by adjustment of reagent concentrations, so that a limited range of pump-tube sizes is used. This ensures that the rollers compress all pump tubes to the same extent.

Three generations of proportioning pump are available for use in the 'AutoAnalyzer'. Marks I and II are basically similar in their functions but Mark II has an increased manifold capacity, 23 tubes compared with 15 for the Mark I. Both are supplied with options of single-speed, dual-speed and variable-speed motors. With the dual-speed pump the lower speed is used for the actual analysis and the high speed to wash out the system between analyses; the high speed is not suitable for analytical use. The Mark III pump incorporates several features not provided with earlier models.

To achieve the maximum accuracy and precision from an 'AutoAnalyzer' the air–liquid bubble-pattern must be regular and reproducible. To ensure exact and reproducible proportioning the Mark III pump is fitted with a device termed an air-bar which is essentially a controlled air-inlet valve. Air is admitted into the system to provide regularity of segmentation. In addition the pump-tube mounting has been simplified in respect of the commonly used clinical methods. Analytical cartridges are pre-prepared and comprise the necessary components, mechanical and chemical, for the particular method. This saves considerable labour in changing from one method to another in that assembly of manifolds and reagent lines is obviated. The capacity of the Mark III pump is 28 tubes so that, provided the manifolds are simple, several methods (up to three) can be accommodated by the pump at any one time.

The use of flexible, elastic pump-tubes is perhaps the most limiting feature of the 'AutoAnalyzer' system. A wealth of experience in many laboratories has demonstrated that the tubes perform reliably for long periods provided that the solutions used do not exert corrosive effects on the tube material. Nevertheless the operator needs to be aware of certain possible causes of difficulty. It is not uncommon for the tubes, when newly fitted, to require a 'running-in' period before the pumping rate reaches uniformity[24]. It is therefore of the utmost importance to include an adequate number of standards with the unknown samples to provide regular compensation for any variability. It is also essential, when using the Mark I and II pumps, to maintain the tubes taut and parallel between the end-block mountings. Ultimately the tubes will suffer some deformation and lose some of their elasticity; there is then a tendency for the tubes to snake and the result is deviation from proportionality. At this stage a new manifold must be assembled and inserted in place of the defective one. Indeed it is preferable to maintain a regular manifold replacement schedule to prevent the incipient tube-deformation stage from being reached. The standard pump-tubes are attacked by strong acids and by a number of organic solvents and cannot be used with these materials. Two resistant types of tubing termed 'Acidflex' and 'Solvaflex' have been developed to improve compatibility with acids and solvents respectively. Nevertheless solvents such as acetone, pyridine and chloroform cannot be successfully pumped by either type of tubing. Pumping by displacement, using liquid immiscible with the reactive solution, can overcome this problem at the cost of some practical inconvenience. Water has been used to displace organic solvents[25] and mercury to displace an acetone solution of pyridine[26].

(*c*) *Separation Techniques*

It has proved possible to incorporate a range of separation techniques into the 'AutoAnalyzer' system, thereby conferring an advantage over the discrete-sample analysers in significantly increasing the range of methods that can be performed. The principles and applicability of these separation methods are discussed fully in other chapters, but briefly they include the following.

(*i*) Dialysis, in which the species of interest is separated from unwanted material by diffusion through a semi-permeable cellophane membrane. In the 'AutoAnalyzer' dialysis unit the segmented sample and recipient streams flow concurrently through two specially-grooved plates separated by the membrane materials. A constant-temperature water-bath held at $37 \pm 0{\cdot}1°$ C accommodates two membrane units.

(*ii*) Filtration, which is achieved by allowing the sample stream to drip on to a moving belt of filter paper.

(*iii*) Distillation, which enables volatile components to be separated from the sample stream and their vapours subsequently re-absorbed into a liquid stream.

(*iv*) Digestion with suitable reagents to break down the sample material and to produce the species of analytical interest in vapour or solution form. The Technicon digestor has proved extremely efficient compared with static digestion methods; it can also be used for distillation, solvent evaporation and photolysis.

(*v*) Liquid–liquid extraction by upward travel through a beaded coil column.

(*d*) *Heating*

In the 'AutoAnalyzer' fixed-period heating is provided by a unit which consists of a glass mixing-coil immersed in a temperature-controlled bath. Two standard units are available, one controlled at 95° C and the other at 37° C.

(*e*) *Detection Systems*

Technicon offer five standard detection units for use with the 'AutoAnalyzer': a colorimeter, dual-differential colorimeter, flame photometer, fluorimeter and UV spectrophotometer.

(*i*) *The Colorimeter.* The Technicon colorimeter employs a twin-beam optical system, both beams being generated by the same lamp. Radiation from the reference beam, which is collimated and passed through an interference filter of appropriate waveband, falls on a photocell detector and generates an electrical signal against which sample and standard absorbances from the sample beam are measured. The voltage difference between the sample and reference outputs generates a d.c. current which is converted into a.c. by a chopper and to a voltage by a transformer. This voltage is amplified and used to drive a balancing motor attached to the wiper arm of the potentiometer measuring the sample output. The arm moves until the voltage balances that of the reference beam. Thus each sample absorbance is presented as a fraction of the reference cell output. Initial setting-up merely involves balancing the two beams under conditions of 100% and 0% transmittance. Narrow-bandwidth interference filters provide the means of wavelength selection.

The flow-cell now in general use with the 'AutoAnalyzer' is of simple tubular design as shown in Fig. 3.13. The sample-stream is pumped past a debubbler, through the flow-cell and thence to waste. The flow-cell is tilted at a small angle to the horizontal to ensure that trapped air is removed.

(*ii*) *The Dual-Differential Colorimeter.* The standard colorimeter does not provide facilities for direct measurement of sample against a blank. In the dual-differential instrument, light from the lamp is split into two complete optical systems so that automatic blank-value subtraction can be achieved. For successful operation the timing of the analytical procedure is controlled exactly to ensure that the sample and blank solutions reach their respective flow-cells concurrently. If the differential mode is not required the instrument can function as two independent colorimeters.

(*iii*) *Flame Photometer.* This is described in detail in Chapter 4.

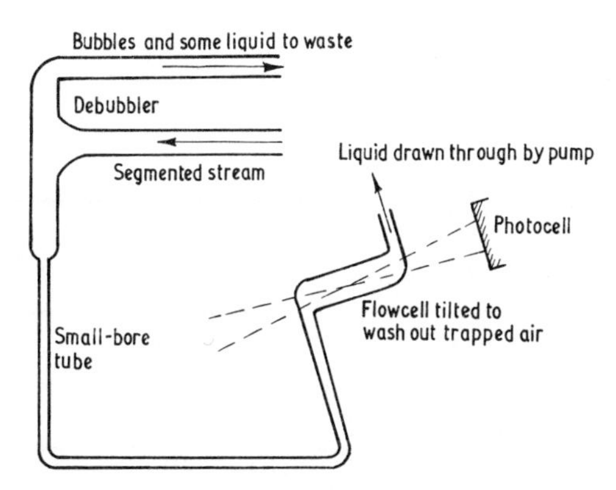

Fig. 3.13 The 'AutoAnalyzer' flow cell.
Reproduced by permission of Technicon Corporation.

(*iv*) *Fluorimeter.* The optical principle of the 'AutoAnalyzer' fluorimeter is similar to that of the colorimeter; it is a dual-beam instrument in which the fluorescence of the sample is compared with the intensity of a fixed reference beam obtained by diverting part of the incident light-beam to a separate photomultiplier. The optical layout is shown in Fig. 3.14. The light source is an 85-W mercury arc lamp which provides sufficient energy for fluorescence measurements to be made in an incident wavelength range of 250–600 nm. The incident beam passes through a rotating wheel having apertures of six different diameters around its circumference; these provide for sensitivity adjustment. Adjacent to each of these apertures are smaller apertures, all of the same size, which permit the isolation of a reference beam at each sensitivity setting. This beam is deflected by a

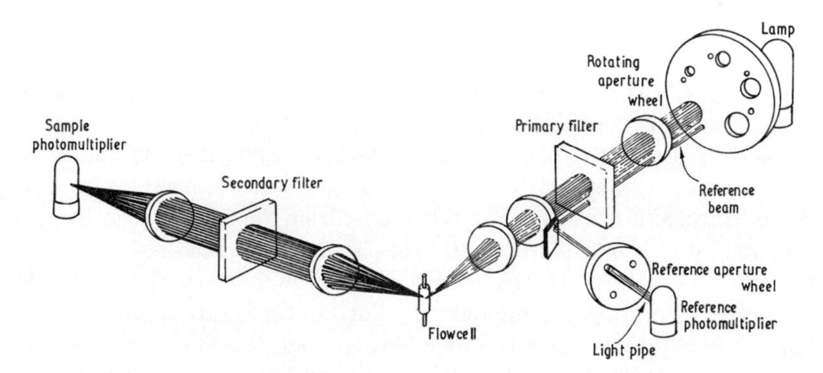

Fig. 3.14 Optical configuration of the 'AutoAnalyzer' fluorimeter.
Reproduced by permission of Technicon Corporation.

mirror on to an IP 28 photomultiplier. The sample-beam is collimated, passed through one of a series of primary filters to select the incident wavelength region and focused on to the sample which flows through a small-diameter flow-cell (Fig. 3.13) which is fabricated as an integral unit with a debubbler. The fluorescent beam, at right angles to the incident beam, is collimated, passed through a secondary filter to remove light of short wavelength, and focused on to an IP 21 photomultiplier. All optical components are of quartz to give the maximum useful wavelength range and to minimize inherent fluorescence. A null-balancing electronic circuit is employed in which the imbalance between the sample and reference photomultiplier signals is amplified and fed to a servo-system which generates the null balance. The long-term stability of the fluorimeter is said to be 1% and mains voltage fluctuations of up to ±10% do not affect its performance.

3.5 Automatic Fluorimetry

Automated fluorimetric methods have been developed in parallel with automated colorimetric ones to meet requirements, largely in clinical analysis, for increased productivity in the laboratory. The automation of sampling, sample pretreatment and the calculation and presentation of results follows virtually identical lines to those detailed in preceding sections for automated procedures based on colorimetry.

Fluorimetric methods are extremely sensitive and sample-volumes required for analysis are frequently very small. This can be advantageous in clinical studies where the volumes of physiological fluids available for analysis are restricted. For example, Ambrose[27] described an automated fluorimetric procedure for phenylalanine, based on its reaction with ninhydrin, which requires only 20 μl of blood, 10 μl each for the sample and peptide control streams. However, not only are fluorimetric methods sensitive in terms of the desired determination, they are sensitive also to interfering effects such as quenching and fluorescence contribution from other components of the sample. Consequently detailed attention must be paid to ensure that these deleterious effects are either eliminated or taken into account through blank and control measurements. Where the 'AutoAnalyzer' is used such effects can arise from the pump-tubes through leaching out of plasticizer, and from dialyser membranes and dust particles[28, 29]. Thorough cleansing of the system with detergent solution before use minimizes the interference. Purity of reagents is important in this context. A dual-channel analyser design, which permits compensation for interference due to chemicals, is necessary to achieve accurate and precise results.

Applications of automated fluorimetry are predominantly clinical in origin and utilize the 'AutoAnalyzer'-type continuous-flow approach. Details can be found in the proceedings of the Technicon Symposia on 'Advances in Automatic Analysis' published by Mediad Inc.

3.6 Kinetic Aspects of Continuous-Flow Analysis

The performance of an analyser which processes discrete samples at intervals is related to the dynamics of the flowing stream and an understanding of the dominant factors is important in optimizing the design of continuous methods. A continuous stream of liquid flowing through a tube exhibits a velocity profile, the flow being fastest at the centre and slowest at the tube surface where frictional retardation occurs. Material at the periphery mixes with that in the centre of the following liquid and is the cause of sample carry-over, referred to briefly above. Segmentation of the stream by air-bubbles reduces carry-over by providing a barrier to mixing, but it does not entirely prevent it, because mixing in the surface layer can still occur. Nevertheless, carry-over occurs mainly in unsegmented streams and in terms of the standard 'AutoAnalyzer' design this implies the initial sample-line before air-segmentation and after debubbling before entering the detector. The need for quantitative correlation of the magnitude of carry-over as a function of the kinetic parameters of

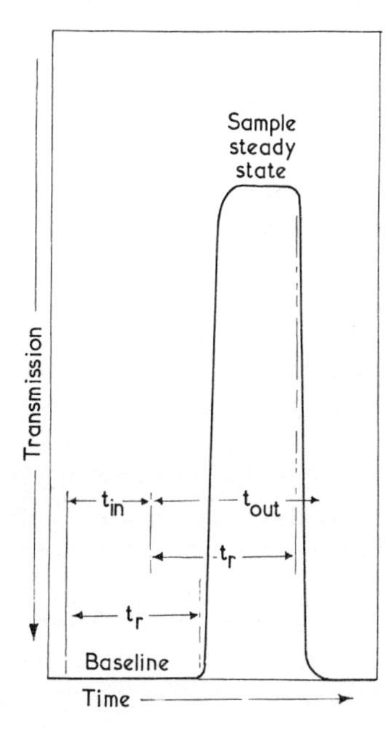

t_{in} = time for which probe is in sample

t_{out} = time for which probe is out of sample

t_r = delay time between sampling and sensing unit

Fig. 3.15 Standard response curve for continuous flow analysis of discrete samples.
Reproduced with permission from Thiers, Cole and Kirsch[30] *and the American Association of Clinical Chemists.*

the analyser has prompted definitive studies by several groups[18, 30, 31, 32]. Two parameters have been demonstrated to be fundamental in calculating the performance characteristics of a continuous analyser, the lag phase and the half-wash time; they afford a correlation between the approach to steady state, fraction of steady state reached in a given time and the interaction between samples. The half-wash time ($W_{\frac{1}{2}}$) is the time for the detector response to change from any value to half that value, the lag phase L is defined in the ensuing discussion.

The standard detector-response curve for a continuous analyser is shown in Fig. 3.15; it is obtained by aspirating a blank solution until a steady base-line is obtained, adjusting the base-line to read zero at the detector, introducing the sample for a fixed period and finally aspirating the blank again. If the sampling time is made sufficiently long, a steady-state plateau is reached; if not, a peak is reached at a fraction of the steady-state reading, the fractional value being a function of the sampling time. The detector response comprises a rise-curve, a steady-state plateau (which may or may not be obtained) and a fall-curve. Detailed studies[30, 32] reveal that apart from an initial lag phase the rise-curve is exponential, thus the measured concentration C as a function of time, t, is given by

$$\frac{\mathrm{d}c}{\mathrm{d}t} = k\,(C_{ss} - C_t)$$

where C_{ss} and C_t are the concentrations at the steady state and at time t.

A plot of log C_t against time takes the generic form of Fig. 3.16. The value of $W_{\frac{1}{2}}$ is calculable directly from the slope of the linear portion of the plot. The initial non-exponential part of the plot is termed the lag phase L and is expressed numerically as the value of the intercept of the linear portion on the time axis. The fall-curve structure is the inverse of that of the rise-curve.

In the continuous processing of discrete samples in the 'AutoAnalyzer' system the reaction-time is held constant by the manifold design, and because the rise-curve is exponential the degree of attainment of steady-state conditions is independent of concentration. Consequently it is not necessary for the analytical reaction to proceed to completion for Beer's law to be obeyed. This confers a considerable advantage upon the 'Auto-Analyzer' approach and one which is frequently emphasized. The relationship between degree of attainment of steady state and $W_{\frac{1}{2}}$ can be generalized in the semi-logarithmic plot of Fig. 3.17 where time is expressed in units of $W_{\frac{1}{2}}$.

The exponential nature of the fall-curve is responsible for sample interaction. Provided the between-sample time is made long enough, or blanks are inserted between samples, the effect is negligible, but if the

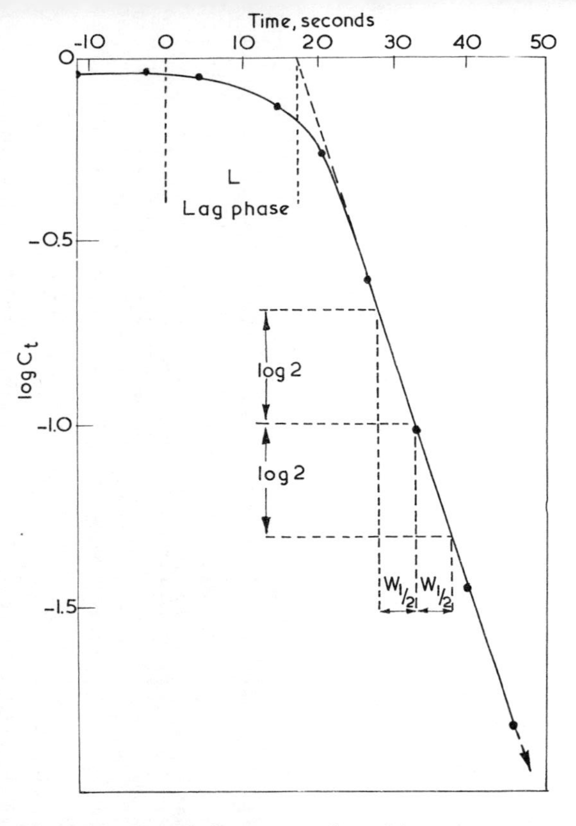

Fig. 3.16 Graphical representation of lag phase L and half-wash time $W_{\frac{1}{4}}$.
Reproduced by permission from Thiers, Cole and Kirsch[30] *and American Association of Clinical Chemists.*

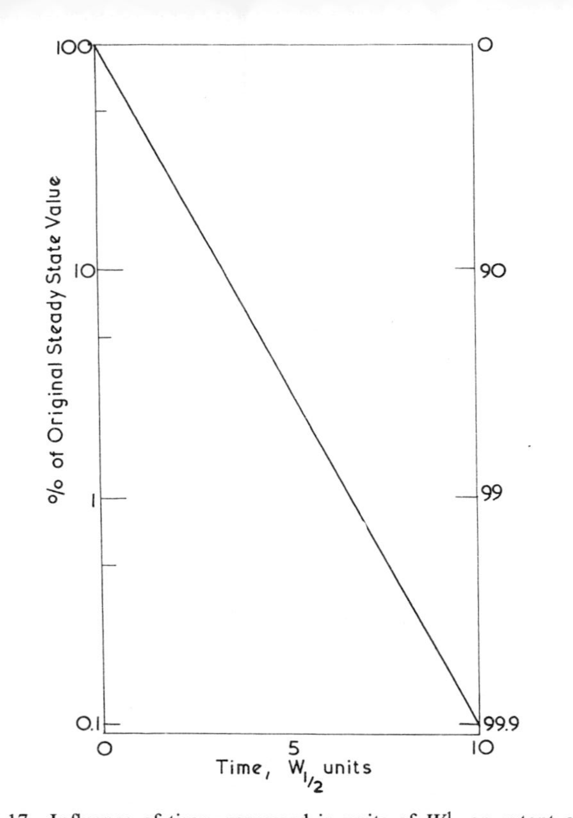

Fig. 3.17 Influence of time, expressed in units of $W_{\frac{1}{4}}$, on extent of achievement of steady state conditions.
Reproduced by permission from Thiers, Cole and Kirsch[30] *and American Association of Clinical Chemists.*

sampling rate is increased then the response to any given sample will be influenced by the tail of the response of the preceding one. The effect can be particularly severe when a concentrated sample preceeds a dilute one; the peak due to the latter may appear as a shoulder on the tail on the one due to the former or in severe cases it may be entirely hidden. Sample interaction can be quantitatively expressed by using $W_{\frac{1}{2}}$ and L. If the between-sample time is t_b, the value of the expression $(t_b - L)/W_{\frac{1}{2}}$ gives a measure of the interaction of a sample with the following one. For values of $(t_b - L)/W_{\frac{1}{2}}$ of 1, 2, 3, 4 – the degree of interaction with the following sample is 50, 25, $12\frac{1}{2}$, 10...% and this interaction appears additively in the response for the following sample. Clearly the smaller the values of L and $W_{\frac{1}{2}}$ the better is the performance of the analyser. For a fixed sampling rate the lower the values of L and $W_{\frac{1}{2}}$ the lower is the degree of interaction, or conversely, for a given acceptable per cent interaction the lower the values of L and $W_{\frac{1}{2}}$, the faster is the allowable sampling rate. The numerical relationship between sampling frequency, $W_{\frac{1}{2}}$ and degree of interaction for the case where L is negligible is plotted in Fig. 3.18. Using 'Auto-Analyzer' methods for sugar, urea and isocitrate dehydrogenase, Walker,

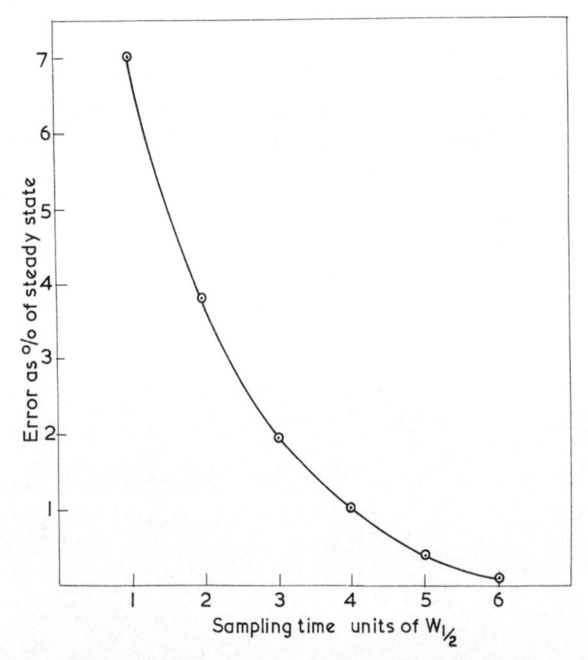

Fig. 3.18 Relationship between sampling time, $W_{\frac{1}{2}}$ and percentage interaction. *Reproduced by permission from Thiers, Cole and Kirsch*[30] *and American Association of Clinical Chemists.*

Pennock and McGowan[32] determined values of L and $W_{\frac{1}{2}}$ as a function of the apparatus components. Significant increases in $W_{\frac{1}{2}}$ were observed when the flow-rate through the colorimeter cell was reduced and when a debubbler with recycling was introduced. In the majority of cases measurable sample-interaction is limited to a sample and the one immediately following it. Wallace[31] demonstrated that a linear absorbance-to-concentration relationship is adequate to correct for the interaction. The absorbance A_n for sample n can be related to its true concentration by the equation

$$A_n = (1-m)\,a\,C_n + m\,A_{n-1} + b$$

where a is the absorptivity and m the fraction of sample $n-1$ in sample n, and b the blank absorbance.

3.7 Drift in Continuous-Flow Analysers

A problem almost universally encountered in continuous-flow systems is that the instrument-response for a given sample assay-value tends to vary with time. This effect, known as drift, affects the accuracy of results. It may be due to several causes, in particular variable performance of analyser components and variations in chemical sensitivity of the method used. It is manifest in two forms, base-line drift and peak-reading drift, which is due to sensitivity changes. The base-line drift may be detected visually if a chart record is kept, provided the magnitude of the effect is sufficient. The peak-reading drift can only be detected if calibration standards are inserted at regular intervals between samples. Experience has shown that there are very few methods in which the effect of drift is negligible over a long period of operation and therefore the processing of periodic calibration standards is mandatory if errors due to drift are to be corrected. Indeed, in 'AutoAnalyzer' usage it has become common to make every tenth reading a standard and to calculate sample results with reference to the new standard response. This approach assumes that the drift is wholly systematic, and cannot take cognizance of random fluctuations in the response to the individual standard. A more meaningful approach is to recalibrate with several different standards from time to time. Where computer techniques are used to process the analytical data it is usually necessary to provide several calibration points each time the calibration is updated. In this context Bennet *et al.*[33] evaluated 16 'AutoAnalyzer' methods and concluded that in all cases the calibration data could be fitted to a third-order polynomial of the form

$$A = a + bP + cP^2 + dP^3$$

where A is the assay value, P the peak reading and a, b, c and d are constants.

3.8 Kinetic Methods Based on Spectrophotometry

Reference is made in Chapter 2 to the use of electrochemical sensors in kinetic analysis. Spectrophotometric techniques have also been successfully employed. There are several well-established methods of deducing concentration from reaction rates but the two most extensively used are the constant-time and variable-time methods. Both involve making two measurements at close to zero reaction-time, the first requiring two absorbance measurements separated by a fixed time-interval, and the second requiring the measurement of the time taken for the absorbance to change from one preset value to another.

3.8.1 Spectrophotometers for Kinetic Analysis

3.8.1.1. *Special Purpose Instruments*

In kinetic analysis it is desirable, in order to simplify interpretation of results, to make both initial and final measurements at close to zero time to ensure that the concentrations of reactions do not change significantly over the reaction period. In practice this frequently means that small changes in absorbance have to be measured and if loss of precision is to be avoided a high-stability spectrophotometer is essential. Several groups of workers have designed and constructed spectrophotometers of excellent long-term stability capable of measuring very small absorbance differences. The main source of instability arises from fluctuations in the photometer lamp, and circuitry for stabilizing the lamp output is a feature of successful instruments.

Loach and Loyd[34] modified a Cary Model 14 R double-beam recording spectrophotometer to achieve high sensitivity of measurement and to enable reaction times between 10^{-4} sec and several minutes to be determined with high precision. The chopper used for normal double-beam operation was replaced by a precision-ground quartz plate. This served as a beam splitter, some 85% of the light from a tungsten iodide lamp being passed through the sample and the remainder into a reference beam. The two beams were detected by a matched pair of multiplier phototubes. The lamp output was stabilized by a proportional feed-back loop operating on the reference beam. The circuit is shown in Fig. 3.19; it achieves stabilization against intensity changes of longer than msec duration. Because the most prevalent form of lamp instability in the

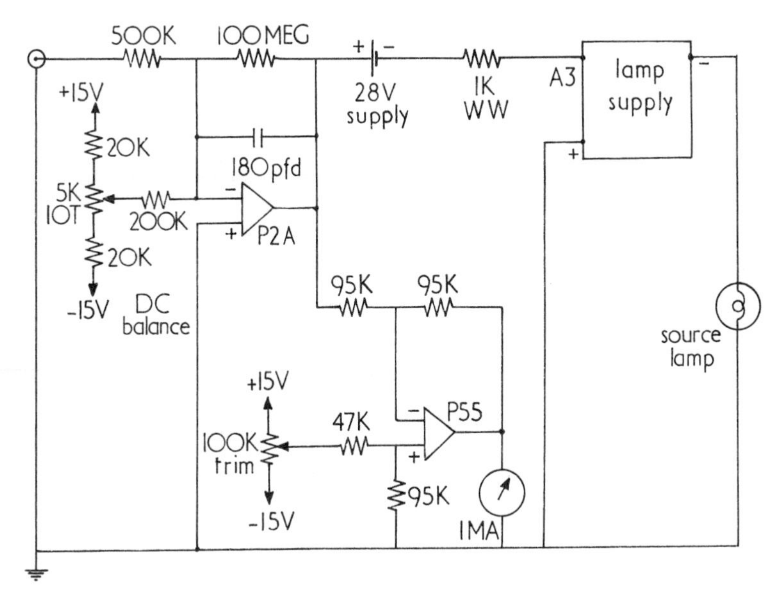

Fig. 3.19 Circuit detail of stabilization of source lamp.
Reproduced with permission from Loach and Loyd[34] and American Chemical Society.

absence of the control circuit was long-term drift, the control proved extremely effective and the sensitivity of the instrument was enhanced almost tenfold. Loach and Loyd were concerned with fast kinetics of light-induced reactions and consequently they did not evaluate the long-term stability of their spectrophotometer. Pardue and Rodriguez[35], in their study of the iodide-catalysed reaction of Ce(IV) with As(III), developed a photometer having emphasis on long-term stability in order that single measurements could be made over a reaction period of 5–500 sec. Lamp stabilization was achieved by using an optical feed-back circuit similar in principle to that of Loach and Loyd. The stability over several hours proved to be within 0·02% transmittance and the repeatability of measurements made over periods of up to an hour was 0·01% transmittance. Over the working range 0·2–0·3 absorbance units the relative error in concentration measurements was less than 0·2%. From a detailed evaluation of the performance of the photometer the authors conclude that the limiting factor in the stability of the lamp output is probably inhomogeneities in the filament; for best results, therefore, both the sample and reference phototubes should view the same portion of the lamp filament.

A continuous semi-automatic reaction-rate measuring instrument is described by Weichselbaum and co-workers[36]. It is modular in principle

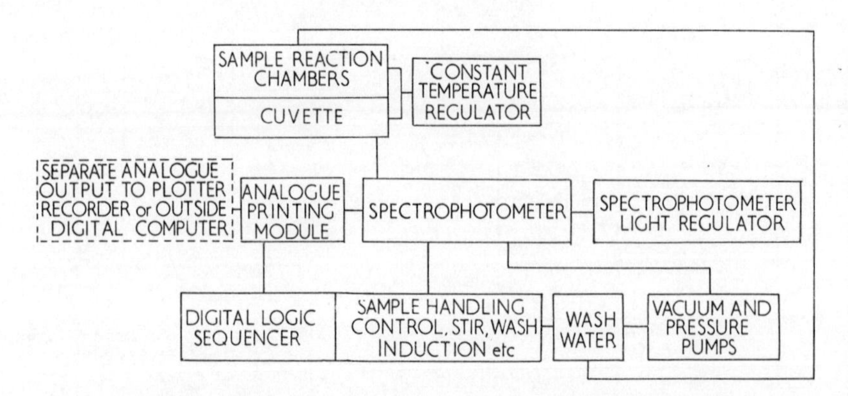

Fig. 3.20 Block diagram of high precision automatic spectrophotometer.
Reproduced by permission from Weichselbaum et al.[36] *and American Chemical Society.*

and is shown in block diagram form in Fig. 3.20. In addition to the spectrophotometer the instrument includes a digital logic sequencer for controlling the chemical operations and print-out of results. The spectrophotometer is designed to perform kinetic analyses by both the variable-time and fixed-time methods in that both continuous and intermittent measurements can be made. In the intermittent mode absorbance or transmittance measurements can be made as a function of time; in the continuous mode reaction rates are measured directly, by using derivative circuitry. The spectrophotometer also incorporates a coefficient multiplier in the servo-system for direct read-out in concentration units. The light-regulator system was designed to the following stability requirements: drift stability

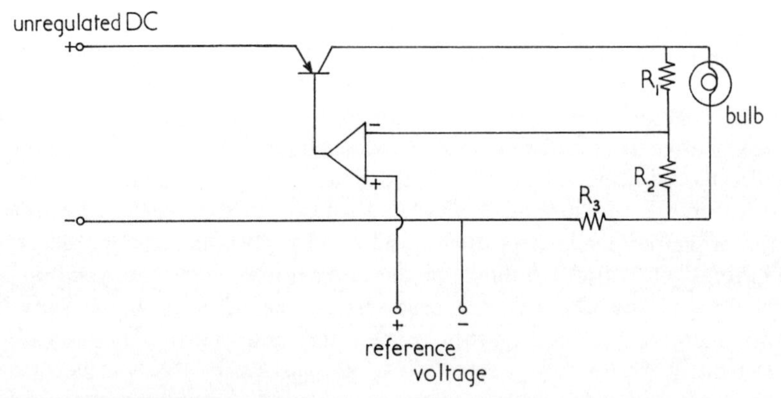

Fig. 3.21 Lamp stabilization circuit.
Reproduced by permission from Weichselbaum et al.[36] *and American Chemical Society.*

0·003 absorbance units per hour, photometric accuracy 0·01 absorbance units at 1·0 and 0·001 absorbance units near zero.

A tungsten filament lamp was used with stabilization provided by a high-gain low-noise electronic regulator, the circuit for which is shown in Fig. 3.21. It enables both voltage and current to be sampled; this is necessary to take into account the change in lamp resistance due to the heating up of the bulb and socket. Referring to Fig. 3.21, the voltage to the inverting amplifier is the sum of the voltage across resistors R_2 and R_3, that across R_2 is a fraction of the voltage across the bulb, and that across R_3 is proportional to the current through the bulb. The ratios across these two resistors can be varied to weight the current/voltage feed-back contributions and so compensate the variations in bulb resistance. Unacceptable variations in light output remained, even with a perfectly regulated and compensated power supply. These were traced to convective heat-flow around the lamp. The effect is serious when the lamp is housed in a chimney fitted with top and bottom vents. Blowing air across the bulb surface reduces the variations but the most effective reduction is achieved by using baffles to prevent convection currents across the optical path. Recorder traces indicate that the baffle-method reduces the variations by an order of magnitude more than vented-chimney method does.

In selecting the most appropriate type of photocell a balance must be struck between signal-to-noise ratio and absolute noise level. Although solid-state photocells can give good signal-to-noise ratios they are not suitable for use at wavelengths below 400 nm. Gas-filled diodes were considered the most satisfactory compromise, particularly for wavelengths in the ultraviolet region, although at wavelengths longer than 400 nm vacuum diodes exhibit lower noise levels.

Reactions were performed in the apparatus shown schematically in Fig. 3.22. It comprises two reaction chambers, used sequentially, connected by a three-way tap to a cuvette for photometric measurements. After manual addition of sample and reagents to the reaction chamber, the subsequent stages of the analysis are controlled automatically from the digital logic unit. This incorporates timers for preselection of reaction-sequence times, circuitry for activation of vacuum and pressure pumps for solution transfer and wash cycles, and the print sequencer and counter.

3.8.1.2 *Commercial Spectrophotometers*

Kinetic methods of analysis are assuming an ever-increasing status, particularly in connection with enzyme studies. Several commercial spectrophotometers are available which have either been designed

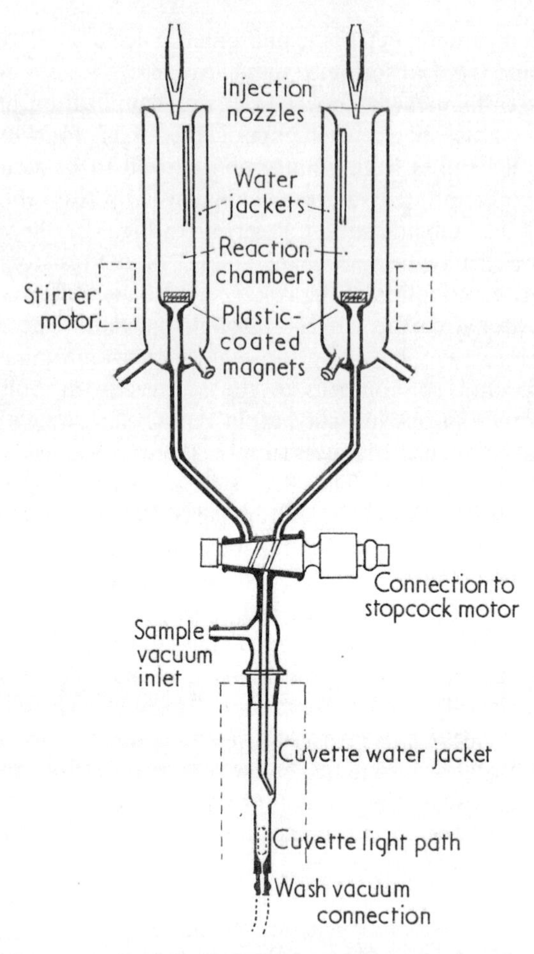

Fig. 3.22 Sequential reaction chambers for kinetic spectrophotometry. *Reproduced by permission from Weichselbaum et al.*[36] *and American Chemical Society.*

especially for kinetic studies or have been adapted from standard spectrophotometers by provision of time-sequencing attachments.

The LKB Reaction Rate Analyser measures rates automatically. Designed specifically for enzyme studies, it is basically an automatic discrete-sample analyser with facilities for automatic sample pretreatment. Up to 100 sample-cuvettes can be loaded into aluminium racks and processed sequentially through the instrument. Each cuvette passes through a dry-heat thermostat and a preheated substrate is added by means of a pump, mixing being achieved by rotating the cuvette several times. The cuvette is programmed to remain in the optical path of the photometer at 340 nm for a period between 1 and 9 min during which time the absorbance change is recorded continuously. The tangent of the angle between the rate-curve and the time-axis is directly proportional to the enzyme activity. Two absorbance ranges are provided, having full-scale deflections of 0·05 and 0·2 absorbance units. If desired, the instrument may be interfaced to an on-line computer programmed to calculate and present results in the desired units and to reject results from non-linear reactions.

The Pye Unicam AC 60 can be fitted with a programme delay unit (AC 61) which controls the length of time (5 sec–5 min) each sample remains in the light-beam. The Cary Model 16 K kinetic spectrophotometer comprises several additions to the Model 16 Manual instrument. The basic additions are an automatic sample-changer, a recorder and a transmittance-to-absorbance conversion unit. Additionally a dual-wavelength drive unit is available, enabling absorbance changes to be measured at two wavelengths simultaneously by driving the dual-prism dispersing unit successively between the two wavelengths. Measurements at two wavelengths are frequently desirable in studying the kinetics of enzyme systems, one being fixed at an isosbestic point and the other at an appropriate wavelength to measure absorbance changes.

Continuous-flow analysers can readily be adapted to perform reaction-rate analyses by the constant-time method; the sample is pumped at constant rate through two photometer cells separated by a coil providing a fixed delay time. The steady-state absorbance difference between the two cells is proportional to the rate of reaction. This approach is illustrated by the work of Blaedel and Hicks[37] who determined lactic dehydrogenase (LDH), using the coupled reaction sequence

$$\text{L-lactic acid} + \text{DPN}^+ \rightleftharpoons \text{pyruvic acid} + \text{DPNH} + \text{H}^+ \tag{1}$$

$$\text{DPNH} + \text{H}^+ + \text{dye}_{\text{ox}} \xrightleftharpoons{\text{diaphorase}} \text{DPN}^+ + \text{dye}_{\text{red}} \tag{2}$$

where DPN represents diphosphopyridine nucleotide and DPNH its reduced form. The dye used was 2,6-dichlorophenolindophenol which is blue in the oxidized form and colourless in the reduced form. The apparatus is shown schematically in Fig. 3.23.

Alternatively a fixed time delay can be achieved by splitting the sample, after mixing with reagents, into two streams each of which passes through a delay coil to one cell of a differential photometer, but one coil is held at room temperature and the other at elevated temperature. Hicks and Blaedel[38] used this technique to determine transaminase enzymes, e.g. glutamic oxaloacetic transaminase (GOT), using the general scheme

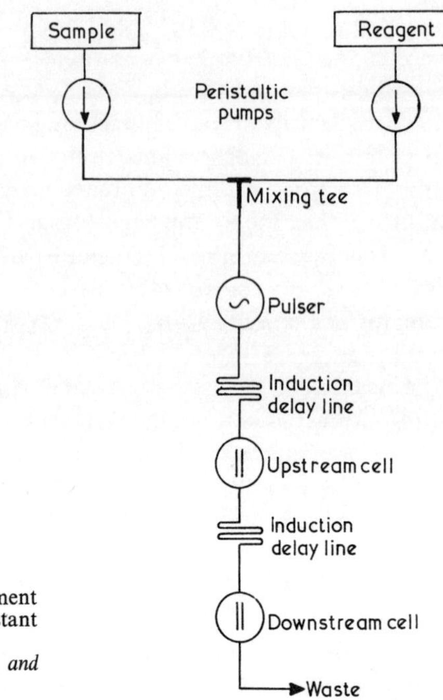

Fig. 3.23 Schematic diagram of equipment for kinetic spectrophotometry by the constant time method.
Reproduced by permission from Blaedel and Hicks[37] and Academic Press Inc.

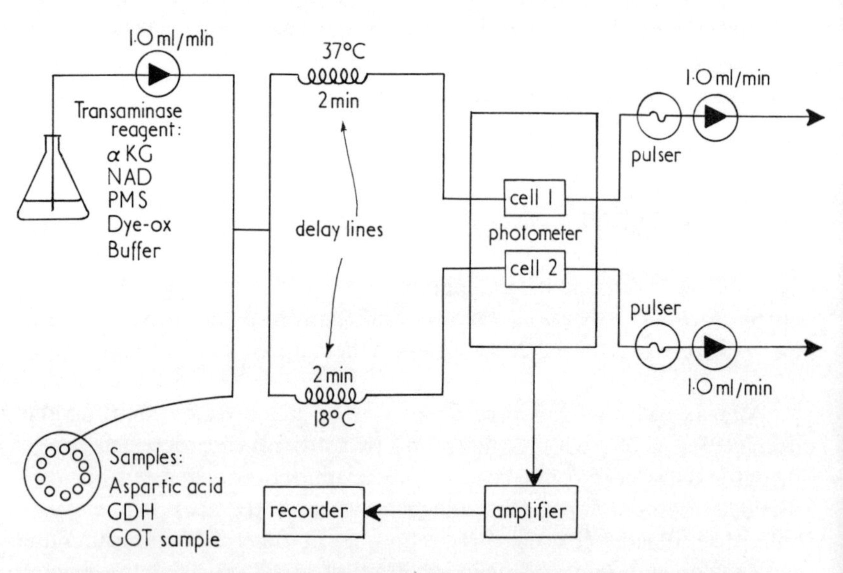

Fig. 3.24 Dual temperature method of constant time kinetic spectrophotometry.
Reproduced by permission from Hicks and Blaedel[38] and American Chemical Society.

$$\text{Amino-acid} + \alpha\text{KG} \underset{}{\overset{\text{transaminase (GXT)}}{\rightleftharpoons}} \text{keto-acid} + \text{glutamate} \quad (1)$$

$$\text{Glutamate} + \text{NAD} \overset{\text{GDH}}{\rightleftharpoons} \alpha\text{-KG} + \text{NADH}_2 + \text{NH}_4^+ \quad (2)$$

$$\text{NADH}_2 + \text{dye}_{\text{ox}} \overset{\text{PMS}}{\rightleftharpoons} \text{NAD} + \text{dye}_{\text{red}} \quad (3)$$

where GXT is a generic term representing transaminase activity, α-KG is α-ketoglutamate, NAD is nicotinamide adenine dinucleotide, $NADH_2$ its reduced form, GDH is glutamic dehydrogenase and PMS phenazine methosulphate. Fig. 3.24 shows the equipment layout. This apparatus has been successfully used to determine enzyme activities separated by using an automatically programmed ion-exchange column[39].

3.9 Automatic Analysis of Microsamples

The two most commonly used techniques for introducing samples into an automatic analyser are aspiration with a peristaltic pump and the use of an automatic pipette or dilutor. Neither approach is applicable when the initial sample volume is very small, say 0·05 ml or less. Both methods involve dipping a capillary tip into the sample, and to ensure that a definite volume is taken there must be a residue of sample in the tube. Natelson[40] has devised a mechanized technique for feeding samples held in capillaries into an automatic analyser. The technique can be utilized with both discrete and continuous analysers and also with a moving tape system.

Samples are taken in disposable capillary tubes of uniform length and bore. Typically the capillaries are 70 mm in length and 25 μl $\pm 2\%$ in volume. Filling of the capillaries is performed manually simply by touching one end against the sample; no attempt was made to automate this stage. The filled capillaries are placed horizontally on a turntable such that they can be picked up individually by a movable arm and clamp. The action of the arm is to lift the capillary, turn it to the vertical position with the end which dipped into the sample uppermost and finally dispose of it into a waste receptacle. The arm also carries a compressed-air line and when the capillary is vertical its contents are blown by the compressed air into a vessel in which the next analytical operation is to be performed. The compressed air is cam-controlled such that the capillaries can only be blown out when they are vertical. Using 25-μl micropipettes, Natelson achieved a reproducibility of $\pm 2{\cdot}5\%$ for the blow-discharge technique. If the sample is to be analysed by a moving tape method it is discharged directly on to the sample tape; alternatively it is expelled into a vessel where it meets a stream of diluent or reagent. When the discharge

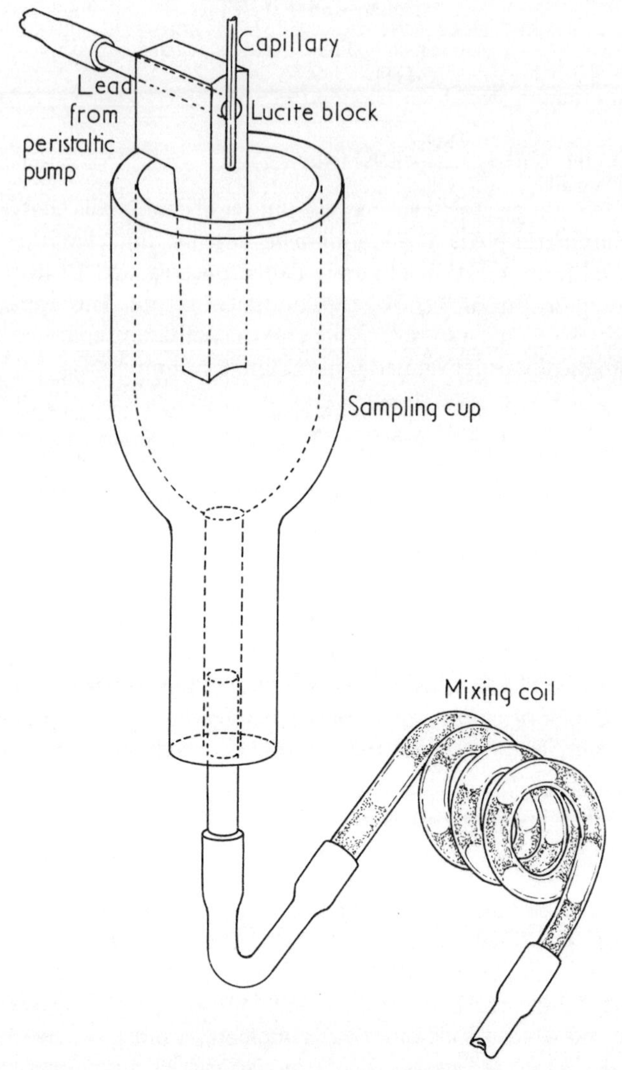

Fig. 3.25 Design of cup for dilution and air segmentation in microsample analysis. *Reproduced with permission from Natelson[40] and Academic Press Inc.*

cycle is complete the arm returns to its rest position and the turntable advances to the next position.

To interface the microsampler with an 'AutoAnalyzer' the cup shown in Fig. 3.25 is used. The Lucite block is drilled and shaped to allow diluent or reagent pumped from the manifold to merge with the sample as it is ejected from the capillary. On leaving the base of the cup the merged stream passes through a mixing coil and is pumped into the 'AutoAnalyzer' for further processing. To ensure that the solution pumped to the 'AutoAnalyzer' is segmented with air, the rate of entry of diluent or reagent into the cup is arranged to be lower than the rate of removal from the bottom of the cup. Full details of the mechanical movements and controls are to be found in the original paper[40]. The sample processing rate can be 36, 72, 90 or 360 per hour. If more than one analysis is required the sample can be divided into several approximately equal portions by the method shown in Fig. 3.26. The appropriate number of pump-tubes is sealed into the base of the cup. A large volume (6–8 ml) of diluent or reagent is metered into the cup when the sample is ejected. The solution is drawn away through the pump-tubes to the next analytical stage. To minimize carry-over between samples a further volume of diluent is added to the cup and pumped out between each microsample addition.

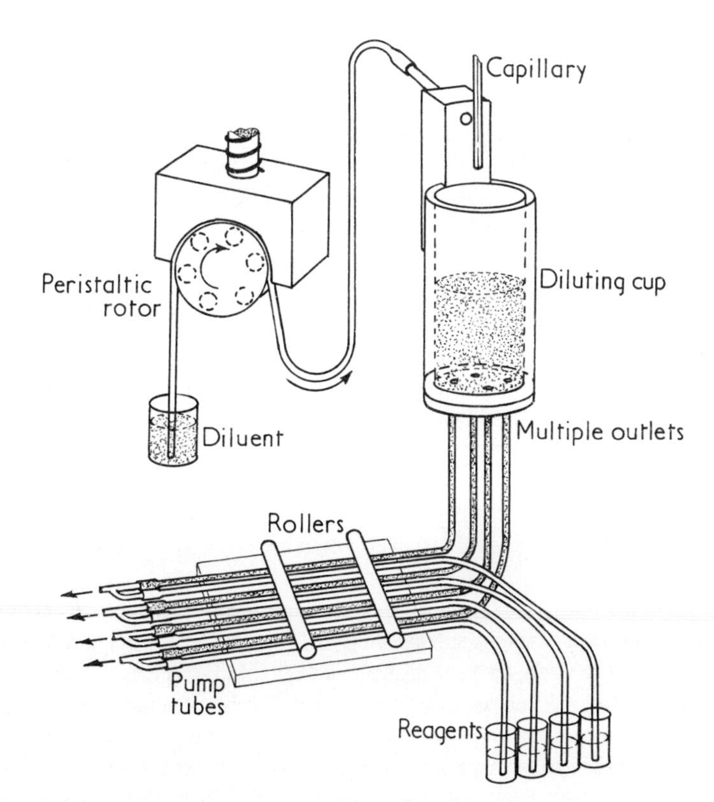

Fig. 3.26 Microsample analysis: method of dividing a sample into four streams. *Reproduced with permission from Natelson[40] and Academic Press Inc.*

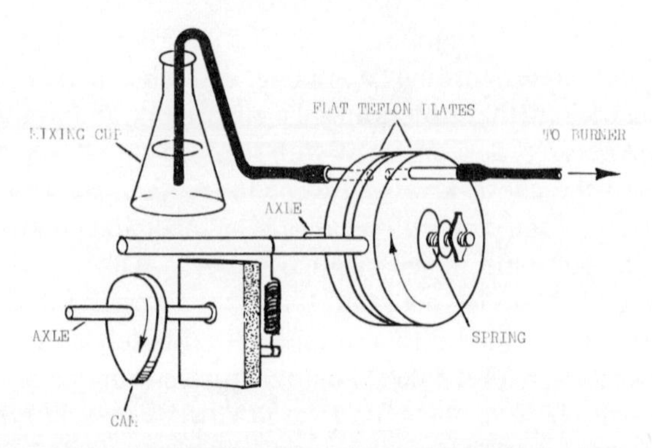

Fig. 3.27 Microsample analysis: method of presenting samples to a flame photometer. *Reproduced by permission from Natelson[40] and Academic Press Inc.*

To enable flame photometry or atomic-absorption spectrophotometry to be utilized as the detection technique the apparatus shown in Fig. 3.27 was designed to provide a suitable interface between the microsample dispenser and the aspirator. Two flat Teflon plates are used as a 'make and break' device. Each plate is drilled with a single hole to carry separate lengths of tubing for transporting the sample. When the two holes are aligned sample can be pumped to the aspirator but when they are not aligned the aspiration is interrupted. The interruption sequence is controlled by a spring-loaded arm attached to one of the plates; the arm is raised by a rotating cam. The operational sequence is as follows: with the Teflon plates in the closed position the microsample is ejected into a suitable receptacle concurrently with the admission of diluent by pumping, after a mixing period (8 sec) the capillary is removed to waste (3 sec), the plate moves to the open position and aspiration to the flame (7 sec) occurs. The excess of sample is pumped to drain, further diluent is admitted and the sequence repeated to flush out the lines. The parallel plate valve is closed and the cycle, which takes 45 sec in all, is repeated. To achieve synchronization between the microsampler and the aspirator device the same motor shaft is used to drive both the sampler and the valve.

The equipment used for analysis of microsamples colorimetrically by a moving tape system is depicted in Fig. 3.28. Three concurrently moving tapes are used, and the sample is deposited and absorbed on discs of Whatman No. 1 filter paper affixed to a nylon backing tape, the reagent is taken up from a rotating ceramic roller on to a tape of Whatman No. 3 paper, and between the two runs an intermediate tape to provide crude separation facilities. For example a tape of cellophane will prevent proteins from passing through to the reagent tape, and cellulose acetate of

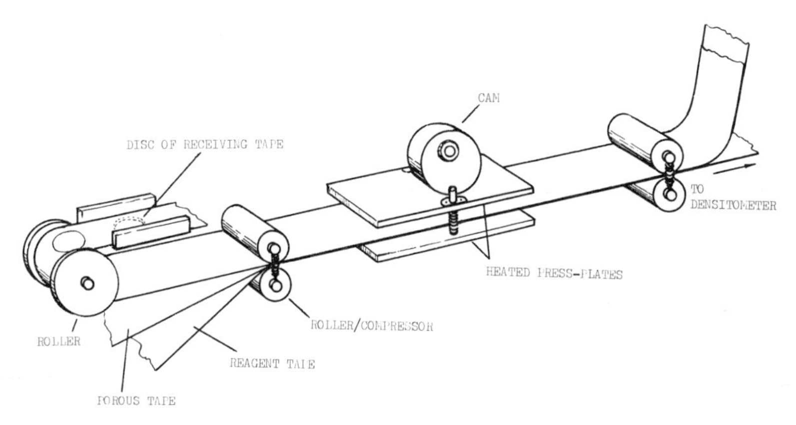

Fig. 3.28 Moving Tape Analyser.
Reproduced by permission from Natelson[40] *and Academic Press Inc.*

0·06-nm pore-size separates proteins from erythrocytes for the determination of proteins in whole blood. The operational sequence is clear from Fig. 3.28; after deposition of the sample in the centre of the absorbent disc the nylon tape passes under a roller and is brought into contact with the intermediate and reagent tapes and passed under heated press-plates at 37° C to complete the colour-forming reaction. The tape, which now carries a series of stained spots, is passed through a drier and then to the sensing device which can be either a colorimeter or a fluorimeter. For colorimetric measurements reflectance proved superior to transmission in that calibration curves gave better linearity, albeit at rather lower sensitivity. The reagent tape is moved across the field of view of a photocell which has a small hole drilled through its centre. The light-source is mounted behind the photocell, incident radiation passes through the hole and is reflected as a divergent beam on to the photocell. For colorimetric studies the appropriate wavelength range is selected by means of an interference filter and for fluorescence measurements an emission filter with a hole drilled through the centre is placed between the photocell and the tape.

References

1. Shapiro, L. and Massoni, C. J. *U.S. Geol. Surv. Prof. Paper, 800B*, 1972, B137.
2. Loebl, H. *Lab. Equipment Digest*, 1966, 48.
3. Anon. *Lab. Equipment Digest*, 1969, 102.
4. Lloyd, P. B. *Spectrovision*, 1971, No. 26, 4.
5. Mills, K. J. *Spectrovision*, 1971, No. 25, 3.

6. Norman, G. A. and Follett, M. J. *Spectrovision*, 1970, No. 23, 2.
7. Johns, P. *Spectrovision*, 1972, No. 27, 17.
8. Welch, H. *Lab. Equipment Digest*, 1967, 73.
9. Anderson, N. G. *Am. J. Clin. Path.*, 1970, **53**, 778.
10. Burtis, C. A., Johnson, W. F., Attrill, J. E., Scott, C. D., Cho, N. and Anderson, N. G. *Clin. Chem.*, 1971, **17**, 686.
11. Jansen, J. M. *Clin. Chem.*, 1970, **16**, 515.
12. Kelley, M. T. and Jansen, J. M. *Clin. Chem.*, 1971, **17**, 701.
13. Tiffany, T. O., Jansen, J. M., Burtis, C. A., Overton, J. B. and Scott, C. D. *Clin. Chem.*, 1972, **18**, 829.
14. Buckley, R. *Lab. Equipment Digest*, 1968, 81.
15. Anon. *Biomedical Engineering*, 1969, **4**, 222.
16. Skeggs, L. T. *Am. J. Clin. Pathol.*, 1957, **28**, 311.
17. *Technicon AutoAnalyzer Bibliography 1957/1967*, Technicon Corporation, Tarrytown, New York.
18. Thiers, R. A. and Oglesby, K. M. *Clin. Chem.*, 1964, **10**, 246.
19. Whitehead, E. C., *Technicon Symp. 3rd, New York 1966*, **I**, 364.
20. Technical Publication No. TA0–0219–00, Technicon Corporation, Tarrytown, New York, Sept. 1970.
21. Marten, J. F. *Technicon Symp., London*, 1963, 24.
22. Holl, W. W. and Walton, R. W. *Ann N.Y. Acad. Sci.*, 1965, **130**, 504.
23. Jansen, R. P., Peters, K. A. and Zelders, T. *Clin. Chim. Acta*, 1970, **27**, 125.
24. Cooke, J. R. and Stockwell, P. B. *Lab. Practice*, 1971, **20**, 125.
25. Taylor, I. E. and Marsh, M. M. *Am. J. Clin. Pathol.*, 1959, **32**, 393.
26. Terranova, A. C., Pomonis, J. S., Severson, R. F. and Hermes, P. A. *Technicon Symp*. 4th, New York, 1967, **1**, 501.
27. Ambrose, J. A. *Ann. N.Y. Acad. Sci.*, 1972, **196**, 293.
28. Ambrose, J. A. *Technicon Symp. Adv. Automat. Anal.*, 1969, **1**, 25.
29. Ambrose, J. A., Ross, C. and Whitfield, F. *Technicon Symp. Adv. Automat. Anal.*, 1967, **I**, 13.
30. Thiers, R. E., Cole, R. R. and Kirsch, W. J. *Clin. Chem.*, 1967, **13**, 451.
31. Wallace, V. *Anal. Biochem.*, 1967, **20**, 517.
32. Walker, W. H. C., Pennock, C. A. and McGowan, G. K. *Clin. Chim. Acta*, 1970, **27**, 421.
33. Bennet, A., Gartelmann, D., Mason, J. I. and Owen, J. A. *Clin. Chim. Acta*, 1970, **29**, 161.
34. Loach, P. A. and Loyd, R. J. *Anal. Chem.*, 1966, **38**, 1709.
35. Pardue, H. L. and Rodriguez, P. A. *Anal. Chem.*, 1967, **39**, 901.
36. Weichselbaum, T. E., Plumpe, W. H., Adams, R. E., Hagerty, J. C. and Mark, H. B. *Anal. Chem.*, 1969, **41**, 725.
37. Blaedel, W. J. and Hicks, G. P. *Anal Biochem.*, 1962, **4**, 476.
38. Hicks, G. P. and Blaedel, W. J. *Anal. Chem.*, 1965, **37**, 354.
39. Hicks, G. P. and Nalevac, G. N. *Anal. Biochem.*, 1965, **13**, 199.
40. Natelson, S. *Microchem, J.*, 1968, **13**, 433.

Chapter 4

Spectroscopic Methods

Flame emission and absorption techniques play a major role in elemental analysis, particularly at the trace level. Simplicity, speed, sensitivity and ability to perform several determinations on a single sample are the principal features which have provided the impetus for their extensive development. Such methods are obvious candidates for mechanization and automation because the step involving the flame is invariably short compared with sample-preparation and data-treatment stages. In this chapter the progress towards mechanization and automation is considered with reference to introduction of the sample into the flame, dispersion and/or monochromation of the radiation to be measured, recording of emission or absorption intensity and calculation and presentation of results. For convenience, emission and absorption methods are considered separately although in terms of approaches to automation it is apparent that there is much common ground between the two.

4.1 Flame Emission Techniques

Early studies using these techniques involved recording the spectrum on a photographic plate and measuring line-intensities with a densitometer. Although recording densitometers are available commercially their use in the analytical field has been limited, largely owing to the inherent shortcomings of the photographic plate for rapid, precise and reproducible measurements and its extensive replacement by multiplier phototubes as the sensing element. Indeed, the advent of multiplier phototubes was largely responsible for the initial stage in the curtailment of manual operation through the development of direct-reading spectrographs. These enable a number of elements to be determined simultaneously or consecutively for a single excitation. Almost all manufacturers of spectrographic equipment include direct-reading instruments in their range of products. Broadly speaking, three design variants have been employed. Perhaps the most widely employed method involves the use of a fixed optical system in which the dispersing element, a prism or grating, is fixed and a fixed slit and a multiplier phototube are appropriately positioned on the focal curve for each element to be determined. Alternatively a scanning technique can be employed. A single slit and detector unit can be scanned slowly across the spectral range of interest and the measured spectrum displayed as a recorder trace. Another approach involves rotating the dispersing element, causing the spectrum to move across a single detector unit.

Spectrographs utilizing the multiple detector approach are inherently inflexible but they are well suited to quality-control analyses where the analytical requirements are fairly rigidly defined, as for example in the metallurgical industry. The scanning instruments offer greater versatility but they are uneconomic in the sense that measurements are made for only a small fraction, often less than 1%, of the total scanning time if several elements are to be determined. Furthermore, excitation conditions frequently cannot be maintained constant throughout the period of the scan.

Dawson, Ellis and Milner[1] describe a rapid scanning spectrograph, using a single fixed slit and detector unit, for the determination of Na, K, Ca and Mg in clinical samples. The optical system is shown in Fig. 4.1; it is based on the Optica CF4 diffraction-grating mounting but modified so that the grating can be oscillated at 6 Hz (1800 nm/sec over the wavelength range 280–770 nm) and supplemented by an auxiliary optical system for wavelength-scale generation. Two trains of electric signals are simultaneously produced, one representing the sample spectrum over a given wavelength range and the other representing the wavelength scale. The latter is a regular train of pulses uniquely locked to the wavelength at the detector; it is produced from a mirror attached to the reverse side of the grating, which when oscillated forms an image of a 200 line/in. graticule on a slit mounted in front of the multiplier phototube. By means of electronic circuitry the reference-pulse train is used to gate the appropriate position of the signal-train for detection and measurement of each required element line.

The grating is driven by an induction motor (2800 rpm) through an idler pulley to the rim of a chromium-plated flywheel. The wavelength-drive cam is attached to the shaft of the flywheel. Signal-counting of spectrum lines is performed only during the forward scan. To inhibit the functioning of the signal-counting circuits during the reverse movement of the grating an auxiliary optical system, comprising a lamp and two photodiodes (Fig. 4.1), is provided. The lamp views the surface of the flywheel, portions of which are painted black to provide signals for recognizing the direction of movement of the grating; one photodiode

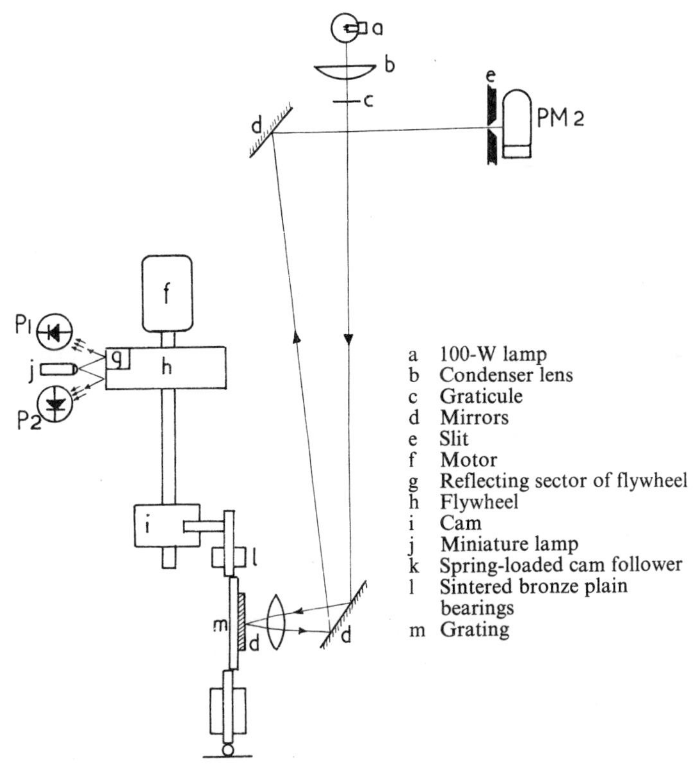

Fig. 4.1 Optical system of rapid scanning spectrograph. *Reproduced with permission from Dawson, Ellis and Milner*[1].

is energized during the forward scan and the other at the end of the forward scan. Signals from these two photodiodes can therefore be used to open or close the counting circuits.

Fig. 4.2 depicts the full function of the instrument in block diagram form. In addition to automatic spectrum scanning the instrument provides automatic dilution and sequential feeding of samples to the flame and automatic calculation and print-out of results. An 'Auto-diluter' (Hook and Tucker Ltd.) automatically dilutes samples and standards which are carried in a modified 'AutoAnalyzer' turntable from which they are aspirated into the flame. Na, K and Ca are determined by flame emission and Mg by flame absorption, with a hollow-cathode lamp. The data-handling equipment comprises a gated integrator for each measurement channel. During the forward scanning sequence the gating circuitry operated by the reference-pulse train opens the photomultiplier–integrator circuitry as each element line passes the exit slit. Accumulated signals for

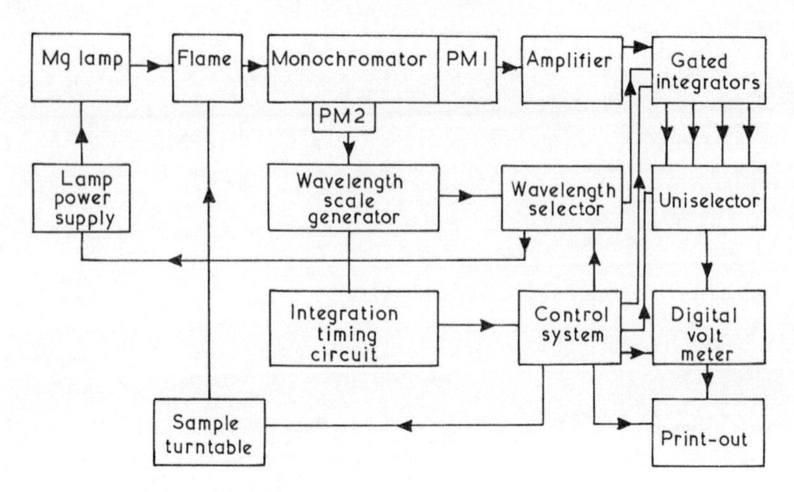

Fig. 4.2 Block diagram showing overall function of rapid scanning spectrograph. *Reproduced with permission from Dawson, Ellis and Milner*[1].

each analytical line are integrated for 45 sec and then the integrators are sequentially switched to a digital voltmeter and electric typewriter. To enable results to be printed out in concentration units an analogue calculator is provided to accept calibration data and correct for any non-linearity in the calibration curve. Zero and instrument-sensitivity correction facilities are also incorporated in the calculator.

The authors comment on the high long-term reliability of the instrument and quote precision values of 3% for measurements on aqueous solutions of 20 ppm Na, 15 ppm K, 1·25 ppm Ca and 0·5 ppm Mg. They note also that the electronic components must be of high quality and it is apparent that the design concept achieves an enhanced degree of efficiency in optical utilization at the expense of appreciable electronic complexity.

Multi-element direct-reading spectrometers can yield large numbers of analytical readings on a single sample and in a busy laboratory the time involved in calculating results is considerable. Where the sample load is large enough automatic processing of the data is economically justifiable. Lowe and Martin[2] describe a computer system for processing results from an ARL 'Quantometer' in a laboratory performing extensive metal analyses on oil samples. The 'Quantometer' has 25 analytical channels and is capable of monitoring 51 analytical lines. Results are presented on a Leeds and Northrup recorder and the analogue-to-digital conversion is achieved by means of a shaft encoder attached to the recorder. The remainder of the equipment comprises a remotely-controlled card punch

(IBM 526), a paper-tape lister and a specially constructed digital read-out console. Data handling, retrieval and computation are performed on an IBM 360/75 computer. By selection of the appropriate switch on the control panel a group of elements to be studied is selected. A number of such groupings are provided to cover the main analytical programs of the laboratory. The digital read-out facility then scans through the preselected combination of elements, digitizes the recorder deflection for each one and writes the values onto data cards, paper tape or both. The scanning mechanism comprises two 26-point rotary stepping-switches. One of these, the channel-sequencer, is wired to step the 'Quantometer' through all of its 25 channels. The other one sequences data read-out. The recorder deflection can be represented as a code and 3 digits (000–999 for 0–100% recorder deflection). The read-out system is controlled by relay circuits and an impulse-counter clock records the reflected-beam time to $\pm 0{\cdot}1$ sec. It is fitted with contact closures to transfer the measured time to the card punch or tape lister. Each analytical channel is read when the recorder is balanced as detected by a sensitive relay connected in parallel to the pen-drive servo-motor. Calibration data are accumulated before the samples are run and the recorder readings adjusted to read in concentration units by operation of the zero and sensitivity potentiometers. Since the quality of the sample results is limited by the accuracy of the calibration data which are stored in the computer, considerable emphasis was placed upon deriving best-fit equations for each element. To this end a separate computer program was written to utilize a least-squares method based upon 9 separate equations ranging from linear to cubic. Separate subroutines are stored in the computer for calculation of results for each preselected group of elements.

The use of an intermediate card punch produces a considerable time-delay between sparking the sample and the production of the analytical report. This delay could be overcome by progressing to an 'on-line' computer system.

Barnett[3] discusses in general terms the possible approaches by which computers can be used to aid the processing of spectrographic data. Four such approaches are distinguished.

1. Data collected by normal methods, manually converted into machine-readable form and processed batchwise.
2. Data prepared in machine-readable form by the instrument and processed batchwise.
3. Data prepared in machine-readable form by the instrument and processed by a time-shared computer.
4. On-line computer operation of the analytical instrument.

To date the first method has been the most widely used.

4.2 Continuous Flame Photometry

Automatic analysis by the continuous flow principle, as exemplified by the Technicon 'AutoAnalyzer', can, in principle, be used with any type of end-measurement technique. The early developments in instrumentation were largely aimed at satisfying the extensive demands for automation in the field of clinical analysis. Requirements for sodium and potassium determinations in blood serum, which are not amenable to colorimetric determination, led to the development of a flame photometer, compatible with the 'AutoAnalyzer'[4]. Basically it is an integrating flame photometer which, by using separate radiation detectors for the sample element and an internal standard element, yields results in the form of sample-to-standard ratios. The ratio-recording approach is advantageous in that errors due to variability in any part of the equipment, such as flame, gas flows and pressures in the photometer, and flow and dialysis rates in the 'AutoAnalyzer', affect both detector units simultaneously, thereby minimizing the overall effect.

A cut-away diagram of the burner is shown in Fig. 4.3. It utilizes a

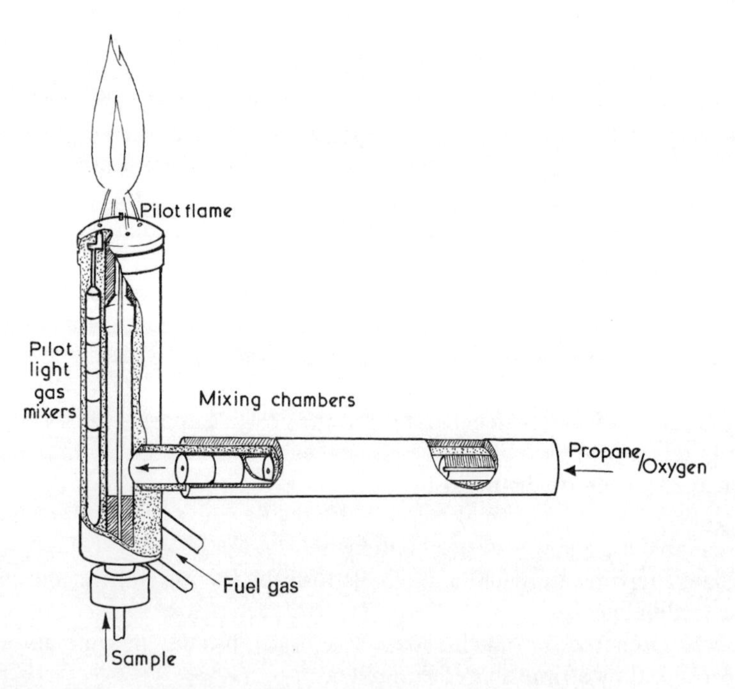

Fig. 4.3 Cut-away diagram of burner for continuous flame photometry.
Reproduced with permission from Israeli, Pelavin and Kessler[4] and New York Academy of Science.

fuel of propane and oxygen and the design principle is essentially that of Vallee and Bartholomay[5] for cyanogen and oxygen. Both a main burner, into which the sample is pumped, and auxiliary burners are used. The flame-front propagation-speed of a stoichiometric propane/oxygen mixture is much less than the gas-stream velocity necessary to aspirate the sample. The function of the auxiliary pilot flames, which are operated at below the flame-front propagation-speed, is to prevent the main flame from being blown away. The fuel gases to both the main and auxiliary burners are premixed and premetered. Mixing is achieved in a mixing chamber which provides alternate elements for expansion and contraction. Needle-valves are used for metering, separate needles being provided for each flame because by maintaining the main flame slightly lean and the auxiliary flame slightly rich the most stable flame is obtained. Once set, the needle-valve settings require only very occasional adjustment. For optimum performance it is necessary to replace the standard cylinder regulators by special-purpose specific range ones; experimental studies indicated that variations in regulated gas pressure cause changes in flame temperature and that such changes affect the emission from Na, K and the Li standard to varying extents. Cooling of the metal surface of the burner is essential, water cooling or heat dissipation by a metal-finned burner are both satisfactory. The burner is mounted in a box surmounted by a chimney, and all air entering the photometer is filtered to remove particulate matter larger than 0·3 μm. The chimney provides a duct for escaping hot gases and also prevents stray light and particles from entering the photometer chamber.

The burner is located in the central part of the chimney, which is made of glass and surrounded by a high-reflectivity (98–99%) layer of finely powdered magnesium oxide. Two sight-tubes are attached to the chimney and each guides the collected radiation from the flame to a photosensitive detector mounted at its end. Photoconductive cadmium sulphide detectors are used because of their small size and high sensitivity in the red region of the spectrum. One detector monitors the sample line (Na or K) and the other the Li standard. The wavelength is selected by means of interference filters mounted in light-tight housings at the end of the sample sight-tube. The electronic circuitry for the detectors provides a ratio null-balance between sample and reference detectors. To ensure optimum cancellation of noise in the detector system the photoconductive cells are matched to give proportionately the same change in response to variations in illumination.

For the determination of Na and K in serum the sample must first be mixed with standard Li solution and then dialysed to remove protein. The aqueous dialysate is further diluted and then a small deaerated fraction of this solution is continuously pumped into the flame. A vertically mounted T-piece vents segmenting air and most of the liquid stream to waste. The residual flow, 1 ml/min, is pumped through capillary tubing into the main flame. The required operating ranges for serum are 100–160 meq/l for Na and 2–8 meq/l for K. The calibration curve for K is linear and for Na it is linear except for slight curvature at the high concentration end. Recoveries of both elements, based on standard additions to serum samples, are essentially complete. Reproducibility is stated to be $\pm 0{\cdot}75$ meq/l for Na and $\pm 0{\cdot}15$ meq/l for K. Sample processing rates of up to 40 per hour are feasible, compared with about 10 per hour by manual methods.

4.3 Atomic-Absorption Spectrophotometry

Atomic-absorption spectrophotometry is now established as the most widely used spectrographic technique for elemental analysis. It is rapid and inherently straightforward and it lends itself to automation more readily than emission techniques, which frequently require the sample to be presented to the flame in a solid matrix. With a manual atomic-absorption spectrophotometer the limiting factor in achieving a high throughput of samples, other than chemical preparation of the samples, is usually the evaluation of results. In addition, constant operator-attention is required to present samples to the excitation system. Where large numbers of samples are to be processed there are clear economic advantages in automating the sample presentation and providing automatic facilities for calculation and display of results. The latter alone is justifiable, but increasing the rate of sample-feed without data-handling facilities serves only to accentuate the computational bottleneck. Considerable progress has been made in the provision of automated procedures for use with basic atomic-absorption spectrophotometers, both in general by the instrument manufacturers, and specifically by analytical research workers to meet their particular needs.

4.3.1 Automated Sample Presentation

The samples can be automatically fed into the burner in one of two ways; either from prepared samples by the discrete method or continuously by using pumping techniques. The discrete-sample approach has been exploited by most of the major manufacturers of atomic-absorption equipment and in terms of design the units are closely similar to automatic samplers for colorimetry. Some typical examples are given below.

4.3.2 Automatic Feeding of Discrete Samples

The Pye Unicam SP 90 atomic-absorption spectrophotometer can be used with the SP 92 automatic sample-changer which is basically a continuous chain carrying holders for 32 polythene sample-cups of 4-ml capacity. Movement of the chain carries each cup in turn beneath an automatic sampling head. The latter aspirates the sample for a fixed time into the burner.

Beckman supply a 24-position rotating tray for use in conjunction with their atomic-absorption instruments. Sample-cups are mounted around the circumference of the tray and as each cup is positioned beneath the sampling probe the edge of the tray is raised by a cam to bring the probe into contact with the sample. The operational cycle can be varied from 5 to 100 sec by adjusting the rate of movement of the tray between samples. The longer times permit the burner to re-establish temperature equilibrium between samples. The aspiration time for sample measurement is continuously variable over the range 5–100 sec. The electronic timing-control circuitry for the sample-changer also triggers the print-out of results by a digital printer. For larger sample throughputs a larger automatic sample-changer having a capacity of 200 tubes is available, based on the same principle.

The Perkin-Elmer Model 3AF incorporates an automatic sample-feed mechanism. It comprises a 200-place sample-table with the tubes mounted in a square matrix and a controller which schedules the operational sequence. The timing cycle can be preset in the range 7–60 sec. Typical coefficients of variation are stated to be better than $\pm 1\%$ for a 10-sec sample rate with standard copper solutions. The controller also incorporates a counter to identify the sample number and facilities for presetting the total number of samples to be analysed.

A versatile automatic sample-changer, the Model 51, is available for use with Varian Techtron atomic-absorption spectrophotometers. Up to 50 sample-tubes can be mounted in a rotatable plate in two concentric rings of 25. The plate is automatically indexed by one position at the end of each sampling operation, thereby bringing each tube in turn into alignment with a dipping probe which aspirates sample into the burner. The control circuitry performs several functions; it not only controls the sampling period (variable between 2 and 20 sec) and the rinsing or settling time (variable between 0 and 20 sec), but it provides synchronization with the output facilities. Thus it provides the command signal for the digital absorption indicator and also synchronizes the operation of a base-line drift corrector between samples. The sample-changer automatically switches off at the end of a run of samples and provides audible indication that the run is complete. This is achieved by inserting a special sample-tube which actuates an automatic stop; this tube can be inserted in any position so there is no limit (within the maximum of 50) on the number of samples which can be examined per run.

Gaumer, Sprague and Slavin[6] developed an automated atomic-absorption technique for determining trace elements in large numbers of samples, for use in medical screening of healthy adults; the elements sought were copper, zinc, magnesium and calcium. The hourly throughput of samples was 200, including the necessary standards for calibration purposes. The sample-feed mechanism was an LKB automatic 'Radi-Rac' fraction-collector which was slightly modified so that samples could be introduced into the aspirator of a Perkin-Elmer Model 303 spectrophotometer. Sample vials of about 8-ml capacity were used and up to 240 could be accommodated in the fraction-collector. Solution was withdrawn through a steel capillary connected by polythene tubing to the atomizer of the spectrophotometer, and a simple eccentric wheel enabled the capillary to be raised and lowered. For the analysis of blood plasma prior dilutions of 1:1 for copper and zinc, and 1:50 or 1:100 for magnesium and calcium, were required. These were performed with an 'Auto-Dilutor' (Scientific Products) designed for repetitive dilutions with the same diluent. It was demonstrated that the precision of automatic dilution was virtually identical to that obtained manually. Test runs on standard solutions of the four elements yielded relative standard deviations of 2–3% over the concentration range 2–700 mg/l. The insertion of a vial containing water between successive samples indicated a carry-over between samples of not more than 3%. The Perkin-Elmer Model 303 spectrophotometer was used in conjunction with the DCR1 concentration read-out unit and the latter provided the signal between samples to index the fraction-collector to the next position.

In a later paper Slavin and Slavin[7] extended the range of application of the instrument to the automatic determination of copper, silver, nickel, chromium, iron, lead, tin, magnesium and aluminium in used aircraft-lubricating oils. Some design improvements were incorporated; the LKB 'Radi-Rac' fraction-collector was replaced by an LKB rectilinear sample-table to give a more compact layout, and samples were diluted automatically by a Fisher diluter. The trace metal concentrations ranged from less than 1 mg/l for silver up to 100 mg/l for iron. In general all nine elements could be determined in this concentration range with a relative standard deviation of 10% or better.

4.3.3 Automatic Continuous Feeding of Samples

This type of approach is based largely on the use of 'AutoAnalyzer' components for sample pretreatment and transport; atomic-absorption spectrophotometry can be used for the end-determination in the same way as flame photometry. Since the range of elements which can be sensitively determined is greater for atomic absorption than for flame photometry more extensive use of this type of automated atomic-absorption spectrophotometry can be expected. The examples given below serve to illustrate the diversity of successful applications.

For the determination of trace metals in aqueous solutions Fishman and Erdmann[8] used a Technicon sampler and proportioning pump in conjunction with a Perkin-Elmer Model 303 atomic-absorption spectrophotometer. No sample pretreatment was necessary and samples were diluted and mixed with an internal standard, using the manifold shown in Fig. 4.4. After mixing in a double mixing coil the samples were pumped

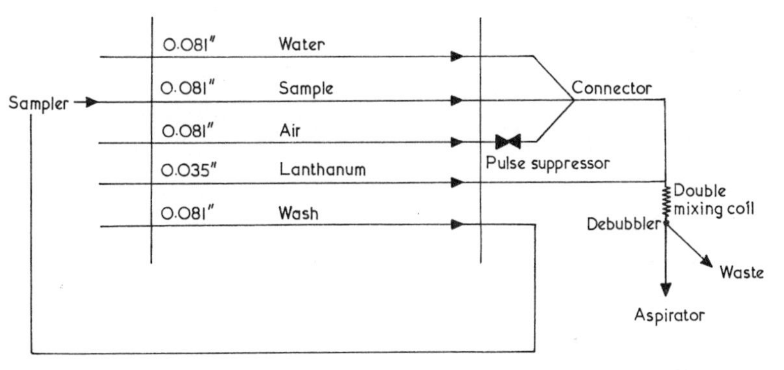

Fig. 4.4 Manifold for continuous determination of trace metals by atomic absorption spectrophotometry.
Reproduced with permission from Fishman and Erdmann[8] and Perkin-Elmer.

to the aspirator and burner of the spectrophotometer. The elements determined were copper, manganese, iron and zinc in the range 0–0·5 mg/l, lithium 0–1 mg/l, strontium 0–2 mg/l, and sodium, potassium, calcium and magnesium 0–50 mg/l. The sample-processing rate achieved varied from 30 to 60 per hour depending upon the amount of noise suppression applied to the instrument. As noise suppression is increased the time taken for the signal to reach a maximum value increases, thereby limiting the rate of sample throughput. For lithium, calcium, magnesium and strontium, comparison data were obtained for both manual and automatic atomic-absorption spectrophotometry; good agreement was obtained in each case.

Klein and co-workers[9, 10, 11] developed automated atomic-absorption spectrophotometric methods for calcium in blood serum, spinal fluid and urine, using a Techtron AA3 spectrophotometer and 'AutoAnalyzer' components. The method is based on earlier manual procedures by Willis[12] and Zettner and Seligson[13] and involves dilution of the sample with 0·5% $LaCl_3.7H_2O$ solution in 0·1*M* HCl and dialysis of the calcium into a recipient solution of 0·1*M* HCl containing 0·5 ml/l of surfactant (Brij). The dialysate was pumped to the atomizer/burner through a C3 debubbler. This was connected to the metal capillary of the atomizer through an N5 nipple with a Tygon sleeve by a length of polythene tube.

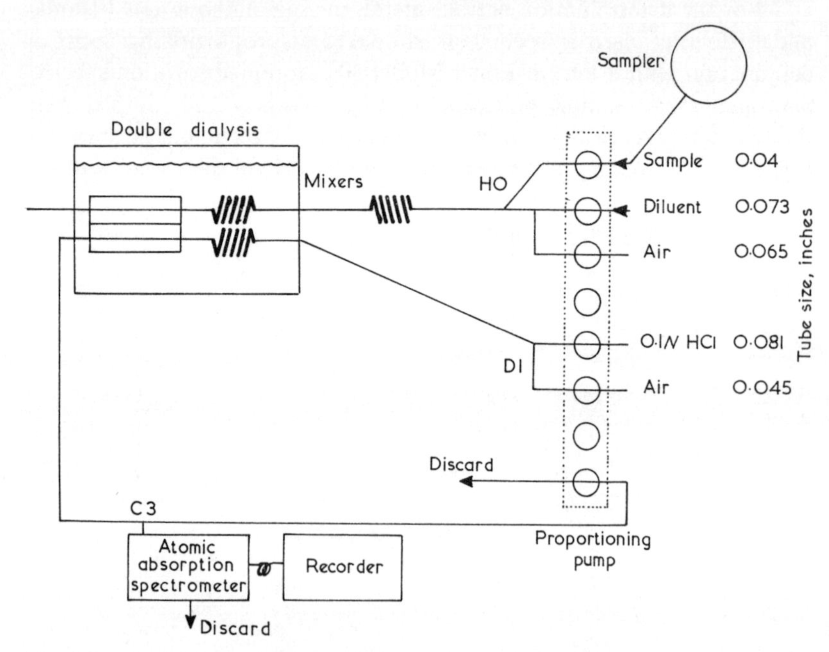

Fig. 4.5 Manifold for continuous determination of Ca and Mg by atomic absorption spectrophotometry following dialysis.

The full manifold detail is shown in Fig. 4.5. The Sampler II was used with a sample-to-wash ratio of 2:1. At a sampling rate of 40/hr and taking 0·6 ml of serum, the recovery of added calcium standards over the concentration range 50–200 mg/l was 99%. Interaction between successive samples was less than 1%. For samples of urine and spinal fluid the protein content is low and the dialysis stage can be omitted.

Gochman and Givelber[14], using 'AutoAnalyzer' components in conjunction with an Instrumentation Laboratory Model 153 atomic-absorption spectrometer, developed a method for simultaneous determination of calcium and magnesium in blood serum. The manifold is shown in Fig. 4.6. It has several advantages over the method of Klein and co-workers outlined above. A small sample volume (30 μl) is used and this obviates the need for a dialysis step; consequently a faster rate of sample processing (90 samples/hr) can be employed. Lanthanum is used as the internal standard for spectrometric measurement and to avoid clogging the burner with solid deposits the standard lanthanum solution is made up in an aqueous solution containing 4% of n-butanol. The performance of the automated method was compared with that of a manual method using a Perkin-Elmer Model 303 spectrometer. For both elements the standard deviations, based on over 100 analyses were almost identical. Similar studies are reported by Griffiths and Raynaud[15] and by Pennacchia *et al.*[16]

The flameless atomic-absorption method of Hatch and Ott[17] for determining traces of mercury in biological samples has been automated by Armstrong and Uthe[18]. It involves an initial manual wet oxidation with potassium permanganate and nitric acid but thereafter the procedure is automatic. 'AutoAnalyzer' components were used as shown diagrammatically in Fig. 4.7, and samples were processed at 30/hr. The mercury vapour was excited in a cuvette 150 mm long and 6 mm in diameter, the sample being swept into the cuvette through a debubbler by a current of air. The concentration of mercury in the cuvette at time t is related to the concentration C of the displacing stream by the equation

$$C_t = C(1-e^{-Rt/V})$$

where V is the cuvette volume and R is the rate of air flow. Using a sample-to-wash ratio of 1:2 and air flow of 27·3 ml/min with a sampling rate of 30/hr, 93% of the steady-state concentration is achieved. Because the removal of mercury from the previous sample is not entirely complete, some 98% being removed, results are recorded on a non-horizontal base-line. The error due to this in measurement of peak heights is stated to be 0·14% of the concentration of the immediately preceding sample. The method, which was used with a Perkin-Elmer Model 403 spectrophotometer, yielded mean recoveries of 94% following standard additions (100–500 ng of Hg) of mercuric chloride or methylmercuric chloride to whitefish samples. Relative standard deviations were in the range from ± 3 to $\pm 8\%$ depending upon the mercury content of the fish (0·10–9·0 ppm). An essentially similar method for determining mercury in organic matrices has been developed by Bailey and Lo[19].

Butler, Brink and Engelbrecht[20] utilized automated atomic-absorption spectrophotometry for the rapid determination of gold at the sub-ppm level in cyanide tailings as an alternative to the time-consuming fire-assay

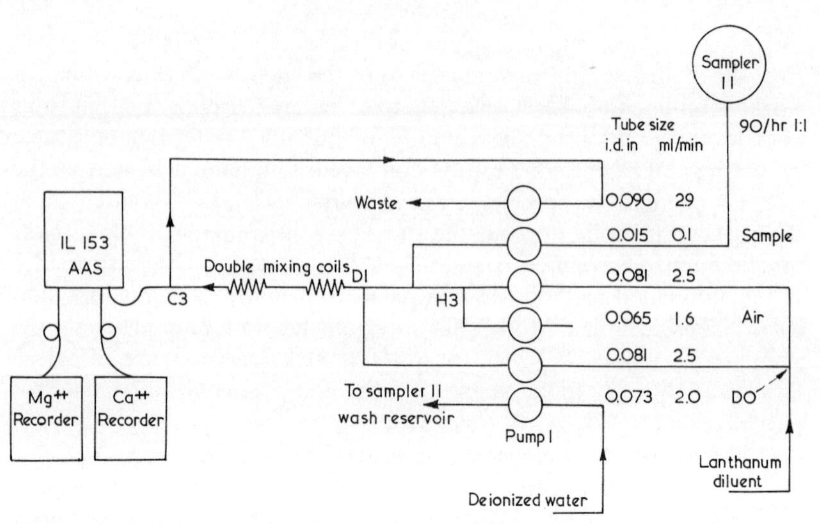

Fig. 4.6 Manifold for continuous determination of Ca and Mg by atomic absorption spectrophotometry without dialysis.
Reproduced with permission from Gochmann and Givelber[14] *and American Association of Clinical Chemists.*

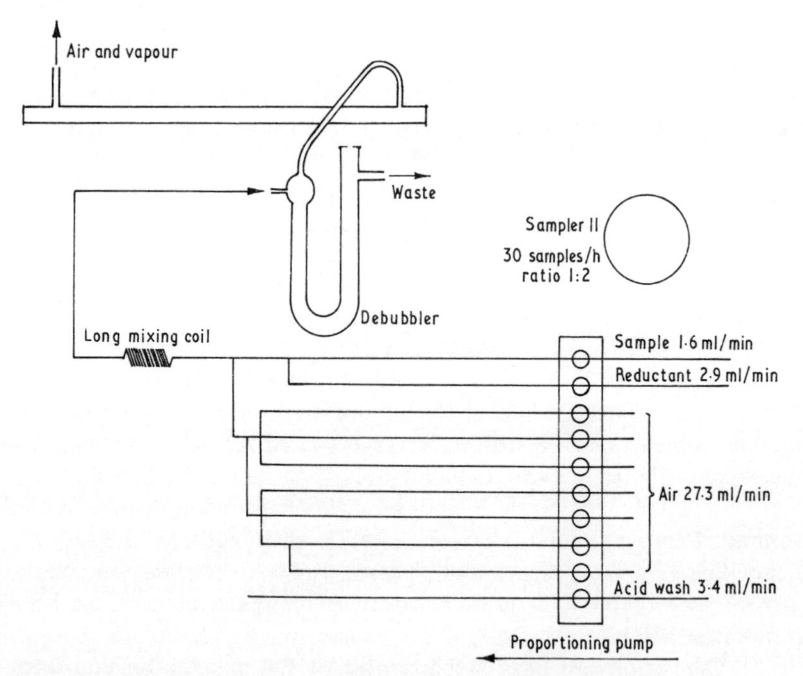

Fig. 4.7 Manifold design for determination of mercury by atomic absorption spectrophotometry.
Reproduced with permission from Armstrong and Uthe[18] *and Perkin Elmer.*

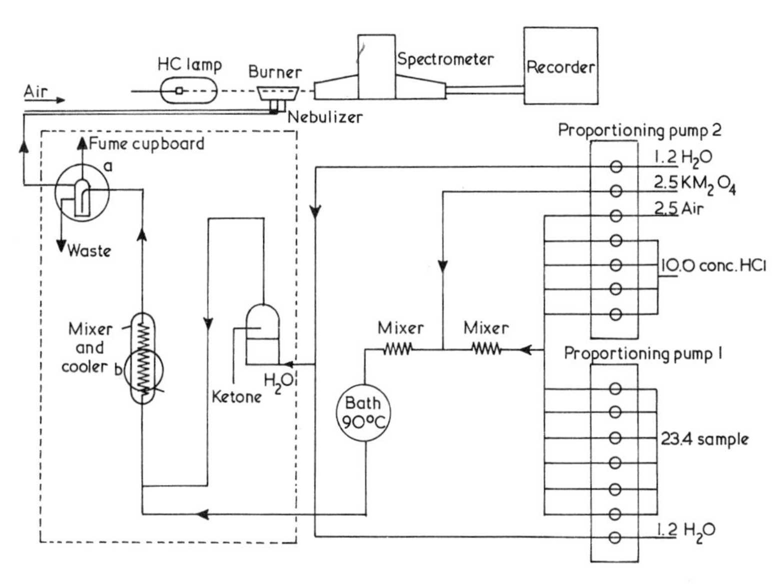

Fig. 4.8 Manifold for separation and determination of trace quantities of gold. *Reproduced with permission from Butler, Brink and Engelbrecht*[20] *and Institute of Mining and Metallurgy.*

procedure. Sample pretreatment comprises oxidation with potassium permanganate and hydrochloric acid to convert AuCl into $AuCl_3$. The latter is then extracted into pentan-2-one which is aspirated into the spectrometer flame. Fig. 4.8 shows the apparatus diagrammatically. The oxidation stage was performed with two 'AutoAnalyzer' proportioning pumps and a heating-bath. The heated solution was pumped to the bottom of a bead-filled mixing coil where it was merged with pentan-2-one introduced by displacement pumping. This was necessitated by the inability of plastic materials to withstand attack by the ketone. This section of the apparatus was constructed entirely of glass. After mixing the phases were pumped to a separator detailed in Fig. 4.9; the aqueous phase passed downwards to waste and the solvent stream was drawn into the atomizer. Dimensions of the separator were critical because the solvent phase tended to be drawn away with the aqueous one. Tests with standard gold solutions indicated that the response time of the system to reveal a concentration change was 2 min and a further 6 min were required for the spectrometer reading to stabilize. Using a sample-to-ketone volume ratio of 10:1 the method could be operated over the concentration range 0·005–1 ppm. At 0·05 ppm the coefficient of variation was 3·5%. With stabilized power supplies the base-line drift was less than 2% over an 8-hr period.

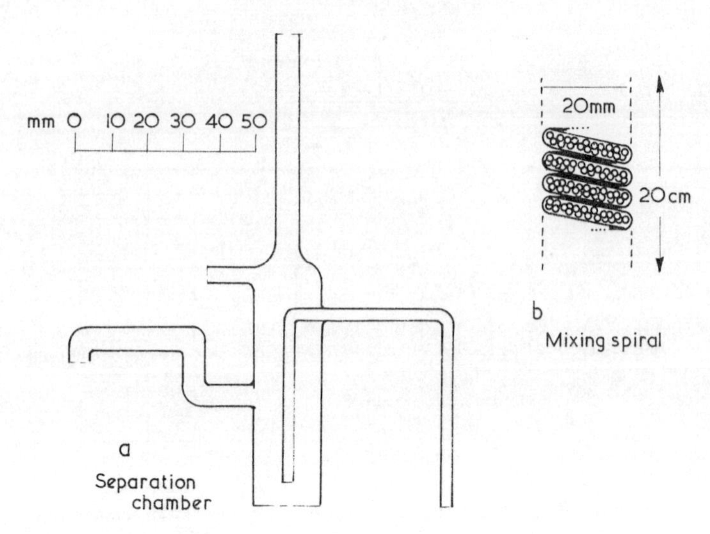

Fig. 4.9. Details of solvent-aqueous separator and mixing coil from Fig. 4.81.
Reproduced with permission from Butler, Brink and Engelbrecht and Institute of Mining and Metallurgy.

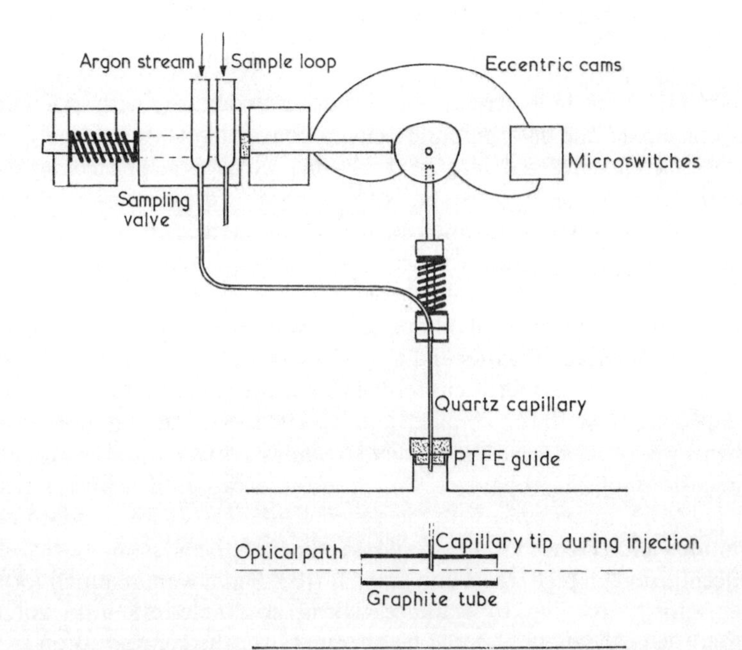

Fig. 4.10 Apparatus for automatic sampling and atomic absorption analysis of a flowing stream.
Reproduced with permission from Pickford and Rossi[21] *and The Society for Analytical Chemistry.*

A method for determining a number of metals at the μg/l level in high-purity water by automatic atomic-absorption spectrophotometry is described by Pickford and Rossi[21]. The flameless method, using a heated graphite tube of the Massmann type in conjunction with a Beckman 1301/DBG atomic-absorption spectrophotometer, was selected. An automatic sampling and injection mechanism was devised which enables 100-μl samples of water, taken from a flowing-sample loop, to be periodically transferred to the graphite tube. The principle is illustrated in Fig. 4.10. The periodicity of the unit is governed by the rotation of two eccentric cams driven by a 2-rpm synchronous motor. The larger of the two performs three operations, the raising and lowering of a quartz capillary through which sample is injected into the graphite tube, switch-on of an oven unit to evaporate the sample from the tube, and stopping the synchronous motor after the quartz capillary has been withdrawn from the tube. The last two functions are initiated by two raised studs attached to the outside of the cam, which operate microswitches. The smaller cam performs the sampling operation; it bears against a push-rod connected to the sampling valve which is of a conventional type fabricated from PTFE and stainless steel, with an enclosed volume of 100 μl. Movement of the push-rod operates a PTFE slider which causes 100 μl of water to be transferred from the sample loop to a stream of argon which carries it as a single plug to the quartz capillary and thence to the graphite tube. When the sample has reached the graphite tube the spring-loaded PTFE slider returns to its rest position, thereby isolating the sample stream from the injection unit. Low flow-rates of sample (20 ml/min) and argon (5 ml/min) are necessary to prevent leakage when the valve is in an intermediate position during sampling and to avoid segmenting of the plug of sample. If the latter occurs the analytical results show a loss of precision.

The sampling and injection unit, together with the power supply and programming for the graphite tube and the read-out system is controlled by a central programming unit comprising a motor-driven 24-position switch. The latter provides some flexibility in the sequencing of the analysis. When the 100-μl sample has been evaporated, further portions of sample (up to 4) can be taken or the program advances to the clearing and volatilization stage.

Avoidance of adventitious sample contamination is critical in corrosion studies with high-purity water; this is aided in the automatic sampling and injection unit by limiting the sample-contact materials to quartz and PTFE.

The performance of the equipment was evaluated by using standard solutions with the sample and injection unit programmed to operate on a 3-min cycle time. Table 4.1 gives the results for 6 elements. The authors state that the analytical precision is limited by the injection stage and that the standard deviations represent an improvement over those obtained when the method is performed manually.

Table 4.1

Element	*Concentration, μg/l*	*Number of replicates*	*Relative standard deviation, %*
Co	50	10	0·6
Cr	100	17	1·0
Cu	50	12	1·2
Fe	30	10	1·1
Mn	10	15	1·7
Ni	50	10	2·4

An automated atomic-absorption spectrophotometric method for the direct determination of lead in air has been described by Loftin, Christian and Robinson[22]. It utilizes a long absorption cell in which the lead compounds present in the air-sample are reduced to elemental lead by carbon monoxide generated from heated carbon rods. The layout of the tube and carbon rods is shown in Fig. 4.11. The quartz absorption tube

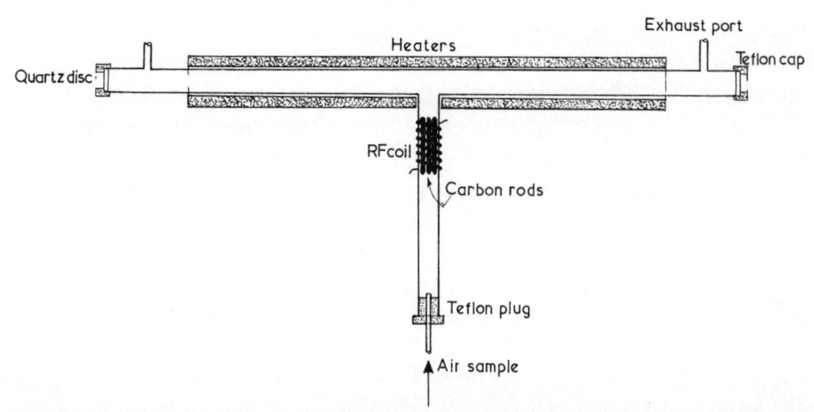

Fig. 4.11 Equipment for atomic absorption spectrophotometric determination of lead in air.
Reproduced by permission from Loftin, Christian and Robinson[22] and Marcel Dekker Inc.

is T-shaped and the air-sample is introduced continuously at the base of the T. It is heated by 1-kW nichrome resistance heaters and the bundle of $\frac{1}{4}$-in. diameter carbon rods by a 5-kW radiofrequency generator. Atomic-absorption measurements are made by using a demountable lead hollow-cathode lamp and a Jarrell Ash 82–500, 0·5-metre monochromator with an 8-speed scanning attachment. To calibrate the instrument the diffusion method of Altshuller and Cohen[23] was employed; this involves diffusion of tetraethyl lead from a tube of known diameter and path-length into an air-stream of known velocity and following the weight-loss of the tube. With an absorption tube having a volume of 160 ml the lower limit of sensitivity of the instrument was shown to be about 3×10^{-11} g of lead. Interferences are minimal; interference from fragments of organic molecules was sought but with the exception of high concentrations of chloroform, carbon tetrachloride and methyl iodide the effects were negligible. The method is rapid and sensitive and clearly applicable to other metallic contaminants in air.

4.3.4 Automatic Read-Out Facilities

Virtually all manufacturers of atomic-absorption spectrophotometers now offer a range of facilities for calculation and presentation of results and as noted above these become indispensable if samples are presented automatically and in large numbers to the instrument.

The read-out facilities relevant to atomic-absorption spectrophotometry are essentially similar to those for any other technique based on the processing of electrical signals generated by a detector sensitive to light in the ultraviolet and visible regions of the electromagnetic spectrum. The principal requirements are a means of calculating calibration factors and of using them to calculate and display the analyte concentrations for the unknown samples. The method of presentation of results can, as in other similar techniques, take the form of a digital indication or print-out, or a computer record. Two aspects of particular significance in automated atomic-absorption methods, considered in more detail below, are first the tendency for the calibration curves for some elements to be slightly non-linear, especially where a wide concentration range is covered, and the tendency for the intensity of the radiation-source (usually a hollow-cathode lamp) to drift. Although this can be compensated to a large extent by the double-beam technique, some base-line drifts nevertheless occur and must be corrected to maximize the accuracy and reproducibility of results.

The Varion Techtron AA-5R instrument is supplied with a range of automatic read-out facilities which typify the range of units which is now commercially available. Samples are automatically fed to the burner from the Model 51 automatic sample-changer described in the preceding section. Automatic base-line correction circuitry is incorporated, which is actuated during the indexing period between the completion of analysis of one sample and the commencement of the next. A digital indicating unit DI 30 enables results to be displayed as absorbance, transmission or direct sample concentration on a four-digit digital voltmeter. Each result can be held indefinitely or the display period can be varied between 0·25 and 1 sec. The spectrophotometer signal-intensity can be averaged over a preselected period within the range 1–10 sec and scale-expansion factors up to 20 can be applied. Deviations from Beer's law can be linearized by using unit DC 31, a digital corrector. This divides the meter scale into 10 segments and curvature is progressively linearized over the entire meter range. Two slope-adjustment channels are provided so that correction curves can be established for two different calibration curves. Data from the DI 30 digital indicator can be displayed as a print-out by feeding them to a digital printer DP 32 or fed to a data-acquisition unit Model 34 for computer processing. Information from the digital indicator in binary coded decimal form is translated to a punched paper tape. The tape code is ASC 11 produced from a teletype or tape punch. This approach is of most value where large numbers of results are to be processed, four input–output channels enabling signals from four spectrophotometers to be processed simultaneously. Computer programs in both Basic and Fortran languages are available and they incorporate correction facilities for sensitivity and base-line drifts and compensation for Beer's law deviations.

4.4 Ultrasonic Nebulization

In the vast majority of methods based upon flame photometric and flame absorption techniques the sample, in solution, is nebulized through

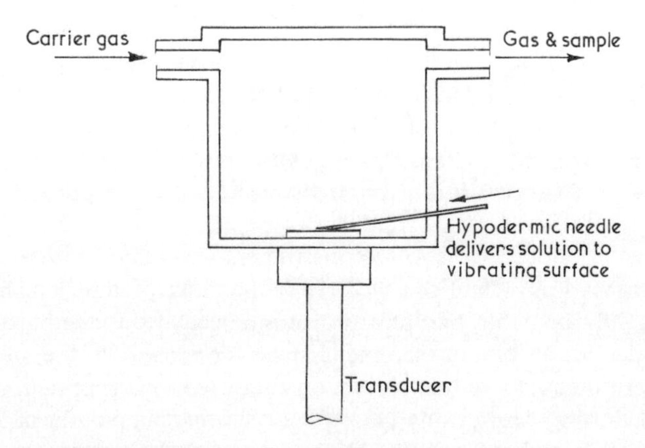

Fig. 4.12 Ultrasonic nebulization of liquid samples.
Reproduced with permission from Kirsten and Bertilsson[27] and American Chemical Society.

a jet into the flame. In seeking to produce an aerosol of finer dispersion and higher sample concentration, ultrasonic nebulization has been examined. In earlier studies[24, 25, 26] solutions were dispersed from their surface into a relatively large vessel with the resulting disadvantages that large sample volumes were required and constant, reproducible nebulization rates were difficult to achieve. Kirsten and Bertilsson[27] describe a technique for continuous and instantaneous nebulization which is of potential applicability in automatic analysers. The sample is fed continuously on to a vibrating surface in a chamber shown diagrammatically in Fig. 4.12. The geometry of the sample introduction system is critical; if quantitative and reproducible nebulization is to be achieved the formation of individual sample droplets must be avoided. Optimum conditions are obtained by introducing the sample on to the vibrating surface through a hypodermic needle mounted at 10–15° to the horizontal with the tip placed about 0·3 mm above the centre of the surface. With sample flow-rates of up to 0·3 ml/min the nebulization is immediate, quantitative and continuous, but at higher flow-rates the sample solution tends to creep along the needle, thereby causing irregular nebulization. The ultrasonic nebulization technique produces an enhanced sample concentration in the aerosol when compared with pneumatic nebulization at the same flow-rate; for example at 0·05 ml/min enhancement factors of 2·8 for sodium (5 mg/l), 4·6 for calcium (5–100 mg/l) and 3·5 for magnesium (50–5000 mg/l) were obtained.

Ultrasonic nebulization offers advantages for continuous control of alloy compositions by facilitating measurements directly and continuously on the molten surface. Fig. 4.13 is a diagram of a nebulizer developed by Fassel and Dickinson[28] for producing an aerosol of molten metal. When the flat tip of the step horn of an ultrasonic generator is placed in contact with the molten metal surface a reproducible metal-dust aerosol is formed which is carried upward in a stream of argon into the burner of the spectrograph. To avoid overheating of the probe when nebulizing high-melting alloys, it is switched off after each nebulizing period of 15–30 sec. Using Woods' metal to test the reproducibility of the technique, 21 repeated scans of the intensity ratio Sn 303·41 nm/Bi 302·46 nm yielded a relative standard deviation of 4·9%.

4.5 Infra-red Spectrometry

Although infra-red spectrometry is widely used as an analytical technique, there have so far been few attempts to use it for automatic analysis of samples. This is probably because sample preparation often requires a measure of manipulative skill (e.g. preparation and mounting

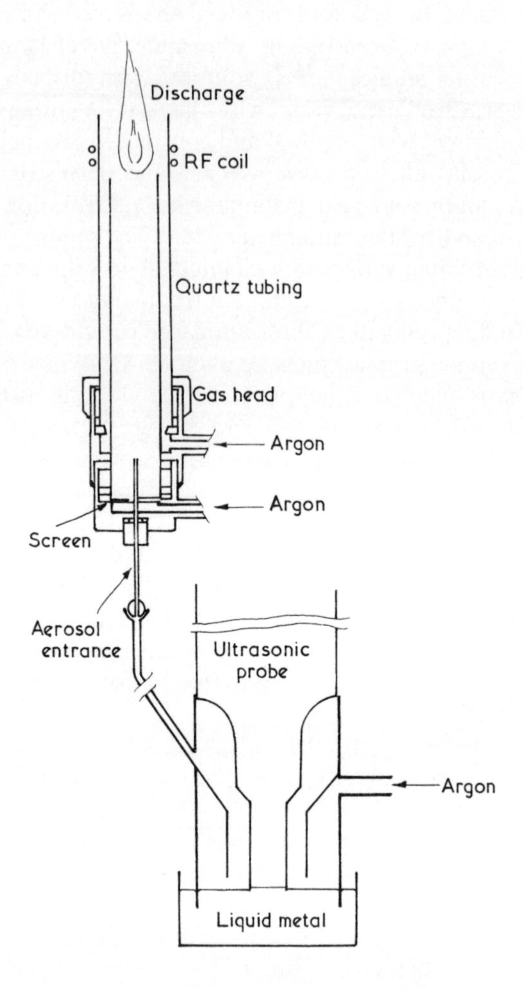

Fig. 4.13 Ultrasonic nebulization from a molten metal surface.
Reproduced with permission from Fassel and Dickinson[28] *and American Chemical Society.*

of mulls) and automation in this context is neither scientifically or economically attractive. However, in cases where samples can be examined direct without special preparation, automatic methods of presenting the sample to the spectrometer are entirely feasible and examples are detailed below. Equally, automatic sample-changing techniques are applicable where the sample can be readily prepared in a rigid, mechanically stable form.

4.5.1 Solid Samples

Infra-red spectrometry is widely used in polymer studies for determination of end-groups, studying degradation, and assessing number-average molecular weights. In such instances the polymer can be examined in the form of a thin sheet or compounded in pellets of material transparent to the desired wavelengths, such as KBr. Early attempts to automate the presentation of samples to the spectrometer involved the use of a modified 35-mm slide changer[29] and mounting the samples on a rotating disc[30].

The rotating disc concept has been extended by Bakker, Frost and Ogilvie[31]. It is limited in that the number of samples that can be examined in a single run can be no more than the sample capacity of the disc. This is 30 in the example under discussion and is sufficient to justify the automation because analyses can be made during non-working hours without operator attention. Furthermore the disc can be mounted in a vacuum chamber if samples are unstable in air.

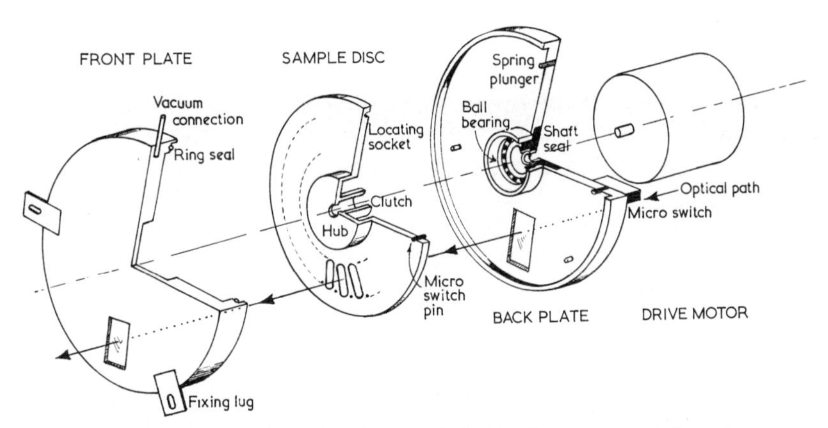

Fig. 4.14 Exploded view of rotating disc sample holder for automatic infra-red spectrometry.
Reproduced with permission from Bakker, Frost and Ogilvie[31] and American Chemical Society.

An exploded view of the rotating disc is shown in Fig. 4.14. The disc has 30 slots cut round its circumference, and samples, in the form of polymer film, are mounted in front of the slots and secured by spring clips. The disc rotates inside a housing which is in two parts, a front plate and a back plate. The former is attached to the spectrometer and the latter to a Crouzet drive motor (0·1 rpm) and a Burgess microswitch (Type V4T4). The bottom ends of each plate are fitted with NaCl windows to provide the optical path and each sample is aligned with it sequentially by rotation of the disc. Accurate positioning of the disc after each sample advancement is important and is provided by a 'ball-and-catch' system; the microswitch is activated by ball-headed pins appropriately mounted at the back of the rotating disc, one pin being omitted so that the end of the sample sequence is recognized and the motor switched off. The control circuit for operation of the sample-changer is built from solid-state logic modules. Using the sample-changer, 30 samples could be examined in 10 hr compared with 10–15 hr if performed manually. However, the automatic unit could be operated during silent hours and saved operator time.

4.5.2 Liquid Samples

4.5.2.1 *Discrete Method*

For the determination of the composition of ethylene/propylene co-polymers Brame, Barry and Toy[32] developed a fully automated infra-red spectrophotometric system for processing samples dissolved in carbon tetrachloride.

The analyser is built round a Beckman Model IR 20 A spectrometer. It is provided with a wave-number programmer which permits automatic selection of three sets of any two wave-number regions, the latter being manually set in accordance with the analytical requirement. Between these two regions the spectrometer scans at rapid speed. Up to three individual programs can be used for up to three analytical methods without operator intervention. The automatic change from one programme to another is achieved through the sample-handling unit which is fitted with a lamp and phototransistor which monitors coding marks on the side of the tray carrying the sample-tubes. To each tray are affixed either one or two black circles in one or both of two positions; this provides three programs identified by black/white, white/black or black/black. The phototransistor system provides the logic circuitry for selecting the desired program.

The liquid-handling system has a capacity of 200 samples and its functions are to extract a small volume of sample from the sample-tube, fill a 1-mm thick NaCl cell with it and subsequently empty it to waste. It comprises a Buchler Fractomette 200 in which up to 200 sample-tubes (8 × 150 mm) can be mounted. To avoid loss of solvent by evaporation the tubes may be covered with aluminium foil or fitted with a rubber seal. As each new sample moves into position for analysis a hypodermic needle pierces the seal and withdraws sample by suction and draws it into the NaCl cell positioned in the spectrometer beam. The flow-paths for filling and emptying the cell are shown in Fig. 4.15. Three solenoid valves containing fluorocarbon O-rings are used to provide the solution paths, and

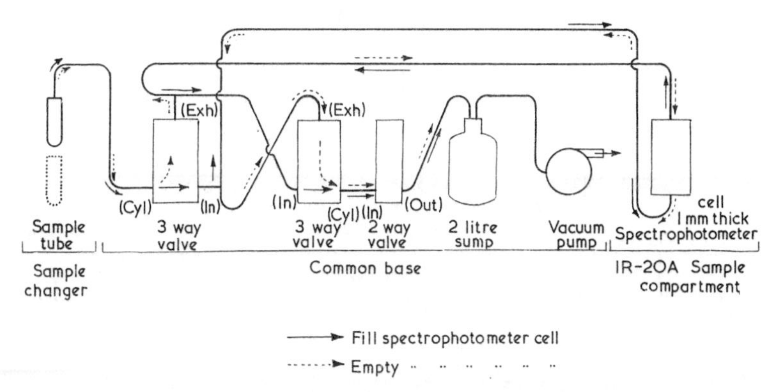

Fig. 4.15 Flow diagram for filling and emptying a flow-through cell for infra-red spectrometry.
Reproduced with permission from Brame, Barry and Toy[32] *and American Chemical Society.*

are operated from the program device described above. Two electric timers provide the selection of the filling and emptying times, and are activated automatically at the sample-change period, the cell being emptied before the next sample is moved into position. To avoid contamination of the NaCl cell a solvent wash is provided by inserting a tube of solvent after each batch of three samples and before a different analysis, as selected by the programmer, is begun.

The read-out from the spectrometer is performed by two digital encoders simultaneously reading wave-number and corresponding sample transmittance. The encoded raw data are stored in a Perkin-Elmer DDR-IC teletype paper-punch unit and the paper tape fed to a computer for calculation and reporting of results.

For the determination of the propylene content of ethylene/propylene co-polymers the automatic spectrometer yielded a relative precision of 1% or less which is better than that obtained when the analysis is performed manually.

4.5.2.2 *Continuous Method*

Under favourable circumstances the 'AutoAnalyzer' method of automatic continuous analysis can be used in conjunction with an infra-red spectrometer detector. This has been demonstrated by Robbins[33] for the determination of D_2O over the concentration range 0·125–1·000% in water and biological fluids. The manifold is shown in Fig. 4.16; it involves the use of an 'AutoAnalyzer' sampler equipped with sample-levelling and constant-volume attachments, a proportioning pump, a dialyser unit, a Perkin-Elmer Model 237 infra-red spectrophotometer and a chart recorder.

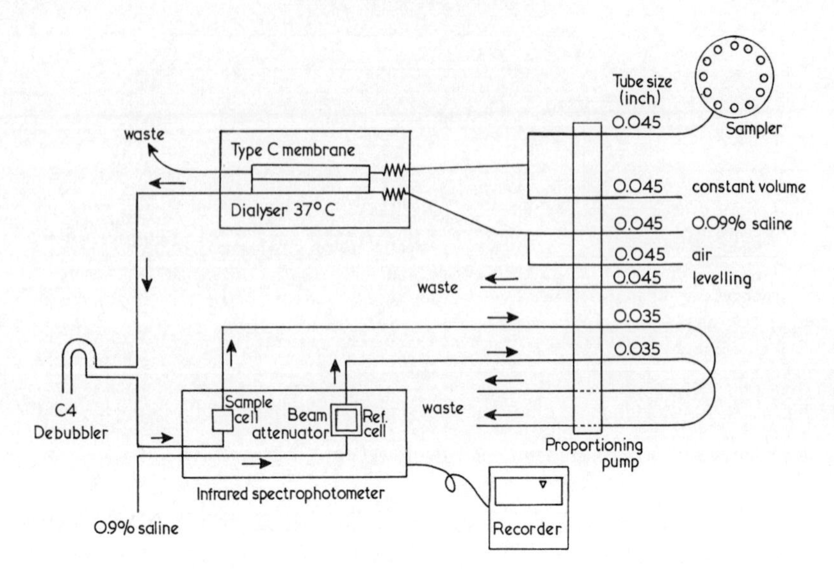

Fig. 4.16 Manifold for continuous determination of D_2O by infra-red spectrometry. *Reproduced with permission from Robbins and American Association of Clinical Chemists.*

The spectrophotometer is a double-beam instrument and the manifold design is such that a blank solution of distilled water or 0·9% saline solution is continuously pumped through the reference cell. The only modification necessary to the infra-red detector is the conversion of the manually filled cells for flow-through operation. This is achieved by fixing polythene tubing to the two filling holes and using the cell at 90° to the standard position so that the filling and emptying holes are vertical with respect to the cell cavity. One of the polythene tubes carries sample from the debubbler (or reference solution from a reservoir) into the bottom of the cell and the other carries the measured sample, or reference solution, through the proportioning pump to waste.

The sampler was operated at 60 samples per hour but the effective analysis rate was 20/hr because each sample was followed by two distilled-water blanks to make sample carry-over effects insignificant. Results were calculated manually from peak height measurements. Recovery values for D_2O when added to water, whole blood, plasma and urine were 99·9, 101·7, 101·1 and 100·2% respectively. The relative standard deviation of peak height measurement varied from approximately ±0·5% at 1·000% D_2O to approximately 1% at the 0·125% level.

4.5.3 Gaseous Samples

Continuous analysis of gases by infra-red spectrometry is well established and widely used as a monitoring method. Instruments for this purpose are available commercially, and are largely based on designs utilizing fixed optics for maximum reliability and simplicity of operation. Their main use is for determining one component of interest in a gaseous sample.

Where a larger number of components of a gas mixture are to be determined the technique of scanning the dispersing element (prism or grating) through each wave-number region in turn may be preferred. An instrument of this type for determining up to 11 components automatically is described by Imbert and Delmas[34]. A grating spectrometer utilizing the Ebert–Fastie mounting forms the basis of the instrument. The grating is rotated by an electric motor, the angular settings for each desired wave-number are converted into a digital angular code which is used to program the grating movement. An auxiliary reversible motor is provided for scanning the grating across each wave-number region of interest. Because the instrument is a single-beam one, samples being examined in a cell 50 cm long, the measured radiation intensity in the absence of absorbing gas varies with wave-number and must be specifically determined for each wave-number used. Each value is related to that measured at a reference wave-number by a correction factor determined experimentally and held in the analogue computer used to calculate results. Calibration factors for each gas of interest are determined by making absorption measurements at known partial pressures; these are also stored in the computer. These conversion factors, together with the total gas pressure, separately measured by an electrical sensor, enable calculations to be made of concentration of the sample constituents. The entire measurement cycle for 11 components takes 1 min. The limit of detection, based on a study of fluorine-containing gases, is about 10 ppm.

References

1. Dawson, J. B., Ellis, D. J. and Milner, R. *Spectrochim. Acta*, 1968, **23B**, 695.
2. Lowe, J. J. and Martin, R. J. *Appl. Spectry.*, 1969, **23**, 587.
3. Barnett, W. B. *Atomic Absorption Newsletter*, 1970, **9**, 6.
4. Isreeli, J., Pelavin, M. and Kessler, G. *Ann. N.Y. Acad. Sci.*, 1960, **87**, 636.
5. Vallee, B. L. and Bartholomay, A. F. *Anal. Chem.*, 1965, **28**, 1753.
6. Gaumer, M. W., Sprague, S. and Slavin, W. *Atomic Absorption Newsletter*, 1966, **5**, 58.
7. Slavin, S. and Slavin W. *Atomic Absorption Newsletter*, 1966, **5**, 106.

8. Fishman, M. J. and Erdmann, D. E. *Atomic Absorption Newsletter*, 1970, **9**, 88.
9. Klein, B., Kaufman, J. H. and Morgenstern, S. *Clin. Chem.*, 1967, **13**, 388.
10. Klein, B., Kaufman, J. H. and Oklander, M. *Clin. Chem.*, 1967, **13**, 797.
11. Klein, B. and Kaufman, J. H. *Clin. Chem.*, 1967, **13**, 1079.
12. Willis, J. B. *Spectrochim. Acta*, 1960, **16**, 259.
13. Zettner, A. and Seligson, D. *Clin. Chem.*, 1964, **10**, 869.
14. Gochman, N. and Givelber, H. *Clin. Chem.*, 1970, **16**, 229.
15. Griffiths, N. and Raynaud, C. *Rev. Franc. Etudes. Clin. Biol.*, 1969, **14**, 315.
16. Pennacchia, G., Bethune, V. G., Fleisher, M. and Schwarz, M. K. *Clin. Chem.*, 1971, **17**, 339.
17. Hatch, W. R. and Ott, W. L. *Anal. Chem.*, 1968, **40**, 2085.
18. Armstrong, F. A. J. and Uthe, J. F. *Atomic Absorption Newsletter*, 1971, **10**, 101.
19. Bailey, B. W. and Lo, F. C. *Anal. Chem.*, 1971, **43**, 1525.
20. Butler, L. R. P., Brink, J. and Engelbrecht, S. A. *Trans. Inst. Mining Metall.*, 1967, **76**, 188.
21. Pickford, C. J. and Rossi, G. *Analyst*, 1972, **97**, 647.
22. Loftin, H. P., Christian, C. M. and Robinson, J. W. *Spectrosc. Letters*, 1970, **3**, 161.
23. Altshuller, A. G. and Cohen, I. R. *Anal. Chem.*, 1960, **32**, 802.
24. Dunken, H., Pforr, G. and Mikkeleit, W. *Z. Chem.*, 1964, **4**, 237.
25. Dunken, H., Pforr, G., Mikkeleit, W. and Geller, K. Z. *Chem.*, 1963, **3**, 196.
26. West, C. D. and Hume, D. N. *Anal. Chem.*, 1964, **36**, 412.
27. Kirsten, W. J. and Bertilsson, G. O. B. *Anal. Chem.*, 1966, **38**, 648.
28. Fassel, V. A. and Dickenson, G. W. *Anal. Chem.*, 1968, **40**, 247.
29. Johnson, D. R., Cassels, J. W., Brame, E. G. and Westneat, D. F. *Anal. Chem.*, 1962, **34**, 1610.
30. McNiven, N. L., Hoffmann, P., and Scrimshaw, G. *Anal. Chem.*, 1965, **37**, 778.
31. Bakker, E. J., Frost, J. S. and Ogilvie, G. D. *Anal. Chem.*, 1970, **42**, 1117.
32. Brame, E. G., Barry, J. E. and Toy, F. J. *Anal. Chem.*, 1972, **44**, 2022.
33. Robbins, R. C. *Clin. Chem.*, 1969, **15**, 56.
34. Imbert, P. and Delmas, P. *J. Chim. Phys. Physicochim. Biol.*, 1970, **67**, 1430.

Chapter 5

Thermal Methods

Analytical methods based upon enthalpimetric changes during a chemical reaction are finding increasing use. To date, however, the automation of such methods has received only limited attention. Nevertheless useful studies have been undertaken in the areas of enthalpimetry, thermogravimetric analysis and differential thermal analysis and these are described below.

5.1 Enthalpimetry

Priestley, Sebborn and Selman[1] demonstrated that enthalpimetric titrations could be carried out by a continuous-flow approach; sample and reagent (in excess of the stoichiometric quantity) are passed at constant flow-rates into a mixing chamber and the temperature change due to the reaction is continuously monitored. This change is proportional to the concentration of the reactive species in the sample. By this means the need to detect a titration end-point is obviated and a basis is provided for continuous automatic analysis.

McLean and Penketh[2] describe the construction of a cell and detector system for performing thermometric analyses on flowing samples. The cell design is shown in Fig. 5.1. The reaction takes place in a single-turn coil of 1-mm bore flexible tubing contained in an effective heat-sink. Since the temperature changes recorded in thermometric reactions are small it is imperative that the heat-sink maintains a constant temperature atmosphere for the reaction to occur in. The authors used a 20-l water-jacket lagged with polyurethane board and controlled the water-temperature manually between runs. For prolonged operation, or for on-line control, use of a heat-exchange coil receiving water from an external thermostat would be necessary.

The cell design depicted in Fig. 5.1 meets the essential requirements for sensitivity and speed of response, namely small heat-capacity, small

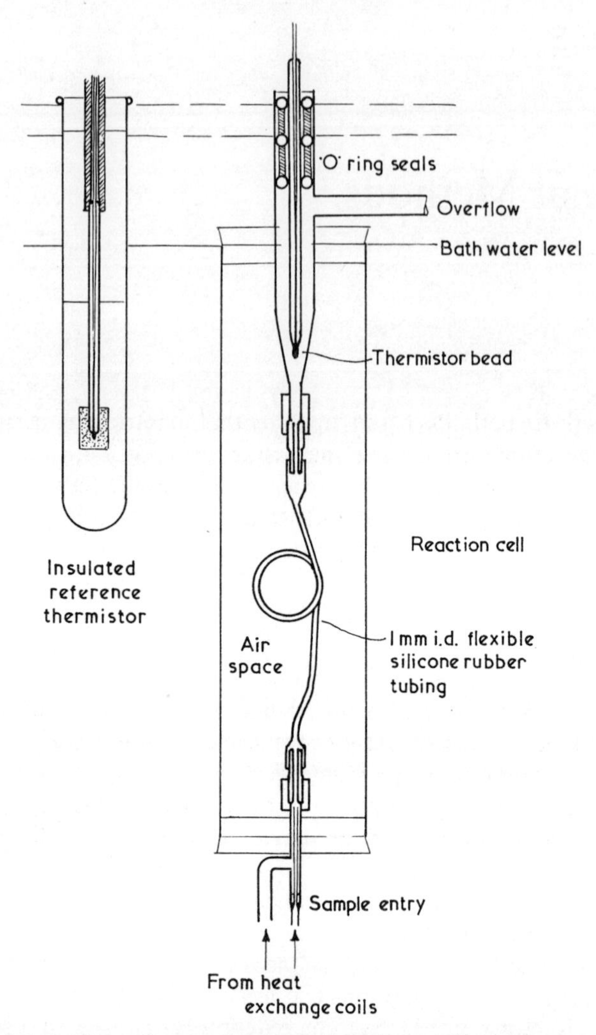

Fig. 5.1 Cell design for performing thermometric analyses on flowing streams. *Reproduced with permission from McLean and Penketh[2] and Microforms International Marketing Corporation.*

internal volume and good mixing. Use of a mechanical stirrer proved undesirable because it provided a source of heat input to the cell. Mixing was therefore achieved by forcing the sample and reagents into the cell at high velocity through a constriction, taking precautions against build-up of back-pressure in the pumping lines.

The temperature change was detected by a thermistor bead with respect to an isolated reference thermistor also contained in the water-bath.

A Wheatstone bridge circuit was used to measure the effect. The sensitivity of the detector could be adjusted by varying the voltage across the cell, the limiting level being set by noise due to power dissipation in the thermistor bead. The equipment yielded a linear relationship between concentration of the thermally reactive component and measured d.c. voltage signal over the concentration range 0·005–0·1 M. A further improvement in sensitivity, up to 20-fold, should be achieved by replacing the single thermistor detector by a double thermistor unit of higher resistance.

Automatic feed of samples to the detector was provided by a Technicon automatic sampler operating at 20 samples per hour. Samples and blanks were processed alternately for equal periods of time.

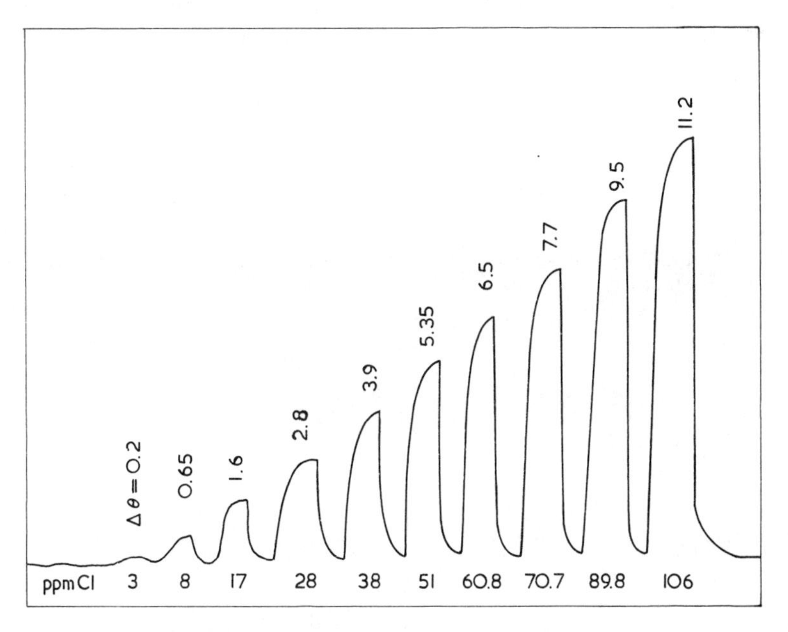

Fig. 5.2 Continuous thermometric determination of chloride; calibration curve for concentration range 0–100 mg/l.
Reproduced with permission from McLean and Penketh[2] and Microforms International Marketing Corporation.

The range of analytical applications for which the apparatus is potentially useful includes acid–base, redox, precipitation and complex-formation titrations; Fig. 5.2 shows a typical response for a precipitation titration, that of chloride, over the concentration range 0–100 mg/l, against silver nitrate. A similar apparatus, in which piston pumps are used to feed sample and reagent to the thermometric cell, is described by Guillot[3].

5.2 Thermogravimetry

A wide range of thermobalances is now available commercially and the history of their development is recorded in a number of books[4–9]. The more modern instruments incorporate a measure of automation in that temperature programming and the recording of weight change are provided by built-in standard facilities. The first fully automatic thermobalance is the subject of a very recent paper by Bradley and Wendlandt[10]; in addition to the aforementioned automatic functions it is also provided with an automatic sample-changer having a capacity of 8 samples. Once the sample-changer has been loaded the instrument performs a thermogravimetric analysis on each sample in turn with no intervention by the operator. The automatic instrument was built around a conventional top-loading balance (Cahn Model RTL). The general layout of the balance, furnace and sample-changing mechanism is shown in Fig. 5.3, and the furnace and sample-holder are detailed in Fig. 5.4. Samples of up to 10 mg in weight are contained in cylindrical platinum containers 5·0 mm in diameter and 2·0 mm in height. Eight such containers can be mounted in indentations cut in a rotating aluminium disc 0·25 in. thick and 8·0 in. in diameter. The disc is driven by a small electric motor.

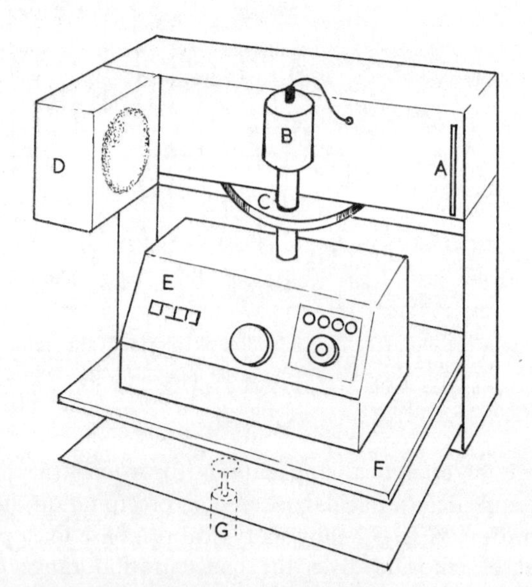

Fig. 5.3 Schematic diagram of automatic thermobalance
A. Gas flowmeter; B. Furnace; C. Sample holder disc; D. Cooling fan; E. Cahn Model RTL recording balance; F. Balance platform; G. Platform motor.
Reproduced with permission from Bradley and Wendlandt[10] and American Chemical Society.

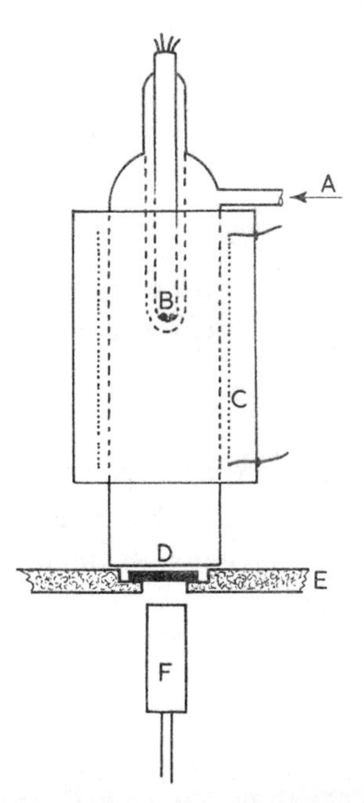

Fig. 5.4 Automatic thermobalance: furnace and sample holder
A. Gas inlet tube; B. Thermocouples; C. Furnace heater windings and insulation; D. Sample container; E. Sample holder disc; F. Ceramic sample probe.
Reproduced with permission from Bradley and Wendlandt[10] *and American Chemical Society.*

Positioning of each sample in turn below the opening of a Vycor tube furnace ($\frac{5}{8}$-in. internal diameter and 4 in. long, wound with Chromel A) is achieved by connecting to the motor a microswitch which is tripped on encountering an indentation in the disc surface. To insert the sample into the furnace it is picked up from below on a ceramic sample probe attached to the balance beam, and the entire balance and balance platform is raised some 2·5 in. by a screw driven by a reversible motor. The limit of movement in each direction is controlled by microswitches. When the sample is positioned in the centre of the furnace at the upper limit of vertical movement of the balance the furnace is flooded with the appropriate gas and the furnace-temperature programmer is activated. On completion of the program the balance assembly is lowered and the sample-container retained by the aluminium disc. The disc is then rotated to position the next sample below the furnace and a cooling fan is switched on to cool the furnace to a preselected temperature for the start of the next analysis. It is possible to change the furnace atmosphere between each analysis when the lower temperature limit is reached. Sequencing of the events described above is carried out by relay circuits controlled by microswitches and preselected temperature values. The thermogravimetric curves are presented on a Leeds and Northrup four-channel multipoint potentiometric recorder. Precision and accuracy of weight recording is estimated at ±2% and for temperature recording ±5%. The upper temperature of the instrument is about 1000° C, the Vycor furnace tube being the limiting factor. The instrument should be of economic value where thermogravimetric analysis is carried out repetitively. The results are amenable to processing by a small digital computer if desired.

5.3 Differential Thermal Analysis

Wendlandt and Bradley[11] describe an instrument which is capable of performing full automatic differential thermal analysis on eight samples sequentially. It is similar in general principles to the automatic thermobalance discussed in the previous section. Fig. 5.5 is a schematic diagram of the sample-changer, furnace and furnace platform.

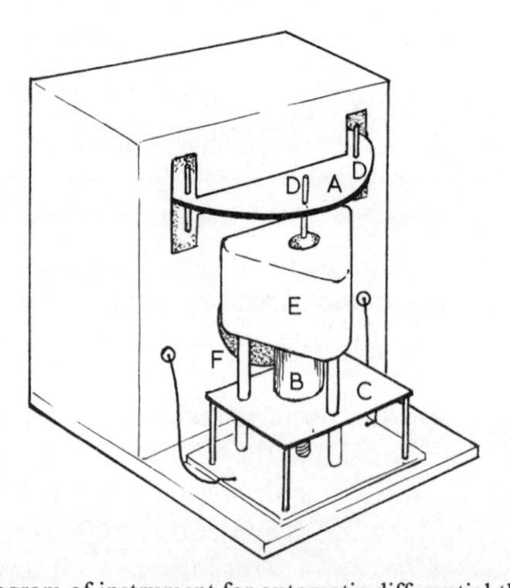

Fig. 5.5 Schematic diagram of instrument for automatic differential thermal analysis
A. Sample holder plate; B. Furnace; C. Furnace platform assembly; D. Sample capillary tube; E. Furnace insulation; F. Cooling fan.
Reproduced with permission from Wendlandt and Bradley[11] and Elsevier Publishing Company.

The sample-changer is an aluminium disc 8 in. in diameter and 0·25 in. in depth. Samples, in powder form, are contained in glass capillaries of 1·6–1·8 mm bore and eight of these can be accommodated around the circumference of the disc, secured by spring clips. The disc is rotated by a small synchronous electric motor fitted with an electromagnetic clutch.

Each 45° rotation of the disc brings a sample in line with the tube furnace. Accurate alignment is achieved optically. Adjacent to each sample a small slot is cut in the aluminium plate and a lamp and photocell are positioned either side of the disc. Interruption of the light-beam through the slot by the photocell provides a signal which stops the motor. The furnace platform is raised by means of a reversible electric motor connected to it by a screw drive so that the sample is positioned in an aluminium heat-transfer sleeve. Microswitches control the upper and lower limits of travel. At the end of the analysis, signified by attainment of a preset temperature for the furnace, the platform is lowered, the aluminium disc rotates to the next position and a cooling fan directs air on to the furnace. Changing from one sample position to the next occupies 15 sec, raising the furnace platform to the upper limit position takes 50 sec and cooling the furnace from 450° C to room temperature requires about 20 min. Once the furnace has cooled to room temperature the entire sequence is set in operation

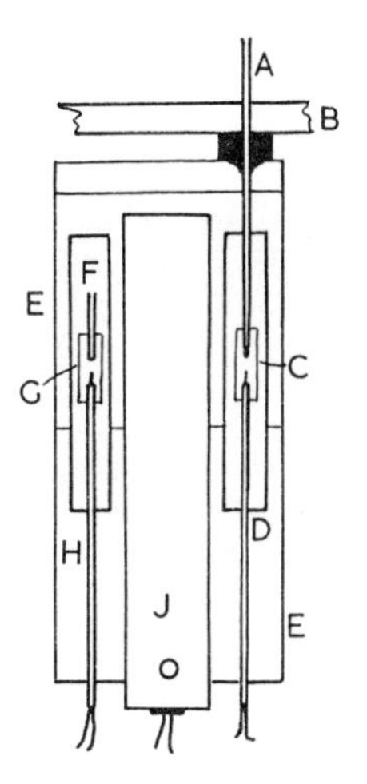

Fig. 5.6 Automatic instrument for differential thermal analysis: furnace and sample changer.
A. Glass capillary for sample; B. Sample holder plate; C. Sample heat transfer sleeve; D. Sample thermocouple; E. Furnace block; F. Reference capillary tube; G. Reference heat transfer sleeve; H. Reference thermocouple; J. Heater cartridge.
Reproduced with permission from Wendlandt and Bradley[11] *and Elsevier Publishing Company.*

for the next sample. The furnace detail is shown in Fig. 5.6. It is cylindrical, 3·3 in. in length and 1·5 in. in diameter, and is heated by a 210-V stainless-steel cartridge heater. The upper temperature limit is about 500° C. Two cavities 1·5 in. long and 0·25 in. in diameter are contained within the furnace tube, one for the sample and one for a reference standard. Aluminium heat-transfer sleeves provide thermal contact for the two cavities. These sleeves are drilled out at the ends so that a close fit is achieved for both the sample tube at the top and the ceramic insulator tube at the bottom.

Sequencing of all operations is achieved by relay circuits responding to microswitches and temperature readings. Results are presented on one of the two channels of a Varian Model G22 chart recorder after amplification of the voltage from the differential thermocouples by a Leeds and Northrup microvolt d.c. amplifier. The sample temperature is recorded on the second channel.

References

1. Priestley, P. T., Sebborn, W. S. and Selman, R. W. *Analyst*, 1965, **90**, 589.
2. McLean, W. R. and Penketh, G. E. *Talanta*, 1968, **15**, 1185.
3. Guillot, P. *Anal. Chim. Acta*, 1970, **50**, 499.
4. Duval, C. *Inorganic Thermogravimetric Analysis*, 2nd Ed., Elsevier, Amsterdam, 1963.
5. Wendlandt, W. W. *Thermal Methods of Analysis*, Interscience, New York, 1964.
6. Keattch, C. *An Introduction to Thermogravimetry*, Heyden, London, 1969.
7. Saito, H. in *Thermal Analysis*, R. F. Schwenker and P. D. Garn, eds., pp. 11–24, Academic Press, New York, 1969.
8. Garn, P. D. *Thermoanalytical Methods of Investigation*, Chap. 10, Academic Press, New York, 1965.
9. Anderson, H. C. in *Techniques and Methods of Polymer Investigation*, P. E. Slade and L. T. Jenkins, eds., Chap. 3, Dekker, New York, 1966.
10. Bradley, W. S. and Wendlandt, W. W. *Anal. Chem.*, 1971, **43**, 223.
11. Wendlandt, W. W. and Bradley, W. S. *Anal. Chim. Acta*, 1970, **52**, 397.

Chapter 6

Radiometric and X-ray Methods

Methods based on the measurement of radioactivity are now extensively used by analytical chemists. They fall principally into two categories, those based on the use of radioactive tracers and those involving the production of induced activities by activation methods. The automation of both approaches has been studied, more especially activation analysis, which has found considerable use in the determination of trace elements[1]. Three distinct areas of automation can be distinguished in the radiometric field; sample-handling, chemical separation techniques, and counting and presentation of results.

The demand for effective and often rapid separation methods for isolating radionuclides of interest following activation has provided a notable impetus for automation in this area, although the development of high resolution α- and γ-spectrometric instruments which facilitate direct analysis of mixed activities has to some extent supplanted them in recent years. Ion-exchange and solvent extraction have proved by far the most effective techniques for this purpose, and progress in the automation of these is fully discussed in Chapters 9 and 10 and is therefore not considered further here.

6.1 **Activation Analysis**

Automation of radioactivation methods begins at the irradiation stage and covers sample-transfer, control of irradiation parameters, discharge of sample either direct to an automatic counter or to a separation facility, and the counting, calculation and presentation of results in the desired form. The initial preparation of samples remains manual because it involves operations such as homogenization of sample and encapsulation in a suitable container for irradiation, and except in special circumstances it is unlikely that automation will offer sufficient advantage to offset the inevitable mechanical complexity and consequent expense which would be incurred. Several examples of automated activation analyses are cited below to illustrate the diversity of approaches which have been studied.

The determination of oxygen is important in controlling the production of steel, and activation methods, using either fast neutrons or photons, have contributed significantly to solving the analytical problem. An automatic method utilizing irradiation with 14-MeV neutrons from a neutron generator is described by Perdijon[2]. The nuclear reaction is

$$^{16}_{8}O\ (n,p)\ ^{16}_{7}N$$

and the count-rate from the high energy γ-emissions from nitrogen-16 is a measure of the oxygen present. No chemical separation step is required.

Samples and standards are introduced into the irradiation and counting facilities by means of a pneumatic transfer unit. This is a two-channel system which enables sample and standard to be irradiated at the same time so that fluctuations in neutron flux affect both to the same extent. One standard serves for a batch of samples and remains in the transfer unit throughout the run. Samples are presented sequentially to the source from a magazine. Operation of the unit is controlled by a programmer based on two electric timers, one for the irradiation time and the other for counting time. One of three irradiation times, 5, 10 or 15 sec can be selected, and the counting time is fixed at 30 sec. The sample-line is fitted with facilities for recovering the sample-capsule and irradiating it once or twice more as well as for discharging it from the system and introducing the next one. The counting assembly comprises two single-channel γ-spectrometers in which sample and standard are counted separately and simultaneously. The method is sensitive to 5 ppm of oxygen in steel. Trials at oxygen concentrations of 50, 500 and 1000 ppm yielded precisions of 10, 4 and 2% respectively. The apparatus accommodates individual samples of about 40 g weight. Perdijon considers that the technique can be extended to the continuous analysis of steel production.

Inhomogeneity of sample and anisotropy of neutron flux are important sources of imprecision in 14-MeV neutron analysis. To minimize these effects it is necessary to both rotate and revolve the sample and standard simultaneously in the neutron beam[3,4]. This aspect has been further developed by Priest, Burns and Priest[5] who describe an irradiation, transfer and counting system for determining short-lived components in samples which are inhomogeneous. They evaluated it with particular reference to oxygen in a range of standard metal samples.

The irradiation unit, which is fabricated from nylon or 'Delrin' to minimize attenuation and back-scatter, accommodates two cylindrical samples (sample and standard) or two in disc form. They are contained in high-density polythene irradiation capsules and during irradiation are subject to three movements, revolution, rotation and traverse. A single synchronous motor and planetary gear-assembly provides the motions of rotation and revolution. Sample and standard rotate on their own axis and they revolve about an axis equidistant between the two sample axes and in the same plane. The rotation is non-synchronous and this ensures that a different aspect of each sample is nearest to the target at each revolution. The speeds of both rotation and revolution are variable, the former from 0 to 180 rpm and the latter from 0 to 600 rpm. In addition a traverse across the target beam is provided by attaching the irradiation head to a traverse table fitted with a rack and pinion drive. The traverse distance is adjustable within the range 0–8 in. and the traverse time between 0 and 10 min. All the variable parameters are controlled from a console which also incorporates safety interlocks to guard against improper sequencing of operations.

Sample and standard are loaded manually and at the end of the timed irradiation period they are picked up by a pneumatic transfer-system which carries them 40 ft through plastic tubing to an uncapping unit. A microswitch-operated solenoid uncaps the capsule. The latter is held in pawls which then swing down to tip the sample into the counting position. A timing system operates both the uncapping and counting sequence; when the counting on one sample is complete the uncapping microswitch opens the next capsule. The counting assembly comprises two 5×5 in. sodium iodide crystals with single-channel pulse-height analysers.

Experiments on a number of National Bureau of Standards standard reference metals showed good agreement with the certified value for oxygen content. It was established that the multiple motion system is unable to correct for gross sample inhomogeneity but this can be detected by repeating the analysis with the sample loaded in the capsule the opposite way round. The mean of the two results is then in good agreement with the certified value. The analytical precision at a concentration of 1 ppm of oxygen is estimated to be $\pm 50\%$.

The application of 14-MeV neutron-activation analysis to the determination of low concentrations of certain elements in flowing streams has been briefly reported by Jervis, Al-Shahristani and Nargolwalla[6]. The irradiation and counting times can be preset to yield optimal results based upon the half-life of the activity to be measured. The decay-time between irradiation and counting, which is needed to permit the decay of unwanted radioisotopes, can be controlled over the range from a few seconds to several minutes by adjusting the length of the delay-line and the flow-rate of the sample solution. To achieve maximum irradiation efficiency a re-entrant sample-loop was used which accommodated the source target. The authors demonstrated the applicability of the method, which is rapid and non-destructive, for the determination of fluoride, vanadium and uranium. The sensitivity of the method depends largely on the neutron-flux of the generator. By use of a commercially available neutron-generator (Texas Nuclear, Model 9900) having a flux of about 10^{10} neutrons/sec each of the three elements could be determined at concentrations of a few ppm with precisions of 5–10%.

Where the half-life of the measured radionuclide is short, the sensitivity of the activation method can be increased by repeated irradiation and

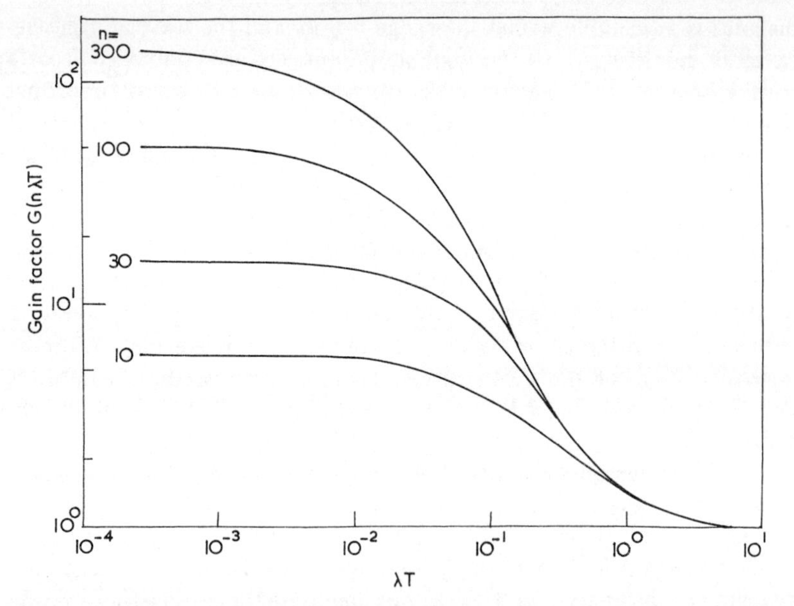

Fig. 6.1 Enhancement of sensitivity of 14 MeV neutron activation analysis as a function of disintegration rate and number of passes.
Reproduced with permission from Ashe, Berry and Rhodes.[7]

counting sequences. Fig. 6.1, taken from the studies of Ashe, Berry and Rhodes[7], shows the predicted sensitivity-enhancement as a function of disintegration constant and number of passes. The authors evaluated the method of recirculating a slurry sample past the irradiation and counting stations. An alternative method of 'cycling' the sample involves holding it stationary and providing circuitry for switching the neutron-beam and counters on and off in sequence. A decay period can then be inserted if required[8].

6.2 Isotopic Tracer Methods

Trace quantities of metals can be determined by radioisotope dilution. An experimental approach which has proved of wide applicability is known as substoichiometric analysis, the principle of which is as follows. The sample is mixed with a known amount y_s of a radioisotope of the element to be determined. Denoting the original specific activity of the standard by S_s and the value after mixing by S, the amount y of non-radioactive element in the sample is given by

$$y = y_s\left(\frac{S_s}{S}-1\right)$$

Because specific activity is related to measured activity a and mass m of the isolated elements by $S = a/m$ it is necessary to determine both a and m in order to calculate y. However, if the experimental conditions are arranged so that the same mass of the element is isolated from both the standard radioisotope solution and from the isotopically diluted sample, then

$$y = y_s\left(\frac{a_s}{a}-1\right)$$

and y can be calculated from measurements of the two isolated activities. This condition can be achieved by reacting both sample and standard solutions with a reagent used in such quantity that it is completely consumed and so extracts the same mass of element in each case. By using appropriate precipitating agents or by solvent extraction of a metal chelate conditions have been worked out for the substoichiometric determination of some 26 elements[9].

The automation of substoichiometric analysis was first considered by Růžička and Williams[10]. They suggested the use of the 'AutoAnalyzer' approach, using a flow-through radiometric detector. A manifold was designed and optimized for the determination of mercury, with zinc dithizonate as the substoichiometric reagent, carbon tetrachloride as the extraction solvent and mercury-203 as the radioisotope[11, 12]. The manifold is reproduced in Fig. 6.2. To prevent attack on the pump tubing, the zinc dithizonate/carbon tetrachloride reagent was pumped by displacement, using the aqueous phase from the preparation of the reagent as the displacing solution. A 55×55 mm well-type thallium-activated sodium iodide crystal served as the detector, and the flow-cell for extracted sample comprised a coil of glass tubing 3·5 mm in outside diameter and volume 0·5 ml, attached to the phase separator. The assembly is shown in Fig. 6.3. The mixture of air and aqueous and organic phases enters the

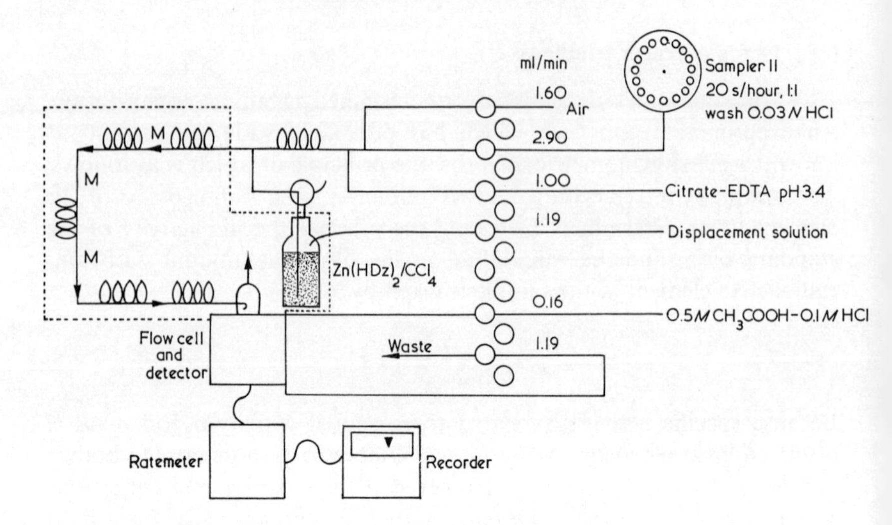

Fig. 6.2 Manifold for automatic determination of mercury by substoichiometric substitution analysis.
Reproduced with permission from Růžička and Williams[10] *and Microforms International Marketing Corporation.*

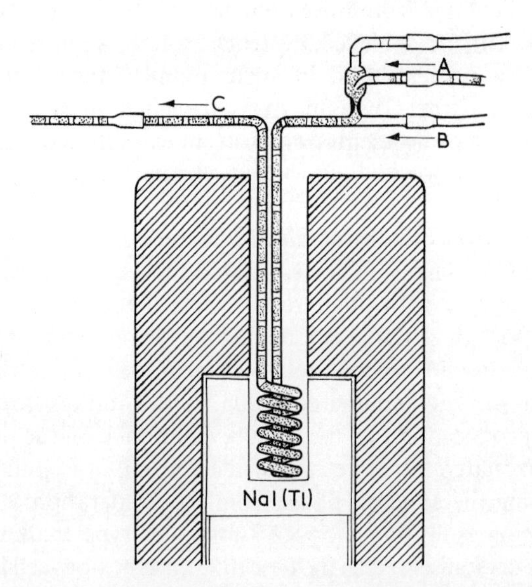

Fig. 6.3 Flow-through detector assembly for automatic substoichiometric substitution analysis.
Reproduced with permission from Růžička and Williams[10] *and Microforms International Marketing Corporation.*

separation trap *A* and the separated organic layer is segmented by non-radioactive aqueous solution *B* and passes through the crystal well for measurement. *C* is a waste outlet carrying the aqueous phase from the separation together with a small, fixed proportion of the solvent phase. The rate at which samples can be processed depends on the volume of the flow-cell and the time taken to clear it of previous sample. With the 0·5-ml flow-cell a processing rate of 20/hr can be achieved, but for the most sensitive measurements it is necessary to reduce this to 10/hr. The absolute sensitivity of the method is dependent upon the specific activity of the mercury-203 solution. With a 300 mCi g solution the limit of determination is 0·005 ppm mercury and the limit of detection is about 0·0005 ppm. Correspondingly higher sensitivities could, if desired, be achieved by using a standard of higher specific activity. Successful use of the automatic

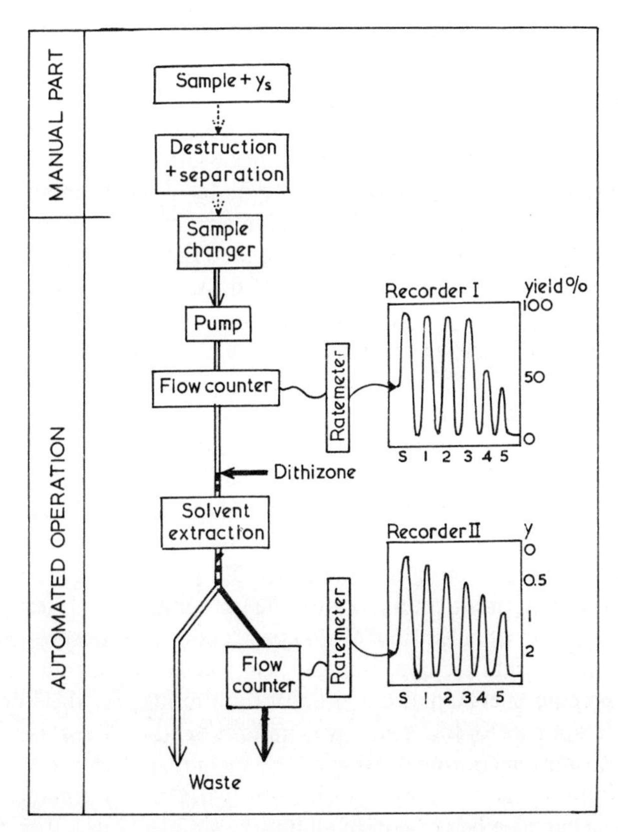

Fig. 6.4 Dual detector method for automatic substoichiometric substitution analysis. *Reproduced with permission from Starý and Růžička*[9] *and Microforms International Marketing Corporation.*

method in conjunction with an oxygen-flask combustion technique for determining small amounts of mercury in a wide range of foodstuffs has been described by Růžička and Lamm[13]. In general, recoveries were 90–105% at the 0·16-ppm level in a standard kale preparation; the relative standard deviation was 2%.

A subsequent modification of the instrumentation for automatic substoichiometric analysis is the addition of a second radiometric detector[9] for which the flow-path is shown in Fig. 6.4. In this configuration the radioactivity is measured both before and after the substoichiometric extraction stage. This serves the valuable purpose of monitoring the manual chemical stages of the method; poor recovery of the analyte over this part of the procedure could lead to quantitative rather than substoichiometric extraction and invalidate the results. Comparison of the two recorder traces provides an unequivocal indication as to whether or not malfunction of the method has occurred.

6.3 Automatic Counting Equipment

In many areas of radiochemistry it is necessary to measure the activity of large numbers of samples. These may be in the form of solutions or mounted on planchettes. Counting equipment for measuring α-, β- and γ-emissions sequentially on batches of prepared samples has been available commercially for some years. The mechanical means of automatically presenting samples to the counter are similar to those described in earlier chapters for automatic discrete-sample analysis using other means of detection. For example, liquid samples, usually contained in plastic vials, can be carried to the detector by a motor-driven endless chain and planchettes can be vertically stacked and transported sequentially on a moving slide.

Undoubtedly the most significant development in recent years has been the extensive development of γ-spectrometry for the analysis of mixtures of γ-emitting radionuclides, with NaI(Tl) or Ge(Li) detectors in conjunction with multichannel pulse-height analysers. Interpretation, and, in particular, quantitation, of a γ-spectrum of a radioisotope mixture can be complex and time-consuming. Each γ-emitter contributes a photopeak of specific energy and a Compton continuum on the low-energy side of it. Responses for lower energy γ-emitters are therefore superimposed upon the Compton continua of higher energy ones and, in addition, peaks due to annihilation of positron–electron pairs and resonance escape phenomena can also be present. In all but the simplest cases, therefore, the quantitative measurement of individual γ-emitters requires a mathematical treatment of some sophistication to isolate the contribution of each one to the total activity profile. The calculations are laborious and it is not surprising that the use of computers has become virtually indispensable in this field. The principal approaches used are matrix inversion, where the component activities are known, and spectrum stripping. The latter involves energy- and intensity-matching of the spectrum consecutively with a series of γ-spectrum profiles obtained from radioactively pure specimens of each possible contributing activity. After matching, each component is subtracted from the composite spectrum until the latter is fully resolved. The accuracy of this subtractive technique depends upon several factors, notably the degree of sophistication of the counting equipment, such as facilities for improving peak-to-background ratio, the purity of the radioisotope standards and the amounts of the individual activities in the sample. For the determination of very low levels of radioisotopes in moon-rock specimens, an anti-coincidence counting system comprising two large NaI (Tl) detectors at 180° orientation with respect to the sample and viewed by 22 photomultipliers was employed and extreme precautions taken to minimize the background activity contribution[14]. The data-acquisition system comprised coincidence–anticoincidence logic circuits and dual analogue-to-digital converters interfaced to a DEC PDP9 computer. The latter was programmed for simultaneous acquisition of data from the individual detector in 255 channels each, a sum-coincidence spectrum in 255 channels and a gamma-gamma coincidence spectrum in a folded matrix of 127×127 channels. A Univac 1108 computer was used for analysis of the data so produced. The performance of the equipment was evaluated by making known additions of radioactively pure standards to moon-rock samples. Recoveries were within 5% of the added value.

6.4 X-ray Methods

6.4.1 X-RAY FLUORESCENCE

X-ray fluorescence spectrometry is an inherently rapid method for elemental analysis which is particularly suited to trace level studies; it is also non-destructive and can be used for both solid and liquid samples. Developments in instrumental aspects of the technique now rest largely with commercial manufacturers, and earlier spectrometers, in which sample introduction and goniometer setting and evaluation of results from recorder traces were manual procedures, are now updated and spectrometers in which these functions are partly or wholly automatic are now available. With the incorporation of this level of automation X-ray fluorescence spectrometry is capable of high rates of sample throughput.

The development of automated means of sample presentation and of processing results has followed along lines having much in common with the automation of other spectrometric techniques. As regards the basic principles of spectrometer automation two distinct approaches can be recognized; these are the sequential and simultaneous methods. In the sequential method a single goniometer is used, driven by a motor such that it scans the range of 2θ angles of analytical interest, whereas in the simultaneous method the spectrometer is equipped with a number of fixed goniometers, each set to diffract X-radiation of an element of interest. To a large extent the advantages and disadvantages of the two approaches are obvious and the choice between them is determined by the nature of the analytical problem. Analysers based on the simultaneous principle are inflexible in the sense that the goniometer settings are fixed and the number of elements which can be determined at a single pass is limited to the number of goniometers provided in the spectrometer. Nevertheless the technique is more rapid than the sequential one in that delays incurred by the scanning time are obviated. The simultaneous method is ideally suited to situations where the sample-load is considerable and constant as regards which elements are to be determined. Not surprisingly, therefore, it has found its principal use in production control of chemical processes. The sequential method, in which the scanning pattern is preprogrammed, offers greater flexibility in terms of the number of elements to be determined and the relative ease with which the initial program can be varied. Indeed, if a digital computer is used to program the goniometer then changes from one set of elements to another can be achieved in a matter of seconds. If a sequential analyser is used in conjunction with an automatic sample-feed mechanism two operating modes are possible; either each sample can be scanned over the required 2θ range in turn or the entire batch of samples can be measured at each setting before the goniometer is moved to the next measuring position. Both approaches have been exploited in commercially available spectrometers.

6.4.1.1 *Automatic Simultaneous X-ray Fluorescence Spectrometer*

The simultaneous measurement principle is exploited in the Philips PW 1270 automatic X-ray spectrometer which provides for analysis of up to 14 elements from atomic number 9 upwards. An earlier model, the PW 1250, has facilities for 7 simultaneous elemental measurements. In addition to the basic spectrometer three other units contribute to the degree of automation provided; PW 1265 is an automatic sample-changer having a capacity of 9 samples, PW 1266 is an automatic sample-changer holding up to 160 samples and PW 1261 is a data-processing unit described in greater detail below. As an alternative to the data-processor results may be evaluated by computer.

The spectrometer comprises the sample-transport mechanism and the analyser. The latter contains the X-ray tube together with the collimators, goniometers, analysing crystals and proportional and scintillation counters. The 14 collimators and goniometers are mounted in a single plane, back-to-back, on the centre plate of the vacuum chamber. Each is directed at the centre of the sample. The co-planar mounting enables the X-ray tube to be mounted close to the sample, thereby maximizing sensitivity. For each channel either a coarse (480-μm spacing) or fine (160-μm spacing) collimation can be selected. Fourteen flow-proportional counters and eight scintillation counters are incorporated, the latter being located outside the vacuum chamber. After coarse adjustment of counter position the final accurate alignment can be performed under vacuum. Independent adjustments are provided for θ and 2θ values. The vacuum chamber, which can also be operated in helium or air, is temperature-controlled to within 1%. The high-voltage generator for the X-ray tube has selector switches to permit operation at 25, 32, 40, 50 or 60 kV and 25, 32, 40, 50, 60 and 80 mA. Voltage and current stabilization is better than 0·02% of the nominal value over a mains fluctuation range from +10 to −15%.

Depending on the number of samples to be processed they can be automatically fed into the spectrometer from a rotating circular tray holding 9 samples or from a rack of 16 trays each holding 10 samples. The tray is mounted adjacent to the spectrometer and a moving arm lifts each sample-holder in turn and places it in the spectrometer. The latter is fitted with an air-lock so that a new sample can be introduced while the first is being analysed. This represents a considerable saving of time when many samples are to be processed. The sample under analysis can be rotated to eliminate effects due to inhomogeneity of the sample surface. If necessary the automatic loading-sequence can be interrupted at any time.

Three types of analysis program are possible with the PW 1270, the appropriate one being preselected before a batch of samples is analysed. They are termed 'absolute', 'monitor' and 'ratio'. The 'absolute' program accumulates counts for a fixed term for each element whereas the 'monitor' program measures the time taken for a predetermined number of counts to accumulate in each monitoring channel. In both modes the spectrometer can accept up to five unknown samples at any one time; while one sample is being measured another is introduced into or removed from the air lock. In the 'ratio' program the sample is measured for the time taken by a previously measured standard to accumulate a fixed number of counts on the same element. Therefore the spectrometer can handle only one sample at a time, the other position being occupied by the standard. Up to three standards can be introduced at the same time. Results from each channel can be presented in tabular form on an IBM typewriter or fed to the data-processor or external computer for further treatment.

The data-processing únit comprises an electromechanical counter which advances one position each time a measurement is made, a timer which determines the time necessary for one measurement and also provides clock pulses to control the calculator and read-out units, and a scaler for each channel. A print-out of the electromechanical counter reading (5 digits) provides sample identification. The timer unit is accurate to better than 10^{-4} sec and permits fixed times of 2, 4, 10, 20, 40, 100, 200, 400 or 1000 sec to be selected.

An optional unit, PW 1281, provides a memory extension for the data-handling facilities. It comprises an arithmetic unit, a switch-memory system and a register unit. It performs the calculation

$$y = ml + b$$

where y is the result, l the measured radiation intensity and m and b are constants. The memory consists of 12 miniature, 10-position switches (4 per programme) in printed circuit form. The information stored consists of values of m and b and their appropriate signs, channel number to which m and b values apply and also the significance of y (concentration results, background or slope correction). If y is a concentration value it is printed out and the register cleared. If y is a background correction (Δb) it is stored in the register and compounded with the following $y = ml + b$ calculation; similarly a slope correction (Δm) is stored and added to the m value for the next calculation.

6.4.1.2 *Automated Sequential X-ray Fluorescence Spectrometry*

Many of the instrumental features of the automatic simultaneous X-ray spectrometer are common to the sequential-type spectrometer. The basic difference lies in the design of the spectrometer itself. Whereas spectrometers of the simultaneous type require an analysing crystal and goniometer for each element sought, the sequential analyser requires only one. In practice, however, several crystals are provided to give a range of lattice spacings so that the full range of elements may be determined.

A sequential analyser which incorporates extensive automation and operational flexibility is the General Electric Model XRD 710. It is designed on a modular basis and allows the operator a wide range of choice of conditions to suit particular analytical problems; samples in pellet, block or liquid form can be examined either in air or in vacuum. It is especially applicable to production control analysis in the metallurgical field and complete programs for a number of important metallurgical analyses are provided. The program control covers the entire analytical sequence including selection of target X-ray tube, analysing crystal and counter tubes, rotation of sample, and data-handling. The programs are in the form of punched paper tape.

The vacuum housing of the spectrometer contains four analysing crystals with associated Soller slits and radiation counters. A high precision of angular measurement is provided. The 2θ protractor rotates from $-2°$ to $+145°$. When an analytical line is being measured the speed of scan is selected by the operator; between element lines the goniometer can be slewed at much higher speed (500°/min) to minimize the time delay. The four analysing crystals can be interchanged under vacuum on a command from the digital controller. Accurate positioning is critical and can be performed to $\pm 0{\cdot}1\%$ of peak intensity, and the change-over time between any two crystals is 1·25 sec. Radiation counters are of the flow-proportional type with an ultra-thin window which allows efficient detection of elements of atomic number as low as 9 (fluorine). The counters are mounted on a cantilever rack and are fitted with integrated circuit electronics. The X-ray tube is a dual-target model and is essentially two tubes in one. Targets of chromium, for lighter elements, and tungsten, for heavier ones, can be selected by program control without releasing the vacuum.

Both solid and liquid samples are presented to the spectrometer in cylindrical containers. The loading mechanism comprises a rotating carousel holding up to 10 containers. It is manually raised to the X-ray tube by a gear-driven crank. The program control provides for rotation of the carousel, and also of the sample at 17 rpm if desired, by two remotely-driven motors.

A digital controller which can be used in either semi- or fully-automatic mode exercises control over the entire spectrometer function. The semi-automatic mode is used mainly for setting-up and testing the spectrometer and for fault-finding. The operator can program single-cycle instructions and observe the spectrometer response. Used fully automatically, the digital controller controls the complete analytical sequence. As many as 49 elements can be determined on a sample in a single run, so that using the full sample-loading capacity of 10 the controller sequences 490 elemental determinations without any intervention from the operator. A teletype is used for data input and output; it comprises an alpha-numeric printer, paper-tape punch and reader and input keyboard. Three data-output options are available, a remotely-located slave teletype, and manual or direct transmission to a remote computer. ASC-11 code is used for manual transmission and direct transmission can be accomplished in any 6 or 8 channel code.

The programming software takes the form of punched paper tapes provided with the spectrometer. They perform the following functions:
(*a*) real time operational control of all spectrometer functions,
(*b*) angular goniometer positioning control to within 0·01° ± 0·002°,
(*c*) setting up the spectrometer and selection of operational modes,
(*d*) tabular storage for setting up and processing 49 elements in each of 10 samples,
(*e*) computation and calculation of results,
(*f*) standardization based on one standard sample and application of corrections to stored conversion data,
(*g*) determination of up to 6 interelement correction constants by multiple linear regression,
(*h*) application of these interelement correction constants to compensate for interelement effects,
(*i*) determination of working curves from standard samples for calibration purposes,
(*j*) determination of peak-intensity wavelengths by a variable 2θ step-scan program,
(*k*) count element peak intensities,
(*l*) instrument check by test programs.

Among the extensive proving trials carried out by the manufacturers are analyses of 10 standard stainless steels from the National Bureau of Standards; the excellent agreement between quoted and analysed figures is illustrated in Table 6.1.

Table 6.1 COMPARISON OF RESULTS FOR PERCENTAGE COMPOSITION OF NBS STANDARD STAINLESS STEELS BETWEEN THE XRD 710 AND QUOTED NBS VALUES

Sample	Mo		Nb		Ni		Mn		Cr		Ti	
No.	*NBS*	*XRD*	*NBS*	*XRD*	*NBS*	*XRD*	*NBS*	*XRD*	*NBS*	*XRD*	*NBS*	*XRD*
846	0·43	0·428	0·60	0·594	9·11	9·20	0·53	0·464	18·35	18·78	0·34	0·340
845	0·92	0·921	0·11	0·116	0·28	0·254	0·77	0·752	13·31	13·00	0·03	0·038
847	0·059	0·044	0·03	0·014	13·26	13·18	0·23	0·214	23·72	23·61	0·02	0·021
848	0·33	0·338	0·49	0·491	0·52	0·514	2·13	2·11	9·09	9·06	0·23	0·214
849	0·15	0·149	0·31	0·308	6·62	6·74	1.63	1.60	5·48	5·46	0·11	0·103
850	—	0·072	0·05	0·049	24.80	24·80	—	0·103	2·99	3·09	0·05	0·049
SS 61	—	0·011	—	0·006	6·26	6·32	0·78	0·798	15·20	15·23	—	0·003
SS 62	—	0·018	—	0·033	12·45	12·40	0·80	0·802	12·80	12·78	—	0·004
SS 63	—	0·003	—	0·004	9·49	9·52	0·79	0·837	18·70	18·71	—	0·003
SS 64	—	0·004	—	0·003	20·60	20·62	0·85	0·958	25·60	25·66	—	0·002

The Carl Zeiss VRA 1 is an X-ray fluorescence spectrometer which utilizes the sequential principle[15]. Operationally it has many similarities to the XRD 710 described above; in particular it is fully programmed, using punched-paper tape, and is consequently capable of completely automatic operation. An important design difference is that the VRA is designed as a two-channel instrument. It has two X-ray tubes operated from a single generator and two spectrometers and associated detector units. Provision of two complete channels imparts a considerable flexibility of operation. In the strict two-channel mode a sample and reference material can be examined simultaneously for the same spectral lines and the results presented as ratios. This not only saves time in relation to a single-channel approach but effects due to primary fluctuations such as generator performance affect both measurement channels equally. Alternatively one channel can be used as a fixed reference by uncoupling the two channels and permanently setting the reference channel to an appropriate spectral line. In addition, normal single-channel operation is possible and if desired the instrument can be operated as two separate single-channel spectrometers. VRA 1 can process five samples automatically and the program can be repeated as often as necessary. Both air and vacuum paths are available, the latter being essential for the light elements from magnesium (atomic number 12) to titanium (atomic number 22).

The two-channel approach is also adopted in the Elliott Model XZ 1030 automatic X-ray fluorescence spectrometer, which has an analytical capacity of 24 elements per analysis and is programmed by punched cards.

The Philips automatic sequential spectrometer, model PW 1212, processes samples in a different manner from the instruments described above. The specimen chamber holds four samples which can be analysed automatically by insertion of a matrix board program. The goniometer is driven to the first programmed angular setting and the first sample is rotated to the measuring position; when the measurement is complete and the results are printed out the second sample moves into position.

Ultimately all four samples (or standards) are measured at the initial goniometer position, then the goniometer is moved to the next selected angle and the procedure repeated. The PW 1212 can be used to determine all elements down to fluorine (atomic number 9) and the program can handle up to 15 elements. Intensities may be presented as a recorder trace or digitally printed as ratios to a preselected standard, or if high precision is not required, as individual readings.

The automatic X-ray fluorescence spectrometers discussed in this section do not represent an exhaustive coverage of those available.

Nevertheless the descriptions emphasize the salient features of this type of instrument. They are complex instruments incorporating components and mechanisms requiring a high degree of precision engineering in their manufacture. Inevitably they are costly, and their use is most readily economically justified in circumstances where a high and recurrent sample throughput exists. Their principal usage is therefore for control of production processes where such control depends on elemental analysis. In this context they offer several important advantages, notably rapidity of analysis, inherently high precision and non-destructiveness. In addition, by correct choice of analysing crystal, windows and detectors all elements except the lightest are amenable to determination and the sensitivity range is wide (from ppm to per cent).

6.4.1.3 *Automatic Sample-Preparation*

In some chemical processes the product to be examined by automatic X-ray spectroscopy cannot be loaded directly into the sample container but requires pretreatment to convert it into a homogeneous physical form. Cement is an example and Bieshaar[16] describes equipment for performing the preparative operations automatically. Dry cement clinker is ground and then pelleted whereas cement slurry is dried, crushed, calcined and fused with lithium or sodium metaborate or tetraborate to give a borax bead.

Dry Pressing and Pelleting

In this technique the powdered material is pressed into aluminium holders and automatically transported into the X-ray spectrometer. Aluminium caps are stacked in a vibrator-type magazine which has a capacity for 400 such caps and is fitted with an ultrasonic detector to indicate when refilling is required. Caps are individually discharged, and with the aid of a pneumatic ram are placed on a die in the pressing unit; a second ram pushes the cap into the die which is then rotated so that it is located beneath the sample-feed mechanism. This is a cone from which sample powder is discharged by volumetric measurement into the aluminium holder. The cone is mounted on a track so that it can be moved aside for periodic cleaning. The measured volume of sample in its aluminium container is moved beneath a plunger where it is compressed for a period which can be set between 1 and 1000 sec. The pressure exerted is continuously variable between 3 and 40 tons. The die is then further rotated until it aligns with a conveyor belt, the pellet is ejected on to the belt and carried into the spectrometer. An automatic sample press of this design is available as the Philips PW 1240 which is compatible with the PW 1270 automatic simultaneous X-ray spectrometer.

Fusion

An integrated unit for converting slurry into a borax bead has been designed[16]. Slurry is transported through a drying oven to remove all free water and thence to a crushing and weighing unit. An electromagnetic vibration-type crusher converts the dry cake into powder which is poured into a small cup connected to an electromechanical weighing mechanism. When the cup contains the predetermined weight of powder a signal from the weighing unit causes the cup to swing away on a movable arm and its contents to be transferred to a platinum crucible. The crucible is moved over a radiofrequency heater which rapidly raises the temperature of the contents to 1200° C. Then a weighed quantity of borax is admitted to the crucible from a hopper and the contents of the crucible are fused at 1050° C. The automatic programming system switches off the heater, and on cooling a glassy bead forms. To prevent the bead from adhering to the crucible a very high potential is applied across it. Beads are tested for breakage and intact ones are fed to the spectrometer for analysis. The total time taken to convert a slurry into a borax bead is 7–8 min. Philips models PW 1234 and PW 1235 are respectively commercially available semi-automatic and automatic units based on the design above. Both are compatible with the PW 1212 sequential and PW 1250 and 1270 simultaneous X-ray spectrometers. Although production of the borax bead is more complex in practice than dry pelleting it has the analytical advantage that effects due to grain size and mineralogical composition are eliminated and matrix effects minimized.

These methods of automatic sample-preparation have been developed to meet the particular needs of control of cement manufacture. Nevertheless they demonstrate that the automation of preparative techniques for X-ray fluorescence is entirely feasible. Both methods are evidently applicable, with modification of conditions as appropriate, to other sample materials encountered in cake or slurry form.

6.4.2 X-RAY DIFFRACTION

Fully automated analysis by X-ray diffraction requires that the diffractometer should be provided with automatic means of introducing samples sequentially, programmed automatic scanning of the goniometer and some form of digital read-out of intensities and angles. An automatic sample-changer, model PW 1170, when used in conjunction with the Philips PW 1050 X-ray diffractometer, enables 35 samples to be analysed without operator attention. It is a mechanical sample-changing unit which carries a magazine loaded with 35 powder samples. The sample changer is fitted with two reversing motors. One moves the magazine so that each sample in turn is aligned with the sample-holder shaft of the goniometer. The second drives a screw-spindle which pushes the sample on to the holder shaft and, when the analysis is complete, reverses to return the sample to the magazine. Analysis conditions, such as angle range to be scanned, scanning speed and ratemeter settings, are preselected by the control unit which is also fitted with a three-position switch so that one of three modes of measurement can be selected, namely, analysis only during scanning of the goniometer in the direction of increasing angle, analysis during both forward and reverse scanning, and analysis of samples alternately in the directions of increasing and decreasing angle. On completion of the analysis of a series of samples the sample-changer is automatically switched off together with the recorder, electronic circuitry and the diffractometer.

References

1. Lyon, W. S. *Guide to Activation Analysis*, Van Nostrand, Princeton, 1964.
2. Perdijon, J. *Talanta*, 1970, **17**, 197.
3. Lundgen, F. A. and Nargolwalla, S. S. *Anal. Chem.*, 1968, **40**, 672.
4. Priest, G. L., Burns, F. C. and Priest, H. F. *Anal. Chem.*, 1967, **39**, 110.
5. Priest, H. F., Burns, F. C. and Priest, G. L. *Anal. Chem.*, 1970, **42**, 499.
6. Jervis, R. E., Al-Shahristani, H. and Nargolwalla, S. S. *Proceedings of the 1968 International Conference on 'Modern Trends in Activation Analysis', Gaithersburg, Md.* (NBS Special Publication 312), p. 918.
7. Ashe, J. B., Berry, P. E., and Rhodes, J. R. *Proceedings of the 1968 International Conference on 'Modern Trends in Activation Analysis', Gaithersburg, Md.* (NBS Special Publication 312), p. 913.
8. Givens, W. W., Mills, W. R. and Caldwell, R. L. *Proceedings of the 1968 International Conference on 'Modern Trends in Activation Analysis', Gaithersburg, Md.* (NBS Special Publication 312), p. 929.
9. Starý, J. and Růžička, J. *Talanta*, 1971, **18**, 1.
10. Růžička, J. and Williams, M. *Talanta*, 1965, **12**, 967.
11. Briscoe, G. B., Cooksey, B. G., Růžička, J. and Williams, M. *Talanta*, 1967, **14**, 1457.
12. Růžička, J. and Lamm, C. G. *Talanta*, 1968, **15**, 689.
13. Růžička, J. and Lamm, C. G. *Talanta*, 1969, **16**, 157.
14. O'Kelley, G. D., Eldridge, J. S., Schonfield, E. and Bell, P. R. *Geochem. Cosmochim. Acta*, 1970, Supplement 1, Vol. 2, 1407.
15. Kramer, L., Mobius, G. and Hasenfolder, E.-P. *Jena. Rev.* (Spec. Issue), 1967, 55.
16. Bieshaar, P. *Philips Bulletin*, 79.177/FS 22.

Chapter 7

Gas Chromatography

Gas chromatography was originally introduced by Martin and James[1] in 1952. The technique developed slowly in the 50's but, with the introduction of the first competitive commercial instruments, the 60's brought such a tremendous acceptance of the technique that today gas chromatography is the most widely used analytical tool of the chemist. In its normal mode of operation a discrete aliquot of the sample is continuously absorbed and desorbed from the column stationary phase and ideally every individual component separates on the column and emerges to the detection system. Automation of gas chromatography will be discussed under three broad headings; (1) automatic preparative gas chromatography, (2) automatic analytical gas chromatography, and (3) multicolumn and column-switching techniques. Other areas such as automatic attenuation-switching and temperature-programming are felt worthy only of passing reference. Attenuation-switching is normally activated directly from a recorder by using either mechanized or optical activators. Recently introduced voltage-comparator circuits offer an alternative approach. Temperature-programming is usually achieved by using a simple timed cycle operation. The subject of automatic data-handling will be considered separately in a subsequent section of the book (Chapter 11).

7.1 **Automatic Preparative Gas-Chromatography**

Preparative gas-chromatography offers several advantages over distillation techniques as a purification technique. The selectivity and the ability to deal with azeotropic mixtures present a solution to the difficulties of the distillation processes. Large chromatographic columns (greater than 3 in. in diameter) do not obey the simple laws applicable to analytical columns, and production by gas chromatography is not a viable proposition. The efficiency of separation decreases rapidly as the sample size increases. Large-scale preparative units have therefore been precluded by this restraint. However, many medium or small-scale preparative units have been designed and automated.

The preparation of a sufficient quantity of pure material for further tests or measurements can be achieved in two ways: either large samples can be processed by a single-shot approach on large columns, or small or medium-sized injections can be made repetitively on analytical scale columns. This involves automatic injections, control of column conditions and sample trapping. The second method offers the best solution since large injections produce poorly resolved chromatograms which are only suitable for simple separations. For very simple separations distillation is adequate and the availability of a gas chromatographic approach has only an academic interest. Large injections also tend to flood the top of the column and strip the liquid phase from the support, causing a rapid fall-off in the column efficiency. An additional problem encountered with large injections is the immediate decrease in temperature, which disturbs the equilibrium of the column. The introduction of vaporization chambers with a high thermal capacity can partially offset this problem, but such chambers are usually made of metal and can cause sample decomposition. Higgins and Smith[2] have made detailed investigations of the effects of sample size on column efficiency and operation. A number of mutually interactive effects have been recognized, such as condensation overloading, feed-volume overload and radial-profile overloading. They are dependent on either the absolute solute concentration or the rate of change of solute concentration at any given point in the column. Neither of these is solely responsible for the change in column efficiency.

Increasing the size of the feed volume may reduce initially the radial-profile overloading effect, but a further increase can lead to feed-volume overloading and ultimately to condensation overloading. Reduction of the feed volume can reduce peak broadening effects but other associated phenomena predominate and result in peaks with a pronounced asymmetry, making the required separation impossible. Good resolution is obtained by injecting with a rectangular profile and by maintaining the sample pressure equal to its saturated vapour pressure at the column temperature. An injection device to meet these requirements has been designed by Higgins and Smith[2]. The injection profile can be studied simply by sampling the gas stream a short distance from the injection port, and passing the sample through a detector. By this method condensation overloading can be recognized as an extended exponential distribution of solute. Flash heaters operated at temperatures above those of the column often produce this effect. Sample is vaporized through the heater on to the column where it condenses and forms a thick liquid film on the support, and this obstructs the flow of gas through the column. These overlapping phenomena make it almost impossible to predict the precise operating conditions required, which can only ultimately be determined experimentally.

Jentzsch[3], developing the ideas of Higgins and Smith[2], has designed a dosing system that has been shown to inject in the way suitable for preparative operation, namely with a rectangular plug profile. This is described in more detail in the section devoted to time pressure-injection systems but in essence, sample from a stock vessel is forced through a capillary on to the column by application of a small overpressure. Samples in the range 50–500 μl are injected by the device.

The system has the following advantages; (*a*) samples from a few μl to several ml can be injected; (*b*) it has good reproducibility of absolute sample quantity; (*c*) it has automatic cycle operation; (*d*) it self-cleans hot parts after sample introduction; (*e*) it is essentially a closed system and the sample is protected against atmosphere and humidity, i.e. sample is only in contact with the carrier gas; (*f*) it injects in a plug-profile manner and (*g*) it is equally suitable for direct on-column injection or flash vaporization.

The precise details of the column configuration and operating conditions for any preparative problem must in practice be controlled by the complexity of the sample mixture and the ease of separation from it of the substance of interest. No real merit exists in a complete separation of the components from a mixture in 6 hr if a simple separation in 1 hr resolves the component of interest in the pure state. A compromise has often to be reached on injection-size and column separating-power in order to provide the most economic operation of the preparative unit. Many examples of automatic preparative units have been described in the literature and of these a number have been tailored to particular requirements. Few have sufficient flexibility to become commercially viable, but on the other hand the commercial designs often become so flexible and expensive that it is difficult to justify their purchase on economic grounds. The various units will be discussed with particular reference to the sample-injection systems, control of column conditions, and the trapping systems used. Most of these systems use moderately sized injections in the 50–500 μl range and columns of similar dimensions to those used analytically. In such systems it is imperative to maintain the operating conditions in the equilibrium state so that the individual fractions are always collected in the correct trap. This can be achieved in one of two ways. Either precise controls on the column parameters can be provided so that no change in separation characteristics occurs during the run time (which is expensive, but sample collection can be reliably maintained solely on a simple time cycle), or the continual shift in column parameters can be observed and the change in conditions used to modify the trapping sequence. A combination of a time-base and peak-sensing system is suitable to trap the appropriate fractions. The second method is more reliable because run times of the order of 3–4 days are often required and column conditions will almost certainly alter during this period. The stream of carrier gas should be diverted via a series of traps and the control sequence should allow the appropriate fraction to be collected in the desired trap. The control of the trapping sequence should be precise and protected against false trippings by spurious noise, since a single mistrapped fragment can ruin a preparative run.

7.1.1 Details of Automatic Preparative Systems

7.1.1.1 *Injection systems based on valves*

One of the first simple automatic gas chromatographs devised for gaseous samples was described by Ambrose and Collerson[4] in 1956. Solenoid valves were used to divert the flow of carrier-gas and to connect the column to a reservoir of the sample-gas. The sample-size is dependent on the sample vapour-pressure and the time for which valves are open. After the injection-period the column-operating conditions are re-established and the chromatograms allowed to develop. One restriction on this system is that for any injection at all to take place, the vapour-pressure of the reservoir must be above the column inlet pressure. The injections made with this apparatus were normally of approximately 1 g. Timing and control of the injection and trapping were obtained by using a London PR/S timer and a 50-position uniselector. A single fraction could be collected by diverting the carrier-gas by a solenoid valve, through a trap. In the particular applications for which this unit was designed, the column conditions were stable for several days; a time-only basis of sample collection proved reliable. The equipment was designed solely for gaseous injections and the requirements of the system are therefore greatly simplified. Only one fraction can be collected and thus most of the sample is lost. Systems based on this design have a limited value for practical problems.

7.1.1.2 *Injection systems based on metering pumps*

A preparative unit described by Atkinson and Tuey[5] specifically for liquid samples is shown in Fig. 7.1. A master controller provides the electrical signals which are used to activate the injection and trapping functions. Trapping of fractions is on a time-only basis and normally a

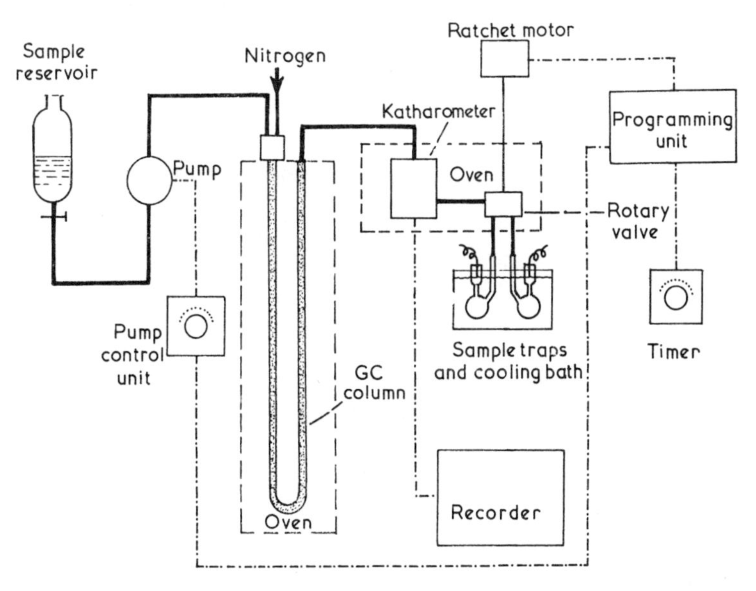

Fig. 7.1 Automatic preparative gas chromatograph.
Reproduced with permission from Atkinson and Tuey[5] *and Institute of Petroleum, London.*

detector is not incorporated into the system but is used initially to program the controller. Even if recorder-drift has been allowed for, two problems affecting the control remain: (1) electrical interferences which produce spurious operations of the valves and lead to contamination of the collected fractions, and (2) the fact that complicated mixtures are difficult to handle if time is the only basis for fraction-collection. Sample is injected by the operation of a piston-pump connecting the sample-reservoir to the column for a fixed period. The quantity of liquid injected is therefore a function of the amplitude of stroke of the pump, the frequency of the stroke, and the time for which the pump operates. A fixed frequency is chosen and, by suitable adjustment of the amplitude and time, volumes of 0·5–3·5 ml are injected. Injection periods of the order of 10 sec are used, and they produce a slight spreading of the peaks but avoid column flooding. There is a tendency for sample-bleed after the initial injection, but the insertion of a ball-valve at the end of the delivery-tube from the pump can overcome this problem, particularly if it is cooled by water. An alternative approach if really effective water-cooling can be maintained is to replace the delivery-tube by a length of stainless steel capillary tubing. Excessive local cooling on injection is avoided by incorporating a vaporization-chamber at the top of the column.

Fraction-collection is controlled by using a custom-built stainless-steel and PTFE rotary switch-valve specifically designed to remain leak-free for long periods. The system allows for four fractions to be collected by diverting the carrier-gas stream through the traps. The emerging carrier-gas stream can either be directly vented or diverted to vent via traps in a strict rotation. Samples must of necessity pass through the rotary switch-valve. The instrument operates in the following manner: the pump is energized and a predetermined quantity of liquid is injected into the column; the controller is programmed so that each fraction of interest is trapped, and the effluent is then passed to atmosphere until the next fraction of interest emerges. As an aid to efficient trapping an electrostatic condensing trap was also described in this paper. The performance of the apparatus was tested with a 50/50 v/v mixture of alcohol and toluene.

The instrument remained in synchronization for 40 replicate injections each of 0·75 ml, and the products were found to be 99·8% pure, by normal gas chromatographic analysis, after a single pass through the column.

A similar injection system was used in the preparative chromatograph of Tenenbaum and Howard[6]. Control of the trapping sequence in this system is based on the detector response, 1% of the carrier gas being diverted through a thermal-conductivity detector. No problems from spurious tripping of the control have been experienced in this system. The solenoid valves are operated by microswitches fitted to the top of the recorder. For the particular types of analysis illustrated no base-line drift was experienced. Problems of sample-bleed were again encountered but were overcome as follows: the sample was contained in a closed vessel, connected to the carrier-gas line via a solenoid valve; after the injection period this solenoid valve was opened to equalize the pressure in the sample vessel and at the top of the column. Three preparative applications illustrated the range and the inflexibility of this type of system. Simple separations of well-resolved peaks were handled adequately, but as the complexity of the chromatography increased so the efficiency of the unit fell rapidly. Crude toluene, ethylbenzene and mesitylene were each chromatographically purified and subsequently analysed cryoscopically. The toluene was found to be 99·95% pure after one pass, 92% of the impurities being removed. For ethylbenzene only 72% of the impurities were removed, because an impurity with a retention time greater than ethylbenzene proved to be difficult to remove. The sample of mesitylene showed many impurities, of which a number were removed after a single pass through the column but a large quantity of an impurity with a retention time close to that of mesitylene itself made complete purification difficult with the column-conditions used. The reduction of the injection-volume to 0·1 ml failed to resolve the impurity from mesitylene. The system described, which uses a simple time-basis of trapping control, has been shown to be suitable for simple separations. Trapping efficiencies of the order of 80% are attainable but these vary according to the type of sample under consideration.

7.1.1.3 *Injection systems based on use of a syringe*

Three systems based on syringes are described in this section. Kauss, Peters and Martin[7] described an automatic preparative system designed to meet the requirements previously set out, limiting the injection-size so that the loss of resolution was minimal. The injection-volumes can be varied from a fraction of 1 ml to several ml by using a specially designed injection-system[8]. The design of the unit incorporates a series of compensating valves which ensure that the entire contents of the syringe are injected. An injection is activated pneumatically, forcing the sample into a flash heater filled with stainless-steel shot. This ensures a rapid evaporation of the sample without local cooling effects at the top of the column. Temperature-programming facilities are provided. A specifically designed detector is described which allows the entire effluent of the column to pass through. This method was considered more reliable than using a conventional detector and splitting device. Trapping of the components is arranged through a rotary manifold valve which diverts the effluent gas-stream and is situated on the outlet side of the trapping vessel; this offers several advantages. The valve does not have to withstand high temperatures, which makes it simpler to make, and it prevents the samples from passing through it, which in turn avoids corrosion problems. The traps are only open to gas-flow whilst carrier-gas is diverted through and the fractions are protected from ambient air and moisture contamination. Problems of cross-contamination between fractions in the traps could arise, but this can be avoided by incorporating ball check-valves into the lines connecting the traps. The particular rotary valve described had six trapping and six venting positions arranged alternately about the circumference of the valve. A strict sequence of trapping and venting must be maintained, but as long as the loss of the majority of sample to waste can be tolerated, this constraint is not too limiting and up to six fractions can be collected.

An injection system designed by Haruki[9] is shown in Fig. 7.2. Liquid samples are injected through a tube (*a*) mounted concentrically inside a piston, which moves horizontally inside a cylinder mounted on the side of the chromatograph injection-port. The injection-tube protrudes beyond the piston so that on injection it extends into the carrier-gas stream. Forward movement of the cylinder (*b*) forces the sealing device (*c*) to open

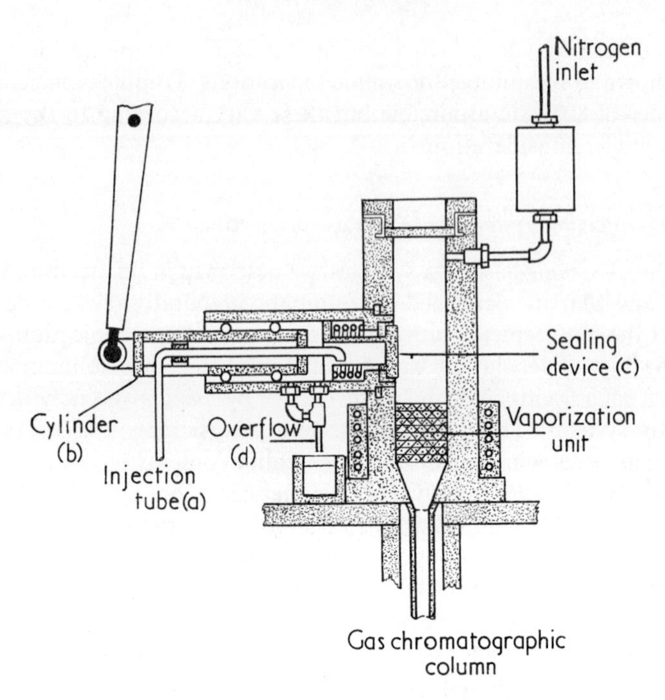

Fig. 7.2 Injection system designed by Haruki[9].

and allows the sample to evaporate, causing an injection. The sealing device is closed by a spring as the cylinder is withdrawn. Excess of sample is removed via (*d*) which is opened by the backward stroke of the piston. As the orifice (*c*) is sealed on the backward stroke, a further aliquot of sample is drawn into the tube from the sample reservoir. A simple control circuit involving the use of two electric motors coupled to a mechanical link operates the device. It is claimed that the efficiency of injection is similar to that for manual injection. There is no sample 'dribble' or loss due to leakage of the injection-device. Injection-size can be varied either by using different tubes or by altering the stroke of the piston.

The system described by Jover and Gastambide[10] uses a mechanical version of a Hamilton PB600 repeating dispenser. Normal operation of the syringe delivers $\frac{1}{50}$ of its total volume each time the syringe is actuated. A sequence of operations designed to overcome the problem of sample-bleed operates in the following manner: the syringe is mounted on a carrier which travels on a fixed framework in a direction parallel to the axis of the injection-port; forward movement allows the syringe to penetrate the injection serum-cap, the dispenser is activated and after sufficient time has been allowed for the sample to evaporate the syringe is withdrawn to its rest position. After an initial loading 50 automatic injections can be performed. The size of the syringe restricts the quantity of sample processed in any one set of operations, but syringe-volumes of 50–500 μl are commonly used.

7.1.1.4 *Injection systems based on time and pressure*

Time–pressure injection-systems were first proposed by Heilbronner, Kováts and Simon[11]. Basically a slug of sample is forced through a capillary restriction on to the chromatographic column by an overpressure of the column carrier-gas. The volume of liquid injected as a function of time, for such a system, is given by the Hagen-Poiseuille law:

$$V(t) = f(p_s - p_c) \cdot \frac{d^4 . \pi . \gamma}{128\, \eta\, L}$$

where d = capillary diameter, L = length

γ = dimension factor, η = viscosity of sample liquid

p_s = applied pressure, p_c = column pressure

$p_s - p_c$ = overpressure

During injection the vaporization of the sample modifies the overpressure. If d and L are maintained constant, then for a fixed time the injection size is dependent on the viscosity of the sample and the overpressure applied. The diameter and length of the injection capillaries are chosen so that the injection-size is compatible with the column dimensions. Three systems using variations of time–pressure injection systems, illustrated in Figs. 7.3–7.5, will be discussed.

The system in Fig. 7.3, proposed by Boer[12], operates in the following manner. *P*1 and *P*3 are opened and *P*2 closed. A single source of carrier-gas is used in the system and a pressure differential between the overpressure and the normal column inlet pressure is arranged by using a variable restriction such as a needle-valve between the inlet and the column. *P*1, *P*2 and *P*3 are pneumatic solenoid valves. After injection, the excess of pressure in the system is vented to atmosphere via the buffer volume and control valves. A continual flow of carrier-gas back through the injection-capillary purges sample from the line. Boer used a vaporizing chamber which improved the resolution of chromatograms for the range of sample- and injection-sizes that he used.

The overpressure system described by Frazer and Morris[13] Fig 7.4, uses two control valves only. A regulated overpressure is fed to the reservoir tank. On initiation of injection, a signal from the master controller

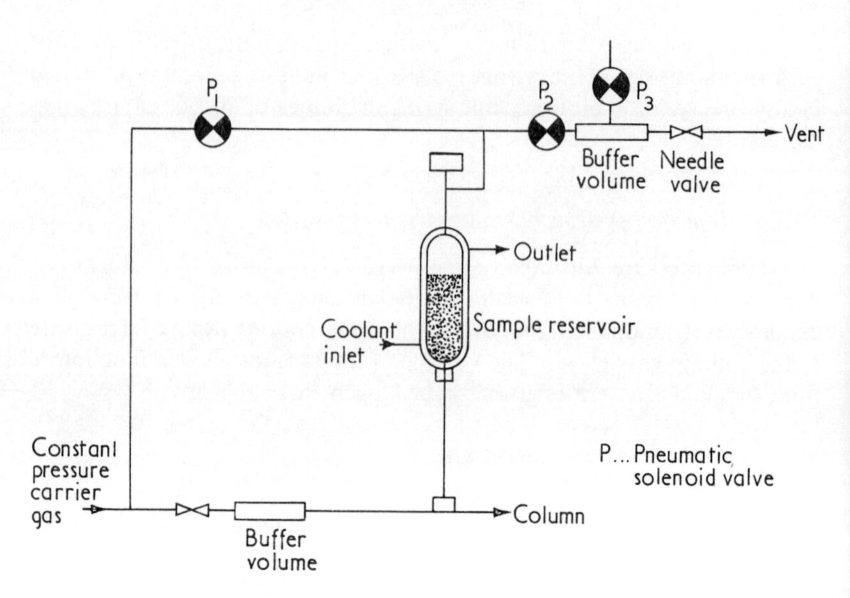

Fig. 7.3 Overpressure system.
Reproduced with permission from Boer[12] and Institute of Physics.

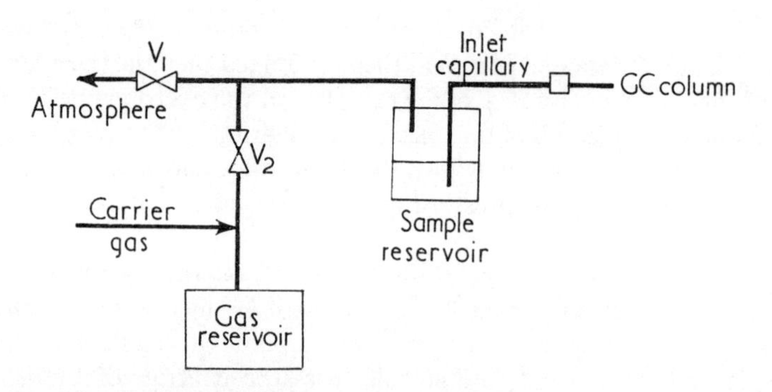

Fig. 7.4 Overpressure system.
Reproduced with permission from Frazer and Morris[13].

opens *V*2 and this forces liquid on to the column. After the injection time has elapsed, *V*2 is closed and *V*1 opened for 0·5 sec; this has three effects – it releases the excess of pressure, gives a sharp cut-off in sample introduction and purges sample from the injection capillary. *V*1 is then closed and the column equilibrium re-established. Suitable co-ordination of the overpressure, capillary-dimensions and injection-time allows an almost ideal plug injection to be obtained. Jentzsch[3] has expressed doubts

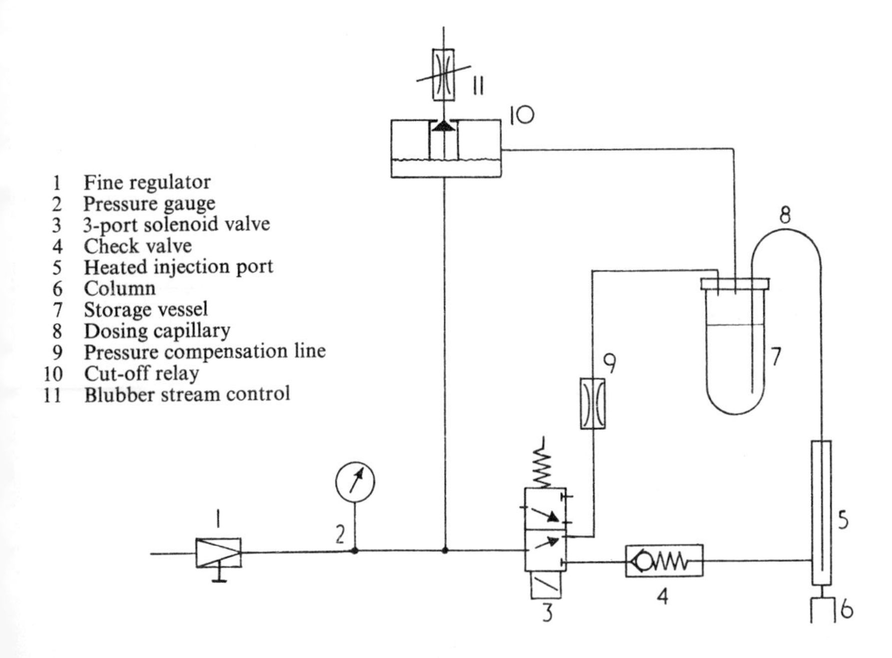

Fig. 7.5 Overpressure system.
Reproduced with permission from Jentzsch[3] and Preston Technical Abstract Company.

about the reliability of both of these overpressure designs. He argues that blowing off the sample directly to the atmosphere could cause sample loss on injection and that alternatively exhausting the excess of pressure via a capillary causes an excessive dosing period. It is also difficult to adjust gas flow-rate and vaporization temperature. Jentzsch has designed a unit, Fig. 7.5, which meets the specification previously given and has a rectangular plug profile.

The unit uses three types of valves, a directional control-valve, a check ball-valve and a 'blubber' release valve. The check-valve ensures that during injection none of the components is directed along the normal inlet of the carrier-gas. If they were, re-establishing the normal column conditions would force them on to the column, causing additional spurious injection. The 'blubber' release valve (11) allows the flow-rate of the bleed-stream to be adjusted so that the overpressure never equals the column-pressure. This avoids any spurious injection which might occur during a cool-down procedure following a temperature-program run. On injection, valve 3 being opened, the column-pressure falls below the overpressure, forcing liquid through the dosing capillary on to the column via the vaporization-chamber. The compensating capillary ensures that

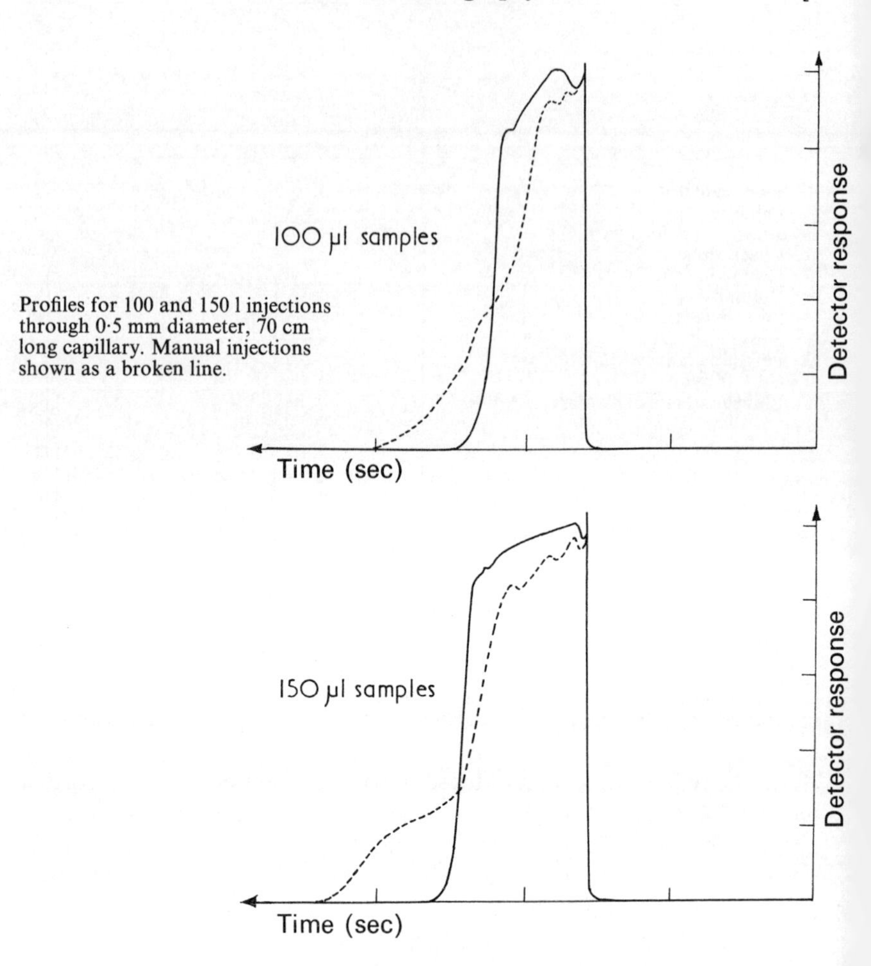

Fig. 7.6 Dosing profiles produced using overpressure system shown in Fig. 7.5. *Reproduced with permission from Jentzsch[3] and Preston Technical Abstracts Company.*

the overpressure never exactly equals the pressure on the top of the capillary, which produces constant injection doses as the sample reservoir is progressively emptied. Sample sizes between 2 and 5 μl are injected with plug-injection profiles. Results of profile-experiments are shown in Fig. 7.6. A manual-injection profile is also shown and it can be seen that the automatic system more closely approaches the ideal state postulated by Higgins and Smith[2]. It should be remembered that each overpressure system has to some extent been tailored to suit the operating conditions and geometry of the particular column. Jentzsch has shown that plug-profile conditions are obtained through using his system. Neither Boer nor Frazer and Morris present any experimental results on this point. They judged the performance solely for the complete unit whereas Jentzsch has only developed an automatic-injection device. The systems devised by Boer and by Frazer and Morris represent the most versatile and flexible systems for preparative gas-chromatography. Both systems use a trapping logic based on a combination of time and peak-sensing and differ only in the detail by which they achieve this. Alternatively, time or peak-sensing bases can be used independently. Morris and Frazer developed their system later than Boer and in many respects it represents a logical development of that system.

The system of Frazer and Morris (and to a lesser extent that of Boer), offers the following attractive features.

(*a*) A wide choice of trapping program.
(*b*) Precise injection-control with a wide dosage-range.
(*c*) Accurate and continuous timing facilities which give precise control of trapping periods and duration of other operations.
(*d*) Infinitely variable peak-sensing threshold levels; more than one level capable of operating within a cycle.
(*e*) Flexibility by interlocking (*c*) and (*d*). This caters for any minor change in column parameters and therefore eliminates the necessity for precise control of temperature and flow-rate, which is expensive.
(*f*) Good trapping efficiency based on control of carrier-gas at the outlet of the traps, thus avoiding contamination by air and condensation.

Boer[14] described a universal programming unit which controls injection, peak-sensing, and the trapping and dumping of separated and unwanted components. Temperature- and flow-programming can also be controlled by this unit. For such a system, time-periods of each separate operation vary greatly from a few seconds for injection to up to an hour for a temperature-programming cycle. The accuracy over this range of time must also be high and this coupled with the need for complete flexibility and ease of change precludes the use of cam-timers, etc. The cycle-time for a cam-timer is fixed by the rotational speed of the synchronous motor, and accurate setting of the desired length of time is almost impossible because of gear-lag and the relatively small circumference of the cams. The controller works on the following principle: a variable synchronous-clocking time-base determines the unit of time, and for each period of the cycle a preset number of pulses is counted. A cycle consists of a number of such periods. After the unit has been switched on, the number of pulses for the first period is counted off and the control of operations passed to the second period, and so on until at the end of the last period control is returned to the first period to restart the cycle. A period can also be jumped by using an external switching device such as a peak-sensor to combine with the time-base sensing. The peak-sensing system described by Boer uses both the rotating cam of the recorder itself and a position-sensing mechanism incorporating a photodiode switch. This requires that the trap-sequence is triggered at the same voltage-level for opening and closing the valves. More than one level of threshold involves a duplication of sensors in the recorder, which often present problems of siting. A voltage-trapping switch controlled directly from the detector signals is preferable. The trapping mechanism devised by Frazer and Morris is also more viable since it does not involve a complicated rotary switch-valve. Individual control-valves on the outlets offer more flexibility. Only one trap has to be used as a dump and therefore if n traps are avail-

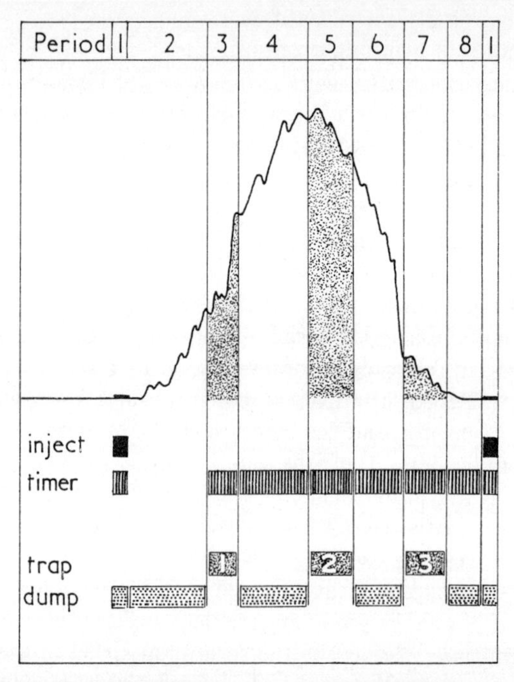

Fig. 7.7a Collection of specified fractions from an incompletely resolved oil sample

This provides a simple example of use of the timing circuits alone. For this application, a pilot run will have indicated the time periods over which traps should be opened to collect the sample 'cuts' of the required boiling range, and the overall period between automatic injections determined. The complete sequence of timing signals is then set up on the matrix board of the Master Programmer, the matrix board of the Trap Controller being set so that the appropriate electro-pneumatic valves for fraction collection or dumping are energized in turn, any signals from the Peak Sensor being ignored. Once the programme is initiated, the entire sequence will be repeated automatically as many times as required.

Fig. 7.7 Use of peak sensing and timing to trap required fractions in preparative gas chromatography.
Reproduced with permission from Pye Unicam Ltd.

able, $(n-1)$ fractions can theoretically be trapped. Trapping in any sequence can also be accommodated, and theoretically all the starting material is recovered.

The systems based on time–pressure injection provide the most flexible approach to automatic preparative gas-chromatography. Three illustrations of separation problems are shown in Fig. 7.7, using (*a*) time

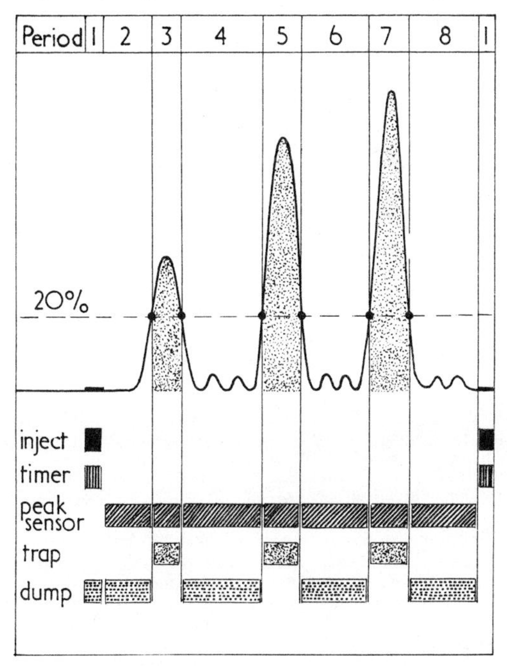

Fig. 7.7b Collection of pure homologues from a wax sample

This illustrates a simple example of Peak Sensor actuation of the collection traps, the timing circuits being used only to control the injection of the sample; these circuits could be arranged to repeat the injection and collection processes automatically if required.

The sequence of operations shown is as follows:

Commencing with a charged sample reservoir, insertion of the 'start' plug pressurizes the reservoir with carrier gas for a short pre-determined period, usually 1–5 seconds. This transfers a defined amount of sample to the preparative column, the timer circuits being disengaged once this period (1 on diagram) ends. From now until a significant component peak emerges, the column effluent is fed through a refrigerated 'dump' trap. When the Peak Sensor (set in this example at 20% full scale on the recorder) registers the emergence of a peak above the preset threshold, Period 2 ends and Period 3 (first component collection) starts, the effluent being diverted into the first trap. As the peak returns to base-line through the 20% position this trap is sealed off and the effluent 'dumped' once more (Period 4).

This sequence is repeated for the collection of the other major components, collected in Periods 5 and 7 in this example. Unwanted minor components, with peak heights below the 20% threshold level, are rejected throughout. For completely automatic operation, the signal generated by Peak Sensor on the tail of the last major peak and used to switch from Period 7 (collection) to Period 8 (dumping) could also be arranged to initiate a time sequence which would result in injection of a second sample, and so on.

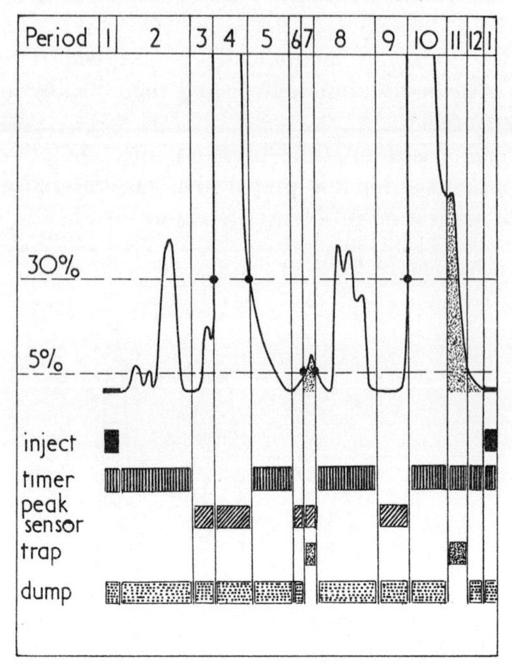

Fig. 7.7c Collection of minor and unresolved components

This example illustrates use of all the facilities of the system, this time 'interlocked' to provide a sophisticated command sequence. Peak Sensors set at two widely different levels are used and an otherwise extremely difficult collection problem is solved without the need for supervision.

The programme sequence is as follows: The sample is injected automatically (Period 1) and, during both this and the preset Period 2, signals from either of the Peak Sensors are ignored by the Trap Controller. In the application being studied, the first minor component of interest follows a major peak, and the relationship between the two peaks is used to control the trapping operation. Between the emergence of the first major peak shown on the chromatogram and the preceding component, the timer switches to Period 3, alerting a Peak Sensor set to 30% full-scale deflection. As the front of the major component passes through this level, Period 4 is started. During this period, the Peak Sensor is set to detect the return of the tail of the peak towards the base-line. As soon as this deflection drops below the 30% value, accurately timed Period 5 commences. After the end of this time (preset to a value derived from an exploratory run), a second Peak Sensor, this time set to 5% deflection, is activated (Period 6). The column effluent, fed to the 'dump' trap up to now, is switched to collect the first required component in Trap 1 as soon as the peak amplitude reaches the 5% value; the trap is sealed off once more when a similar level on the tail is reached. Period 8 has been preselected to run from the recommencement of effluent dumping to a convenient point in the base-line just before the emergence of the second major component, on the shoulder of which is the unresolved peak produced by the second component required to be collected. During Period 9 the 30% full-scale Peak Sensor is energized, awaiting the emergence of the front of the second major component. From that point onwards, Period 10 commences, terminating in the opening of Trap 2 after an accurately preset time (once more based on the exploratory run). Trap 2 remains open for a fixed time (Period 11), after which the column effluent is dumped (Period 12) until all components have emerged from the column; at this point, a new sample is automatically injected and the whole process repeated until sufficient quantities of the desired components have been collected in the two traps.

It will be seen that this technique allows the Peak Sensor to 're-align' accurately set timed periods and permits narrow bands of column effluent to be located *with respect to the current performance of the column*; loss in collecting efficiency due to the minor changes in retention time which frequently occur between runs is thus avoided.

only, (*b*) peak-sensing only, and (*c*) a combination of both for trapping-logic. These systems will handle virtually any analytical problem. The various other systems described have some merits for the particular problems for which they have been tailor-made. If analysis needs are likely to change often, then the high cost of a more sophisticated preparative unit in the laboratory becomes a more economic proposition. If only a single application is considered, a custom-built system based on the principles set out in this section will prove reliable and economic.

7.2 Automatic Analytical Gas-Chromatography

Automatic analytical gas-chromatography, unlike preparative work, does not require control of the effluent gas-stream, the prime consideration being the optimum precision of the analytical results. In the manual approach precise injections and ultimately reliable results can be achieved only by highly skilled and qualified staff. Where large blocks of routine work are involved, the staff can be completely occupied in preparing samples, injecting them into the chromatograph and interpreting the results. From the viewpoint of job-satisfaction this is both undesirable and uneconomic; moreover the instruments are only utilized for about a quarter of the available working time. These disadvantages, coupled with the high cost involved in equipment, make automation an attractive and economic advance over manual analysis. Automation allows the output of an analytical gas chromatograph to be effectively increased without the necessity for shift work or the excessive duplication of equipment. The cost of automation, including the initial cost of the gas chromatograph and recorder and other equipment must be balanced against the savings achieved by the deployment of skilled operators for more useful and potentially more satisfying tasks.

An example is the fatty-acid composition of foodstuffs, determined by methylation followed by gas-chromatographic determination. The complete range of esters of acids from C_{10} to C_{28} is normally found. Chromatograms can take up to 5 hr for complete resolution and running automatically in the 'silent hours' would achieve up to five-fold increase in output for each instrument. A different type of approach is that in which medicinal preparations, e.g. tinctures and essences, are examined for their dutiable spirit content. The original method and indeed the statutory analytical method (Thorpe and Holmes[15]) involve various distillation and clean-up procedures which are both laborious and time-consuming. Harris[16] has described a gas-chromatographic method using a column of porous polymer beads (Porapak Q). A flame ionization detector is ideal for determining the ethanol content since it has a low response to water. Propan-l-ol is added to the sample, as an internal standard, and the mixture injected into a column. A calibration graph of proof spirit strength against the ratio (area ethanol peak/area propan-l-ol peak) is drawn. An unknown sample can then be analysed by reference to this graph. By this method an operator can analyse approximately 20 samples per day. Automation of this process could free him for more worthwhile tasks.

A fully automatic gas chromatograph is a potentially valuable instrument in any laboratory carrying out repetitive analysis. An outline specification for such an instrument requires the inclusion of the following features: it should be capable of handling discrete samples; should mix them with internal standards where necessary and do such manipulations as solvent extraction; it should control the gas-chromatographic conditions (i.e. temperature-programming, column-switching, etc.), and finally should present the computed results in a manner suitable for reporting. This chapter will not attempt to cover the data-processing aspect involved (this will be covered in Chapter 11). Bowman and Karmen[17] discuss the salient features required from an injection device for precise analytical gas-chromatography. The following features are required: (*a*) the sample should be delivered to the inlet of the column as a vapour over a short time interval, in order to keep dilution with the carrier-gas to a minimum; this is, however, a critical parameter and a compromise must be used since excessive flooding of the top of the column can be as damaging to the chromatogram as overlong injection periods; (*b*) quantities of sample injected should at least be reproducible, and better still, accurately known; addition of an internal standard avoids the need for precise repeatability of injection; (*c*) the injection should not require the stopping of the gas-flow; a rapid return to equilibrium flow-rates resulting in a minimal base-line disturbance is a critical requirement for reliable results; (*d*) the quantity of air introduced should be minimal. Bowman and Karmen described an injection device which met these requirements, using glass capillaries filled with sample which was injected by crushing the sealed ampoule with a stainless-steel piston. The device was not automatic but this technique of encapsulation forms the basis of many known and commercially possible instruments.

The systems available for automation vary tremendously in the extent to which they have been automated. They will be discussed in this chapter according to the type of injection system used. A schematic diagram of the requirements for automatic gas-chromatography is shown in Fig. 7.8.

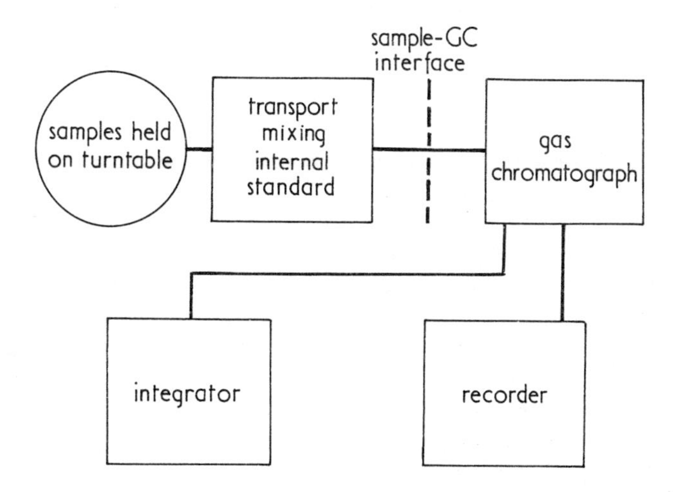

Fig. 7.8 Schematic diagram of the configuration of an automatic analytical gas chromatograph.

7.2.1 Detail of Automatic Analytical Systems

7.2.1.1 *Systems using mechanical syringes*

Early attempts at automatic instruments were simply mechanical devices for operating a series of hypodermic syringes. In 1964 Jarrell and Allison[18] proposed a system using a bank of twenty separate syringes, and even in the 1970s designs based on a hypodermic syringe are still prominent (notably the Hewlett Packard[19] system). Jarrell and Allison designed a system which used twenty syringes mounted on a turntable, where each syringe was presented in turn over the injection-port and a pneumatic mechanism forced the syringe through the serum cap. The liquid was displaced by actuating a plunger. The plunger was drawn back and the syringe withdrawn after evaporation of the sample. All syringes had to be preloaded and then held on the turntable; a cam-timer was used to control the sequence of operations. After the initial loading sequence, each injection was carried out in a similar, reproducible fashion and none of the errors normally associated with manual injections such as inter-operator errors, occurred. Visual peak-inspection was used to check the reproducibility of injection. Up to 40% more samples could be analysed automatically than manually. Measurements for the determination of water content, using a 5% water/ethanol mixture, based on peak height, showed a 4% repeatability of injection.

In 1969 Jover and Gastambide[10] patented a device for continuous operation of a Hamilton PB600 repeating dispenser, described above in section 7.1.1.3. The automatic analytical gas-chromatograph (HP model 7670A) marketed by Hewlett Packard uses similar principles. Currently this system most closely meets the specifications set out previously. Samples are held in containers mounted on a turntable. The syringe can travel in both horizontal and vertical planes. During operation the syringe is lowered into the sample and the plunger withdrawn; sample is drawn into the barrel. The syringe is withdrawn and the sample flushed out into a drain reservoir by forcing the plunger downwards. This wash-out cycle can be repeated for a single sample many times. After the required wash-out the syringe is filled with sample and then withdrawn; the carriage moves the needle so that it is directly in line with the injection port. The system then operates similarly to that of Jover and Gastambide. After injection the syringe is withdrawn and then positioned for the next sample. At the end of the chromatogram the turntable rotates one position and then a new cycle is begun. An integrator/computer interface is also available to evaluate the results and to prepare reports. Response factors and normalization techniques are readily incorporated and, if retention data are fed into the computer, peak-identification and retention-indices can be compared. Reports can be produced in any desired format. The design is well engineered and, although primarily built as an attachment to Hewlett Packard gas chromatographs, can be fitted to any gas chromatograph with minimal engineering effort, correct alignment of the syringe being the prime consideration. Recently two other instrument companies, Pye Unicam and Varian Associates, have introduced similarly automated syringes.

7.2.1.2 *Encapsulation techniques*

With the general improvement of instrument design and column configuration, the injection has become a limiting factor. Operation at high inlet-pressures and temperatures causes volatile components to be driven off from the serum caps, and this affects the distribution of the absorption in the column and modifies the detector-response by causing base-line drift and noise. These will affect the quantitative accuracy of the system, particularly when temperature-programming is used. Encapsulation techniques obviate the need for serum caps, but despite this advantage the technique still requires a prior injection of samples which, whenever considerable time is spent on sample preparation, leaves much of the work in the hands of the manual operator. Automation of these time-consuming processes is far more worthwhile. The requirements for injection into an analytical chromatograph have already been discussed. Nerheim[20] has developed an injection technique using samples sealed in indium capillaries. The technique was introduced to combat the difficulties met in injecting volatile liquids via a syringe, since the problems of variation in the volume injected and of selective vaporization were encountered, particularly when flash vaporization was used. Samples often vaporize on storing, and encapsulation can overcome this since accurate weighing of the sample capillary before and after filling is a simple procedure. This also avoids the use of an internal standard. Capillaries sealed in this way remained stable to within $\pm 0{\cdot}5\%$, over a 4-week period. A heated injection-block was designed to allow the removal of the indium capillary after the injection. The internal volume of the block was less than 5 μl and this minimized peak-broadening effects. Hudy[21] and von Rudloff[22] have also used this method of sample introduction, with different types of capillary for varying sample types.

Walker[23] introduced an automatic feed-system using a soft tin alloy. Capillaries of this are dropped into a channel milled into the shaft of a probe which moves into the vaporizing zone of the injection port. The vaporized sample is swept on to the column and the components separated. At the end of the chromatogram the probe is withdrawn from the injection-port and turned through 180°, and the globule of solder is then discarded. A further revolution through 180° allows the next sample to fall into the channel ready for the next injection. Solids, liquids or gases may be analysed in this way. The system can accommodate 36 samples of varying composition.

A more sophisticated and well-engineered capsule-injection technique has recently been presented by Otte and Jentzsch[24] and is available through Perkin-Elmer. It requires samples to be placed in small aluminium capsules which can be crimped to ensure closure. Solvents can be evaporated before the crimping and a storage life of the capsules greater than 3 months is attainable. Each tube is presented in turn to the inlet of the gas chromatograph and pierced with a sharp needle. The sample is completely flushed with hot carrier-gas, thereby ensuring good injection-characteristics and complete vaporization. An identification system is also available which uses a photoelectric device operated in conjunction with the sampler. Accurate results with this technique have been obtained even with capillary columns. However, such a system can only be maintained by a well-equipped engineering section and, since it is a complex system will require careful attention to detail for fault-free running.

Encapsulation techniques have found many uses, particularly in the analysis of pesticides and steroids. Often in these applications only small samples are available and in order to obtain meaningful results the solid sample is injected. Podmore[25] devised a system in which samples are deposited on to a small ferrous metal cylinder and the solvent evaporated off before the analysis. The technique allows the sample to be concentrated by using multiple additions to the cylinder. A manual injection still has to be made and the accuracy of the method depends on the injection precision. More accurate results are obtained by weighing the cylinder before and after sample is deposited on it. The metal cylinders are fashioned around a suitable glass former so that a slit 1 mm wide remains along the length of the cylinder. The cylinders are silanized and conditioned at 300° C before use. A number of prepared samples can then be loaded in the injection device shown in Fig. 7.9. The small cylinders are fed to the flash-heater and then removed electromagnetically, a cam-timer being used to control the sequence of the operations. The injection-port introduces a large dead-volume on to the top of the column, allowing back-diffusion of the sample, which produces poor resolution. Samples are in

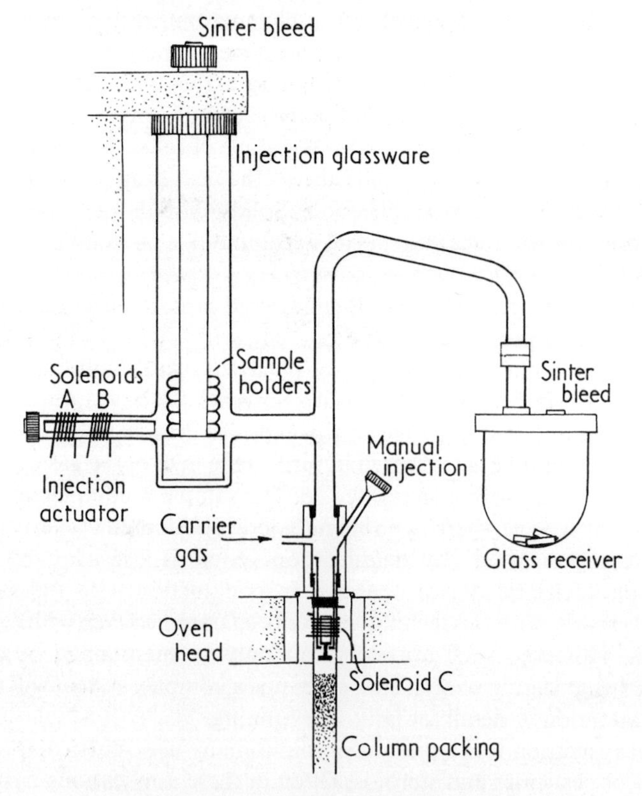

Fig. 7.9 Encapsulation system designed by Podmore[25].
Reproduced with permission from Pye Unicam Ltd.

contact with each other on the feed system and this causes intersample contamination. Results of analysis of steroids evaluated by simple peak-height or triangulation techniques were, however, quite satisfactory.

A commercially available system using capillaries based on this design is available from Pye Unicam Ltd. (No. 12006 series 106 model 6).

Table 7.1 STANDARD MIXTURE OF TRIMETHYSILYL ETHER DERIVATIVES (RESULTS FROM 18 REPETITIVE ANALYSES)[26]

Androsterone	$33{\cdot}8 \pm 0{\cdot}8\%$
Etiocholanolone	$19{\cdot}0 \pm 0{\cdot}6\%$
Dehydroepiandrosterone	$27{\cdot}7 \pm 0{\cdot}5\%$
Coprostanol	$19{\cdot}4 \pm 0{\cdot}3\%$

Ruchelman[26] also described a system for handling solid samples, using wire gauzes (Dixon rings). The rings are carefully cleaned and then placed in the cavities of an evaporation-plate; solid samples are dissolved in a volatile solvent and an aliquot of this solution is dropped on to the gauze from a syringe. Solvent is evaporated and the sample drawn by capillary attraction on to the wire gauze. The sample gauzes are then carefully transferred to a Teflon wheel and placed in slots in the wheel, situated over the column. When the wheel is loaded the cover can be replaced and the chromatograph carrier-gas turned on (the carrier-gas flow is also used to purge the system). This is a disadvantage since it takes time to re-establish equilibrium conditions. Once conditions are stabilized, the Teflon wheel rotates and a single gauze is allowed to drop into the flash-heater zone of the gas chromatograph. The sample vaporizes on a basket in the heater zone with the ring retained in the basket. The Teflon wheel rotates at a constant speed, and when the next slot is aligned with the column-inlet another sample-gauze is dropped into the heater-zone and a new cycle begins. No attempt is made to remove the gauzes, and this, coupled with the large dead-space above the column, could cause peak-tailing and sample-interaction. However, for the types of analysis discussed, good chromatograms are obtained. Results for a standard mixture of keto steroids and coprostanol are shown in Table 7.1.

Sample decomposition could take place because of storage at high temperatures above the column-oven. Applegate and Chittwood[27], using a modified Ruchelman system, have shown that pesticide solutions are stable for periods of up to 12 hr in the sample container used in their system. Difficulties were experienced with the Dixon rings, caused by (*i*) loosely wound rings being prone to stick in the Teflon seats and (*ii*) the rings catching in the lugs of the Teflon wheel and therefore requiring trimming. Applegate and Chittwood also found that the instrument needed recalibration every few hours, with the result that reliability for several days could not easily be guaranteed.

Tinti[28] and Appleqvist and Melin[29] have also produced a variation of this type of design. Both have eliminated much of the dead-volume, which remains an undesirable feature of the Ruchelman design. Tinti concludes that spiral-type holders have better performance than capillaries and that, for substances affected by hot metallic surfaces, glass spirals are preferable. Appleqvist's design has the advantage over other systems of this type in that the empty sample-container is withdrawn and sample-holders do not accumulate at the top of the column. A build-up at the top of the column can produce band spreading and any high-boiling components may slowly decompose to cause irregularities in the response of subsequent samples. To avoid this, systems have to be dismantled frequently to remove the sample-holders.

Borth *et al.*[30] and McGregor[31] have also presented similar approaches which differ from those already discussed in the method of sample-transport. Neither uses a sample-wheel and, in the system described by Borth, samples are placed in a glass side-arm to the injection-port and separated from each other by lengths of silanized glass rod, and a soft-iron rod placed beneath the last sample. A solenoid loosely encircling the tube mechanically transports the iron rod at a fixed velocity along the side-arm, tripping samples and spacer into the injection port. McGregor impregnates porous balls with the sample and these are transported into and out of the vaporization chamber pneumatically.

7.2.1.3 *Systems based on valves*

Kipping and Savage[32] described an automatic analytical system based on a Loenco valve (LSV-220) and used a modified laboratory fraction-collector to hold samples which were transferred from the sample-tubes pneumatically via a double probe system. Air is forced down one of the tubes and the sample forced up through the second tube and into the valve. When the valve is activated a slug of sample is forced into the flow of carrier-gas and on to the top of the column. This system produces a relatively high carry-over effect from one sample to another, often to the order of 1%. Higher-boiling components are not transferred quantitatively because of condensation in the connecting line from the valve to the column. Well-resolved peaks without any noticeable tailing are obtained by use of this system for moderately volatile liquids.

Wicks[33] has devised a system using a sandwich-type valve which is suitable both for liquid and vapour injections. For optimum performance the sample-valve is located outside the oven and the sample transferred first to a vaporizer and then to the column. Siting the valve in the oven causes bubbles to form in the flowing stream and results in erratic injection-volumes. The presence of bubbles in the valve also causes gas/liquid hold-up on the valve-plates and produces a slow dribble of sample into the vaporizer after an injection. The injection device also incorporates a by-pass valve which maintains the pressure in the vaporizer below that in the sample line, allowing sample to be forced into the vaporizer. The valve described is constructed from a gold-lined Teflon slider sandwiched between stainless-steel plates. It is actuated by compressed air and operates reliably on samples boiling in the range 50–500° C.

Evrard and Couvreur[34] designed a valve-system to overcome the various objections discussed above. A solvent wash-out step is incorporated into the injection-cycle. Injection of very small samples and wash-out of the assembly are achieved by the addition of a volatile solvent. A solvent-venting technique using a precolumn to trap the sample[35], and a schematic diagram of the automatic injection device are shown in Fig. 7.10. The valve comprises a Teflon body and a stainless-steel plunger with a groove cut in the side. Sample-transfer lines of stainless-steel capillaries are fitted to the valve, and connect it to both the turntable and the vaporizer.

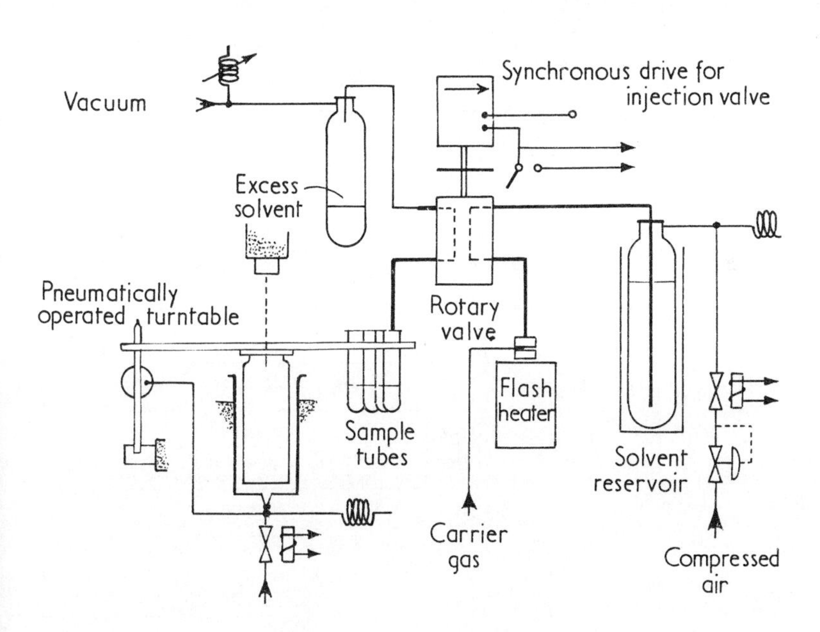

Fig. 7.10 Automatic analytical gas chromatograph.
Reproduced with permission from Evrard and Couvreur[34] and Elsevier Publishing Co.

A sample-tube is located under the capillary probe, and when the turntable is raised the capillary pierces the aluminium cap. Sample is drawn through the valve by a vacuum pump. The valve is rotated 180°, the pressurized solvent-reservoir is connected through the valve to the vaporizer, and this forces the sample in the groove and a quantity of solvent onto the column. The turntable is lowered and moved to the next tube which contains solvent only. The valve is turned a further 180° and the sample-groove fills with solvent so that the lines are cleaned. A further injection of solvent is then made. The sample and solvent are absorbed on a pre-column and the volatile solvent is allowed to vent to atmosphere. When most of the solvent has been removed the valve is operated and equilibrium is established in the column. The sample is then injected on to the separating column by heating the precolumn. In this manner multiple injections can be concentrated on the precolumn before analysis on the second column. However, the system does need attention from time to time if it is to operate for long periods. An internal standard is necessary, since the aluminium-foil caps do not prevent sample evaporation when long cycle-times are involved. No intersample contamination is observed, owing to the good wash-out characteristics of the system.

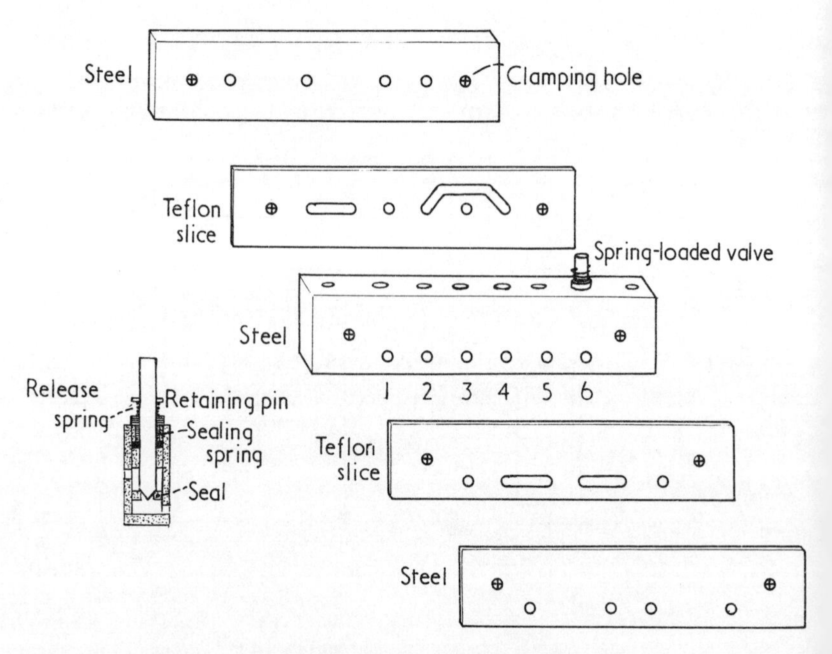

Fig. 7.11 Detail of needle valve injection system.
Reproduced with permission from Dugdale and Jones[36].

Dugdale and Jones[36] describe the construction of a simple valve for gas-injection, comprising six needle-valves. The design is shown in Fig. 7.11. Ports 1, 3 and 4 and ports 2, 5 and 6 are ganged together. In the normal position the carrier-gas is directed into the analyser and the sample-gas flows through the sample-loop. Carrier-gas flow is momentarily stopped as the sample-loop is switched into the carrier-gas stream. This system is not suitable for use with liquid samples. Smith and Harris[37] utilized a 16-port rotary-valve assembly coupled to the sample-loop. The system was developed for analysis of soil-gases. The sample-loop is pumped out, and as each sample is connected in turn, the sample-gas is partitioned between the sample and the loop. The volume of sample in the loop is then swept on to the gas-chromatograph column by the carrier-gas stream. Table 7.2 shows the results for 16 replicate samples. A relative standard deviation of 1·9–2·7% based on peak-height measurements, was found.

Table 7.2 RESULTS OF ANALYSIS OF SOIL GASES BY AUTOMATIC GAS CHROMATOGRAPHY

	Nominal concentration	*Peak height*	*Relative standard deviation*
Methane	6·1	66·1	1·4
Ethane	5·9	82·3	1·7
Ethylene	5·8	65·4	1·4
Propane	6·9	71·5	1·9
Propylene	7·9	41·1	0·8

The Perkin-Elmer head-space analyser (Multi Fract F40) also uses a system of valves for sampling the volatile components from perfumes and similar flavouring materials. Samples are sealed in containers and held in a turntable rack at an elevated temperature. Each sample is then drawn from the head space and diverted into the carrier-gas stream, and thence on to the gas-chromatographic column.

7.2.1.4 *Hamilton Automatic Injection System*

Hamilton have recently introduced an automatic system designed complete with an injection-port. The unit will not, however, fit all types of chromatographs. Addition of an automatic-injection unit system to an existing model is extremely costly and often impossible. The Hamilton automatic sampler is built around a new type of injection-system based upon capillary forces and uses a piston-type action to force the sample on to the column. The system is able to handle viscous samples held in custom-made sample-holders, but is complex and relies on absolute alignment for fault-free operation. A sample-identification arrangement is also included.

7.2.1.5 *Systems based on time–pressure injection*

Stockwell and Sawyer[38, 39] make provision for some pretreatment of the sample under test before presenting it to the gas-chromatographic column. The system is designed to meet the particular requirements of the authors' laboratory and the sample is reasonably plentiful (more than 500 ml available). Fig 7.12 shows the schematic diagram of the system used

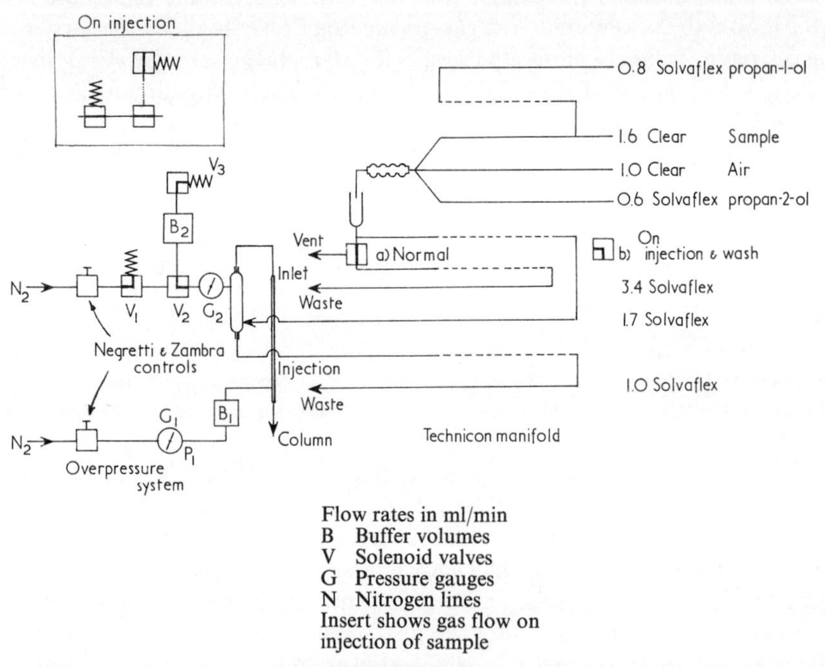

Fig. 7.12 Automatic gas chromatographic system for determination of ethanol in tinctures. *Reproduced with permission from Stockwell and Sawyer[39] and American Chemical Society.*

for sampling, dilution and addition of internal standard to the sample. The mixed sample-stream is then resampled and subsequently injected from the injection-vessel by the application of an overpressure. The dilution stage is incorporated to swamp out any wide viscosity variations and to ensure that the volume of sample-mixture injected on to the column is reproducible.

Samples 0·2–2 μl in size are injected. The liquid streams are transported by means of a 15-channel Technicon peristaltic pump which uses a mixed manifold of Tygon and Solvaflex pump-tubes. With the sample-probe in a sample-tube, sample, air and internal standard are pumped through a mixing coil into a vessel. This vessel, which has two outlets, is controlled by a three-port two-way valve. In the normal mode of operation the vessel is drained to waste but when the valve is operated the sample-mixture is drawn into the injection-vessel.

Differential pumping rates into and out of this vessel cause a build-up of liquid which is subsequently drained as the liquid-control valve is opened. The cycle is repeated to build a representative sample to a fixed level in the vessel, then injection is initiated by the application of an overpressure of carrier-gas. The probe is then removed from the sample-tube and repositioned in a solvent-reservoir to initiate a wash-out cycle of the apparatus. Wash-out liquid is not injected through the capillary tubing, which is continuously flushed by the back-flow of a hot stream of carrier-gas from the top of the column. The system can easily be adapted to any gas chromatograph. The injection-port is replaced by a T-coupling incorporating the column-feed gas-line, the column itself and the capillary restrictor being mounted in a configuration with minimum dead-volume.

Experience has determined the overpressure configuration which achieves a rectangular plug profile. Carrier-gas is fed into the column-inlet from a pressure-regulator and the overpressure is allowed to vent through a 500-ml buffer vessel, the valve *V*1 and the restrictor. A back-pressure is maintained in the injection-vessel through *V*2 and *V*3 and controlled by the restrictor. On injection, *V*1, *V*2 and *V*3 operate for a timed interval and this has two effects; first an overpressure is applied to the vessel, forcing a slug of sample-mixture on to the column and secondly the vent-line is opened to atmosphere so that the overpressure can be quickly released after the valves revert to the standby position at the end of the injection-period. This automatic system, complete with off-line data-handling equipment, has worked reliably in our laboratory for several years. Modifications can be made to the system so that it will handle a number of sample-types. Combination with other techniques such as heart-cutting or back-flushing, discussed in the next section, greatly extends the usefulness of a system which can already handle such pretreatment techniques as solvent extraction and some elementary chemistry.

7.3 Multicolumn and Column-Switching Techniques

In the 1960s it was a widely held view that the application of computer data-processing should be used to quantify chromatograms which have unresolved components or components appearing on a solvent tail. The use of multiple columns for obtaining complete separations in such cases offers a better solution to the analytical chemist. Flow-switching is common practice in process chromatographs but is not widely used in analytical work. Two types of approach are of interest; (*a*) maintenance of a continuously flowing stream, and (*b*) use of a trapping device which selectively absorbs part of the stream before further analysis on a different column. The transfer of one or more selected groups of components eluted from one column on to a second or series of further columns is usually referred to as heart-cutting or cutting. There are two major areas of interest. (*i*) The analysis of complex mixtures in which complete resolution cannot be achieved on a single column in a reasonable time. Groups of compounds are selectively transferred to a second column. A simple example is the analysis of saturated hydrocarbons, where a non-polar column resolves the components according to carbon number and a further separation on a polar column separates naphthenes from paraffins. (*ii*) Determination of trace components. In this type of analysis it is usual for the detector to become overloaded with the major component, and this either swamps the trace peak or causes it to come out on the tail. A marked improvement can be made by including a cutting device some way down the column so that only a small fraction of the major components is transferred on to the second column. This often eliminates overloading and tailing. Flow-diagrams for two systems which automatically accomplish 'heart-cutting' techniques are shown in Fig. 7.13*a* and 7.13*b*. In system (*a*) the flow-switching employs a multiport valve. Operation of this valve diverts the gas stream selectively to vent or on to the second column. The shortcomings of a valving system have been discussed in other parts of this chapter. The second system (*b*) has been discussed by Deans[40] and sets out to overcome the limitations of the valve. The automatic switching-valves are not critical in this second scheme since they are located outside the boundary of the oven. Taps *A* and *C* are set open throughout general running and are used to set up the system. The flow of the components from columns 1 and 2 is controlled by operation of the valve *B*. When *B* is shut the two columns are used in series, and sample-cutting is effected with *B* in the open position. The system proposed by Deans can easily be adapted to meet further analytical requirements. Where long-retention components are of no analytical interest, they can be purged from the system by coupling it with a back-flushing system, thereby reducing the analysis time. The inclusion of microreactors such as a hydrogenation or pyrolysis chamber can also extend the usefulness of this approach.

The overall advantages of system (*b*) can be summarized as follows. (*a*) The dead volume associated with the use of taps in the sample-path is

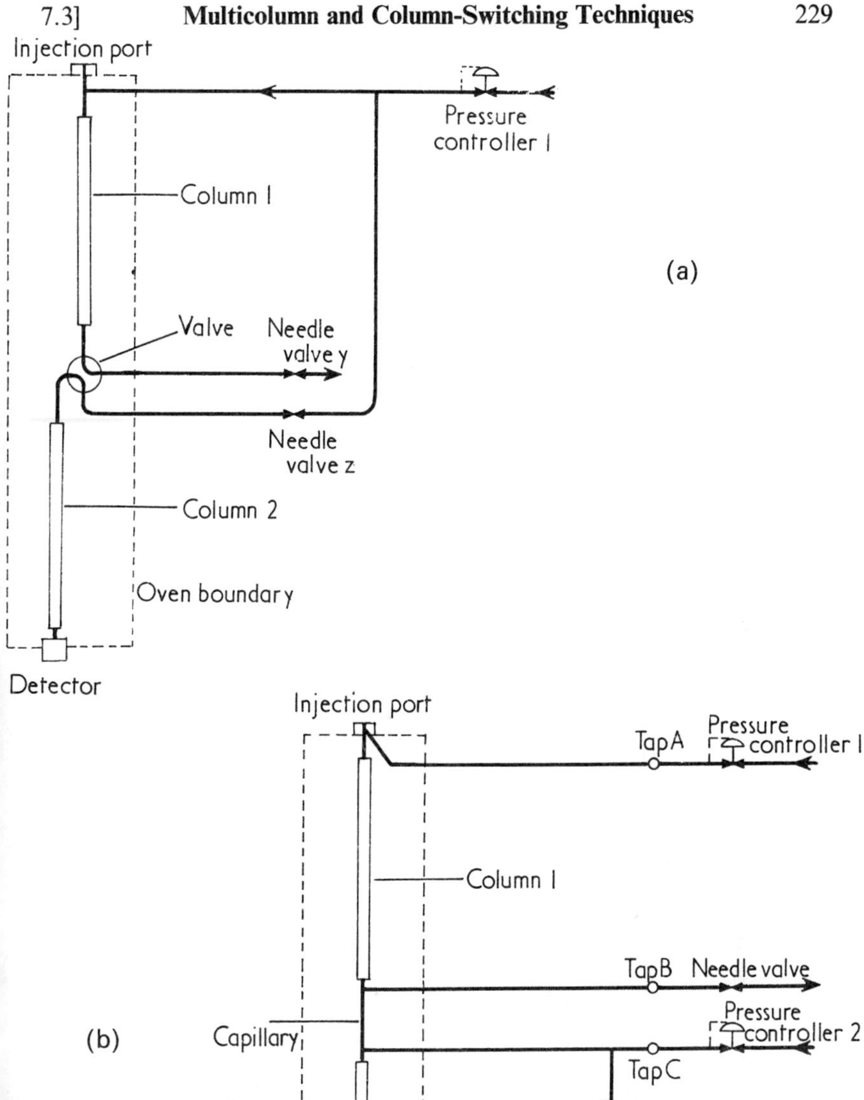

Fig. 7.13 Flow diagrams for heart-cutting using (*a*) Multiport valve and (*b*) external valves.
Reproduced with permission from Deans[41].

eliminated. Consequently the cutting techniques may be applied to capillary and other high-efficiency columns.

(*b*) Absence of valves eliminates the possibility of absorption of sample components in tap lubricants or low-friction plastics.

(*c*) Taps are used outside the oven at an ambient temperature; therefore, no special design features are necessary and it is easy to add to the existing gas chromatograph. Such a system is, according to Deans, simple to set up and to operate. In practice the flows have to be precisely balanced when working at maximum sensitivity. Many authors have found in some cases that the disadvantages of valve systems have minimal effect, and their ease of operation makes them preferable to Dean's system. Graham[41], in his work on the analysis of cigarette smoke, has described a fraction-trapping and transfer device which has many applications in this type of analysis. A two-stage gas-chromatographic method has been developed in which a primary

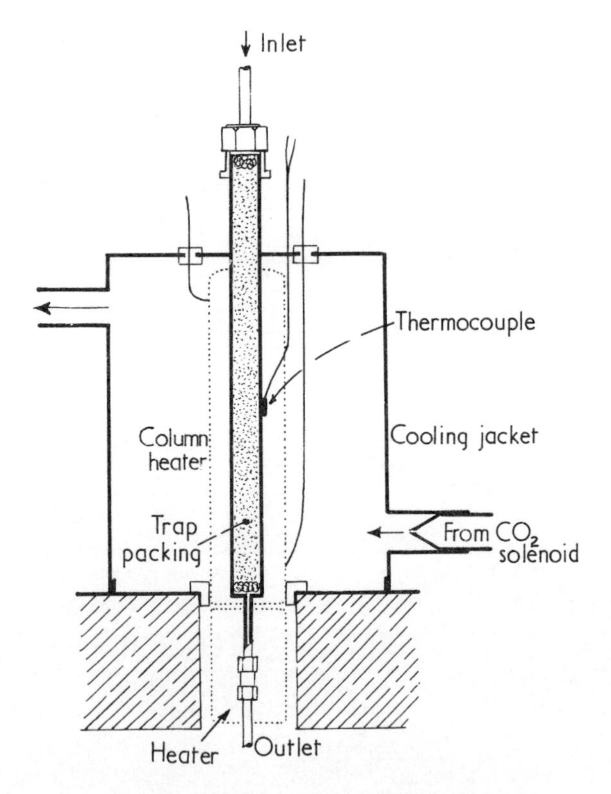

Fig. 7.14 Fraction collection trap designed by Graham[41].
Reproduced with permission of Cigarette Components Ltd.

separation is achieved on the first column and part of the effluent is trapped, and a further separation is performed on a second column of different polarity. Systems described for fraction-cutting entailing switching the gas-flow by valves can cause undue band-spreading, leading to poor separation. The trapping device described by Graham eliminates this adverse effect. Fractions are trapped in the packed-column trap and cooled by carbon dioxide. The trapped sample is evaporated by an electric heater and diverted on to the second column via a system of valves. The trap is illustrated in Fig. 7.14. This system differs from other trapping devices described previously, in that the trapping and transfer are in-line. Removal of the traps, with a resultant loss of material, before injection on to the second column is not required. The trapping procedure is efficient, and good-quality chromatograms are obtained even with a capillary column in line. Green, Albert and Barker[42] described a similar system for the determination of paraffin and hydrocarbon types in gasolines. It is expected that the use of multiple columns and automatic column-switching will increase rapidly in the near future.

References

1. Martin, A. J. P. and James, A. T. *Biochem. J.*, 1952, **50**, 679
2. Smith, J. F. and Higgins, G. M. C. *Brit. Pat. Spec.*, 1055091,1 (1967) and *Gas Chromatography 1964* (ed. Goldup, A.), p. 94, Institute of Petroleum, London, 1965.
3. Jentzsch, D. *J. Gas Chromatog.*, 1967, **5**, 226.
4. Ambrose, D. and Collerson, R. R. *Nature*, 1956, **84**, 177.
5. Atkinson, E. P. and Tuey, G. A. P. *Gas Chromatography 1958* (ed. Desty, D. H.), p. 270, Institute of Petroleum, London, 1959.
6. Tenenbaum, M. and Howard, F. L. *J. Res. Natl., Bur. Stds.*, 1962, **66A**, 255.
7. Kauss, J. M., Peters, U. J. and Martin, A. J. *U.S. Pat. Office*, 3267646 (1966).
8. Kauss, J. M., Peters, U. J. and Martin, A. J. *U.S. Pat Office*, 3155289 (1964).
9. Haruki, T. *German Pat.*, 1203501 (1965).
10. Jover, L., and Gastambide, B. *French Pat.*, 1464867 (1967).
11. Heilbronner, E., Kováts, E. and Simon, W. *Helv. Chim. Acta*, 1957, **40**, 2410.
12. Boer, H. *J. Sci. Instr.*, 1964, **41**, 365.
13. Frazer, J. W. and Morris, C. J. *UCRL–14359 (Microfilm)*, Dec. 6, 1965 (California).
14. Boer, H. *J. Sci. Instr.*, 1963, **40**, 121.
15. Thorpe, T. E. and Holmes, J. *J. Chem. Soc.*, 1903, **83**, 314.
16. Harris, J. R. *Analyst*, 1970, **95**, 158.
17. Bowman, R. L. and Karmen, A. *Nature*, 1958, **182**, 1233.
18. Jarrell, J. E. and Allison, A. W. *J. Gas Chromatog.*, 1964, **2**, 192.
19. Hewlett Packard, Model H.P., 7670A.

20. Nerheim, A. G. *Anal. Chem.*, 1964, **36**, 1686.
21. Hudy, J. A. *J. Gas Chromatog.*, 1966, **4**, 350.
22. Von Rudloff, E., *J. Gas Chromatog.*, 1965, **3**, 390.
23. Walker, J. Q. *Hydrocarbon Proc.*, 1967, **46**, 122.
24. Otte, E. and Jentzsch, D. *Gas Chromatography 1970* (ed. Stock, R.), p. 218, Institute of Petroleum, London, 1971.
25. Podmore, D. A. *J. Chromatog.*, 1965, **20**, 131.
26. Ruchelman, M. W. *J. Gas Chromatog.*, 1966, **4**, 265.
27. Applegate, H. G. and Chittwood, G. *Bull. Env. Contamin. Tox.*, 1968, **3**, 211.
28. Tinti, P. *J. Gas Chromatog.* 1966, **4**, 140.
29. Appleqvist, L.-A. and Melin, K.-A. *Lipids*, 1967, **2**, 351.
30. Borth, R., Canossa, A. and Norymberserski, J. K. *J. Chromatog.*, 1967, **26**, 258.
31. McGregor, R. F. *Clin. Chim. Acta*, 1968, **21**, 191.
32. Kipping, P. J. and Savage, C. A. *Gas Chromatography 1968* (ed. Harbourn, C. L. A.), p. 276, Institute of Petroleum, London, 1969.
33. Wicks, K. *Brit. Pat. Spec.*, 1016461, 1966.
34. Evrard, E. and Couvreur, J. *J. Chromatog.*, 1967, **27**, 47.
35. Evrard, E. *J. Chromatog.*, 1967, **27**, 40.
36. Dugdale, N. and Jones, K. *Chem. Ind. London*, 1965, 1460.
37. Smith, K. A. and Harris, W. *J. Chromatog.*, 1970, **53**, 358.
38. Stockwell, P. B., Sawyer, R., Bunting, W. and Ingram, P. H. C. *Gas Chromatography 1970* (ed. Stock, R.), p. 204, Institute of Petroleum, London, 1971.
39. Stockwell, P. B. and Sawyer, R. *Anal. Chem.*, 1970, **42**, 1136.
40. Deans, D. R. *Chromatographia*, 1968, **1/2**, 18.
41. Graham, J. F. *Beitrage Tabakforsch.*, 1969, **5**, 43.
42. Green, L. E., Albert, D. K. and Barber, H. H. *J. Gas Chromatog.*, 1966, **4**, 319.

Chapter 8

Thin-layer and Paper Chromatography

The techniques of thin-layer and paper chromatography are considered together; both are widely used in routine and research applications and are of great value to the analyst. Each of the techniques is basically simple but attention to detail is vital for the best results. Complete automation of a system from sample application to development and measurement has received little coverage in the literature and has a low priority in terms of pure cost-benefit amongst instrument manufacturers. Generally the technique can be considered as three discrete stages: (*a*) sample application, (*b*) development of the chromatogram and (*c*) quantitation of results. Many automatic and mechanical aids have been described for these operations and are discussed under these individual headings. Finally, the more complete systems designed primarily for preparative applications are discussed.

8.1 Sample Application

One of the most critical stages in thin-layer or paper chromatography is the application of the sample. For analytical work the sample is generally applied as a small, regular spot and for preparative methods a thin streak is applied to the starting line of the plate or paper. It is essential that the surface of the absorbing media are not damaged in any way and that the dimensions of the spot or streak are not too large. Where dilute solutions are used, it is preferable to apply the sample in several small portions rather than as a large single injection which produces a large irregular spot. The application, which requires a high degree of skill and dexterity, must be performed with the utmost care for reliable results.

8.1.1 SPOT-APPLICATORS

Three types of spot-applicators, using glass pipettes, hypodermic syringes and a method based on a Technicon pump, have been described in the literature. An example of the first of these is in the form of an extremely simple device, described by Clarke and Sowter[1], which consists of a glass capillary drawn out to a U-shaped tube which can be filled with sample solution and held slightly below the horizontally clamped chromatographic paper. A draught of warm air from a blower mounted above the paper, depresses the paper on to the tip of the primed capillary to effect the application. Solution is drawn on to the paper by capillary attraction and rapidly evaporates in the stream of warm air. Switching the blower off causes the paper to spring away from the capillary and eliminates flooding. Samples of about 0·2 ml can be spotted in approximately 5 min in regular shaped spots.

A device developed by Radin[2] also makes use of a glass pipette, held vertically in a holder. The pipette can rest in two positions – either directly above the thin-layer plate or paper, and in contact with it, or at 90° to this position above a sample container. In order to effect the application, the filled pipette is lowered by a swinging motion to the appropriate position on the plate. Liquid is expelled from the pipette and applied to the plate by pressure on an attached bellows, at a rate proportional to the evaporation of the solvent. The pipette is then removed, washed, and filled with another sample, or a further aliquot of the same sample. Further samples can be applied after manual movement of the thin-layer plate but the device forms a mechanical aid which is both reproducible and simple to operate, and greatly increases the precision and rapidity of sample-application. A simple modification to the technique would be to couple the device to a sample-turntable and plate-travel mechanism as the basis of a worthwhile automatic spot-applicator.

Simultaneous multiple-spot application, by using a series of capillary pipettes, is the subject of further papers[3, 4]. In each case, pipettes are filled with sample and clamped into position close to the plate or paper, and the sample is applied by physical contact with the tips of the pipettes (either mechanically or pneumatically) so that the sample is drawn on to the plate. A return spring terminates the spotting. A new plate and reloaded pipettes are needed for a further application. In practice the small dent in the plate produced by the physical contact aids the identification of the start-point.

The adaptation of the continuous infusion apparatus, described by Metzner and Volcsik[5] for simultaneous preparation of 6 paper chromatograms, is particularly suitable for temperature-sensitive and radioactive solutions. Samuels[6] has devised a similar system which uses a number of microsyringes to apply (simultaneously) samples to the thin-layer plate, as shown in Fig. 8.1. A modification has been made to the hypodermic needle used in this system in order to quantify sample applications; a reverse bevel is cut symmetrically to the original bevel (Fig. 8.1*a*) to produce an effect which is superior to that produced by grinding the syringe

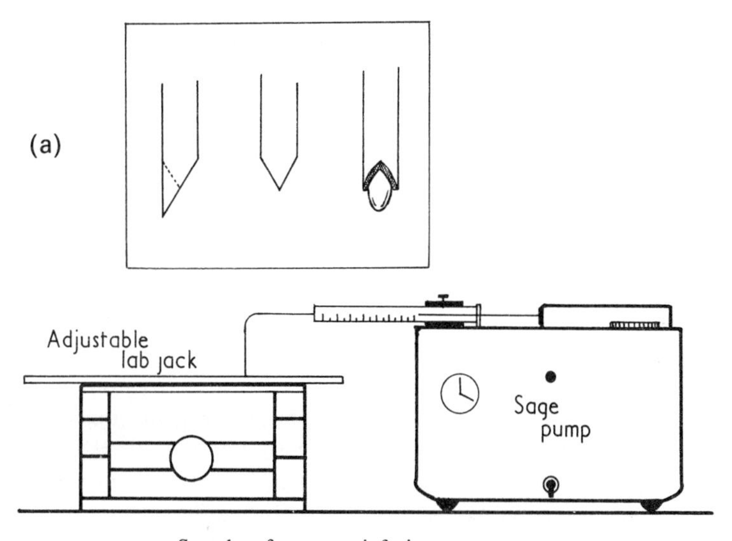

Samples from an infusion pump are applied to the thin-layer plates mounted on an adjustable lab jack. Inset (a) shows needle modification to improve sample application, reverse bevel being formed along dotted line.

Fig. 8.1 Arrangement for sample application.
Reproduced with permission from Samuels[6] and Elsevier Publishing Co.

square (the syringe then tends to pick up absorbent and also become blocked). There is no retention of liquid on the side of the needle and minimal damage is caused to the surface layer. Use of a small Teflon sleeve over the end of the needle[7] is rather less satisfactory. In the operation of this system the syringes are filled and clamped into position (normally at least 4 syringes are used) and accurately aligned by a series of manual adjustments. Spotting is actuated by forcing fluid from the syringe on to the plate which is simultaneously raised to a position close to the needle. Samples with a wide range of viscosities and of widely differing types can be handled but, despite this versatility, the applicator is not fully automated.

A further type of spot-applicator has been designed by Musil and

Fosslien[8], based on a peristaltic pump to dispense samples from eight disposable Oxford pipette tips fitted to the input side of the pump. The output from each pipette is directly connected to a hypodermic needle, mounted in an adjustable Teflon block, hinged to allow easy insertion and removal of the thin-layer plate in order to prevent needle-damage. The applicator is placed on a thermostatically-controlled hot-plate which maintains the preheated thin-layer plate at a constant temperature, and speeds up solvent evaporation. Evaporation can be additionally accelerated by a stream of hot air directed to the points of application. A limiting feature of this applicator is that the use of organic solvents must be avoided since traces of organic components extracted from the pump tube distort the analytical results.

8.1.2 STREAK-APPLICATORS

These are used primarily for purifying samples on a micropreparative scale. Two types of technique have been used, repetitive application of a number of spots side by side, and continual application of the substance along a moving plate. Vandenheuvel[7] described a semi-automatic device which provides sufficient pure samples for a subsequent analysis either by infra-red or by gas chromatography. To produce a sharp separation individual spots should not contain more than 1 μg of substance. Repeated spotting at intervals along the starting line ensures a sharp separation and also the required increase in sample-loading for micropreparative work. A low loading density of the sample on the starting line is an important aspect of successful operation. A series of neatly separated horizontal bands should be obtained from the developed plate, but irregular loading characteristics, whilst not affecting separation, result in distorted bands which are difficult to remove individually from the plate. Vandenheuvel has mechanized a Hamilton PB600 repetitive syringe to apply spots continuously on to the thin-layer plates and produce streaks 4–5 cm long, containing up to 10 μg of substance. As each impulse is passed to the 'Autospenser', which is mounted vertically above the thin-layer plate, $\frac{1}{60}$ of the contents of the syringe is expelled on to the plate. The plate is then moved fractionally in a horizontal direction before the next impulse. Three 'Autospensers' are normally coupled to put sample on either a single plate or on three separate plates. Standard hypodermic needles are used and in this case a small section of Teflon tubing is slipped over the end to prevent damage to the needle-tip. This modification also aids the formation of liquid droplets by preventing the liquid from creeping over the needle. Development of plates prepared in this way produces precise horizontal band separations.

A similar system, which forms the basis of the Desaga 'Autoliner', was originally described by Stahl and Dumont[9] and has similar facilities to the instrument mentioned above. A preloaded syringe is mounted in a vertical column above the thin-layer plate, and the sample expelled by a pneumatic plunger. Size of sample, length of streak, and width can easily be varied. A timed interval can also be allowed between two successive applications of the sample, and provision is made for the use of an electrostatic field in which to apply the sample. This system is extremely versatile and is equally suitable for analytical, qualitative, micropreparative and quantitative thin-layer chromatography and is also of value in paper chromatography and electrophoresis.

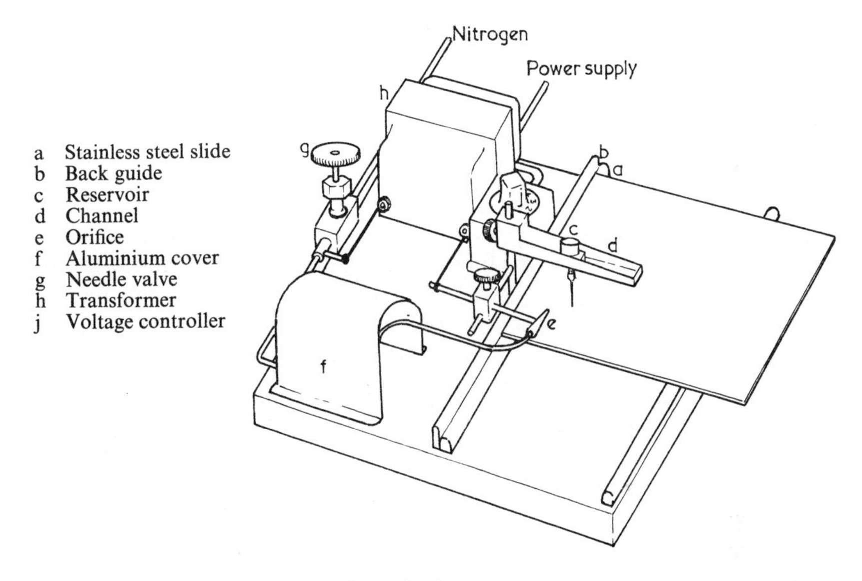

Fig. 8.2 Arrangement for streak application.
Reproduced with permission from Scott and Beeston[10] and Journal of Lipid Research.

The system described by Scott and Beeston[10] (Fig. 8.2) for streak-application does not make use of a hypodermic syringe but relies instead on the capillary flow from a head of liquid to control the streak-application. A sample-reservoir is mounted on a moving assembly with a protruding capillary drain touching the plate. The height of the reservoir can be adjusted to compensate for varying plate thicknesses. The reservoir is filled with sample from a device such as a syringe and the sample flows through the capillary on to the plate. As contact is made between plate and capillary drain the sample is applied by capillary attraction and the solvent is evaporated by a stream of hot air. As the flow commences from the needle, the plate oscillates along the starting line. The stroke of this movement can be adjusted so that a number of samples can be applied on the same plate by a multiplicity of dispensers.

It is of interest to note that the combination of thin-layer and gas chromatography, as exemplified by Janák[11, 12] in his method of streak-application of an effluent gas-chromatographic stream, enhances the overall potential of both techniques. The outlet from the gas chromatograph is split between the conventional detector and a further capillary outlet which diverts the gas stream on to a thin-layer plate held on a small flat carriage. A continuous screw driven by an electric motor moves the plate slowly across the outlet. Carrier-gas flow-rates of about 60 ml/min produce starting lines on the thin-layer plate approximately 3 mm wide. This type of two-dimensional chromatographic separation exploits to the maximum the separation potential of gas chromatography and thin-layer chromatography.

8.2 **Development**

It is essential for accuracy that each individual sample component is fully resolved and at the same time to ensure that each is retained on the plate itself. Since the rate of solvent movement can vary enormously from one plate to another, it is necessary for the operator to be in attendance for a considerable time to control the analysis. R_f values can only be ascertained by experiments requiring visual observation. Fig. 8.3 shows an apparatus designed to automate the development[13].

A trough containing the eluent is placed in a vacuum-sealed container and the prespotted plates or papers are hung in the tank. Essentially this system is designed to fulfil the function of terminating development when the solvent front reaches a preset level and this is achieved with the aid of an electronic sensing device. The remaining eluent is withdrawn by a vacuum pump triggered by the signal from this device, and other functions, such as acid or alkali spray during development, hot-air drying or application of spray reagents, can be incorporated into the cycle. Equilibrium is not set up in such systems and the flow characteristics may therefore not be very precise. Brodasky and Griffith[14] describe a similar but less versatile development tank which uses a thyratron to sense the end of development. The adjustable bias on the thyratron allows variation to suit a wide range of solvents. Powell[15] and Blondeel[16] describe systems which are more suitable for quantitative work because equilibrium conditions are set up as part of the automated procedure. Powell described a completely valveless system which can be programmed for multiple runs, each with independent timing and which is equally suitable for

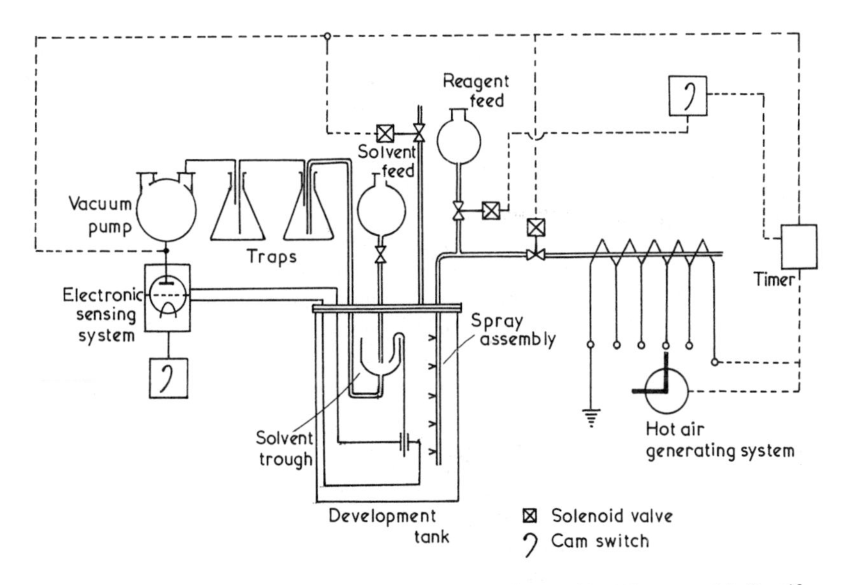

Fig. 8.3 Automatic development tank configuration designed by Rizescu and Pelloni[13]. *Reproduced with Authors' permission.*

ascending and descending chromatography. Once the liquid/vapour equilibrium has been established, a portion of liquid is allowed to flow into the trough and a constant head of liquid can be maintained throughout a run. This produces a gradual flow which avoids surging effects and provides optimum flow characteristics so that the system can be easily controlled on a time basis. Blondeel, however, uses an electrode-clamp to bring about the conclusion of development and thus attains excellent repeatability without the need for a precise form of flow control. This system is probably simpler in design than that described by Powell, and, in both systems, hot air drying can be carried out immediately following development of the final plate or paper.

8.3 Quantitation of Results

In the majority of cases quantitative results are estimated by visual comparison with known standards. Accuracies are consequently limited to the order of $\pm 10\%$. There are few satisfactory alternatives to this procedure which offer increased precision and generally the separated sample zone can only be analysed accurately by conventional wet chemical techniques. Such an approach has been automated by Snyder and Smith[17], using an automatic zonal scraper in combination with a collection system[18] coupled to an automatic scintillation counter with a direct data-output printer. In this instance narrow zones of 1, 2 or 5 mm are scraped from the plate, collected and automatically analysed by scintillation techniques. Results are visually presented in graph form, showing variation of component concentration across the plates.

Three approaches have been made. The first two make use of the standard thin-layer plate, the third utilizes a specially prepared narrow rod on which the chromatogram can be developed.

In the first system, developed by Hamman and Martin[19], an automatic scanning assembly with a spectrophotometer for obtaining quantitative results from thin-layer plates or paper chromatograms (Fig. 8.4) is described. The plate-scanner consists of a mobile arm attached to a constant-speed motor situated in a light-tight box. Flexible-fibre optics transmit

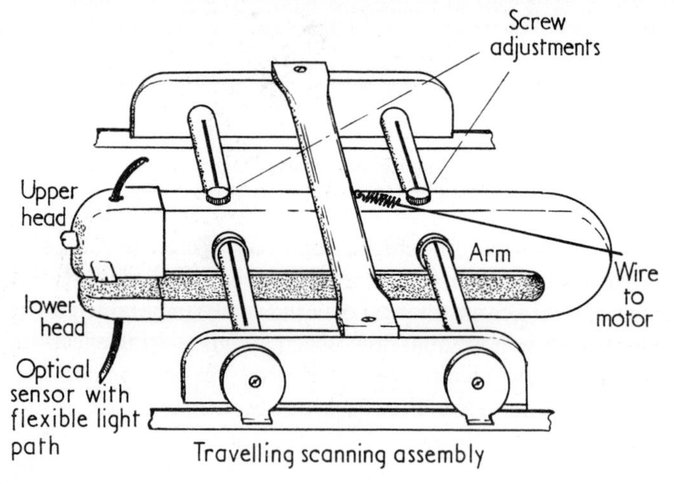

Fig. 8.4 Spectrophotometric scanning device for measurement of thin-layer plates. (Device travels over thin-layer plate.)
Reproduced with permission from Hamman and Martin[19] and Academic Press Inc.

light to the spectrophotometer, allowing automatic scanning of all areas of the thin-layer plate. This apparatus was tested with a mixture of steroid standards (less than 1 μg each) with 10% ethanol in chloroform as eluent. For detection a colour reaction with Tetrazolium Blue was used and the plate was subsequently scanned, using a 560-nm filter. Results were obtained faster and with more reliability than with conventional visual techniques. In the second system Khalameizer[20] has developed an automatic recording and integrating densitometer for use with paper chromatograms. A series of control analyses of known fats was used to assess this instrument and the precision of measurement found to vary between $\pm 1{\cdot}5$ and 2·6% a considerable improvement on results from normal visual techniques. Padley[21] describes the application of a flame-ionization detector for quantitation of thin-layer chromatograms. A narrow rod is coated with silica and the components are separated by normal elution techniques. After development the rod is passed directly through the hydrogen flame of a flame-ionization detector. Fig. 8.5 shows the resulting chromatogram obtained, as a plot of the detector-response against the movement of the rod through the flame.

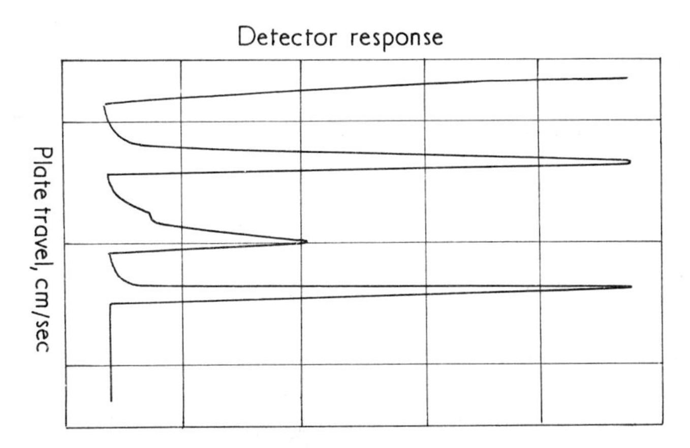

Fig. 8.5 Chromatogram produced by thin-layer measurement system. *Reproduced with permission from Padley*[21].

Packard Instruments produce a device similar to that described by Padley. However, a heated zone in the Packard instrument is moved along the rod and the evolved gases are swept into the analysis gas-stream by a carrier-gas, for subsequent analysis in a manner similar to conventional gas chromatography. Both systems are susceptible to noise on the output signal, due to irregularities in the silica coating and glass spirals but distinct advantages again present themselves over conventional visual matching techniques.

8.4 **Completely Automatic System**

Completely automatic systems have received only limited attention in the literature, and the more useful of these contributions are discussed in this section. Vissor[22] has devised a complicated arrangement for thin-layer chromatography, based on an endless-belt system using two revolving drums (Fig. 8.6). The separations are carried out on the upper portion of an impermeable belt of PTFE-impregnated glass fibre (*A*) which passes around two horizontal cylinders (*B*) and (*C*). The belt is drawn over a flat

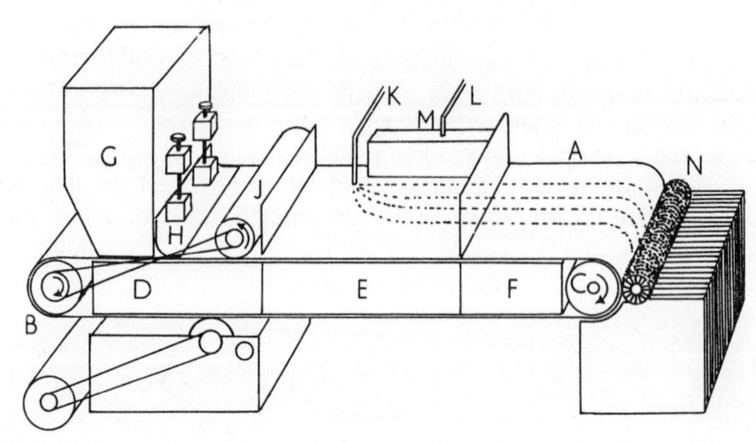

Fig. 8.6 Automatic preparative thin-layer assembly. (Characters explained in text.) *Reproduced with permission from Visser*[24] *and Elsevier Publishing Co.*

plate consisting of three separate horizontal compartments (*D*) (*E*) and (*F*) situated between the upper and lower portions. An absorbent-filled tank, (*G*), is placed at the starting point on the upper surface of the belt and a continuous layer is applied through the adjustable slit and compressed by the cylinder (*J*). Electric heaters mounted in compartment *D* are used to activate the adsorbent. The sample is applied continuously as a thin strip at the leading edge of compartment *E*, through a stainless-steel capillary, (*K*). Separation is achieved in this section by the application of a constant stream of eluent fed through tube *L* into a container (*M*). from where it is passed by means of a thick strip of filter paper on to the adsorbent layer. Both sample and eluent streams are delivered from a dosing pump. Separation at constant temperature takes place perpendicularly to the movement of the belt. The effective volume of compartment *E* is minimized to facilitate saturation of the eluent vapour, and *E* is connected to a circulatory thermostatic bath. In the third section (*F*) the solvent is evaporated and allowed to vent, and the adsorbed layer is subsequently removed by a rotating brush (*N*) and segmented into a series of containers for analysis on collection. The underside of the belt is cleaned on the return stroke by a foam cylinder (*O*) which brushes over the belt and rotates in a trough of water.

The scope of the instrument can be extended to include preparative electrophoresis by the addition of electrodes situated parallel to the edges of the adsorbent layer. However, the performance of this instrument is closely related to the precise control of the liquid flows relative to the belt-rotation speed. Two other systems have been described for automatic preparative-scale paper chromatography. In the first, Pavliček, Rosmus and Deyl[23] use centrifugal paper chromatography operating on a continuous basis. The sample is continuously applied on to a rotating disc of chromatographic paper adjacent to the mobile phase inlet by means of a capillary tube. This tube undergoes only a relatively slow rotary motion and each component traces a spiral path towards the edge of the paper as it is separated. The slow rate of rotation of the capillary, together with the retardation factor R_f determines the relative positions at which the separated components are eluted from the paper. The fractions are collected from the front of the paper by a series of collection cups rotating at the same speed as the disc. The method offers a time saving of some 85% in processing samples of the order of 700 μg. This type of system could easily be modified to cope with quantitative work by providing a sample-loading and collection system. An approach based on a Technicon sampler requiring only limited engineering effort would be simple to construct and could have considerable uses in this field. The preparative system described by Solms[24] uses a cylinder of paper as the transport medium. The paper is rotated in a horizontal plane at 90° to the travel of the eluent front. The mixture to be separated is continuously fed at a fixed point on the top of the paper from a capillary tube and the separated components trace a spiral path. The discrete fractions are collected in receptacles at the base of the cylinder and removed for analysis. Separations of lithium and potassium chlorides, xylose and galactose, and Methylene Blue and Fuchsin mixtures by this approach are described by Solms.

8.5 Conclusions

Completely reliable total automation of thin-layer or paper chromatography is not easily justified by the cost of the equipment which would be needed. However, the information presented above reveals several possible areas that can usefully be automated or mechanized with substantial savings in staff-time, provided operation can be carried out on a routine basis with samples of a similar nature.

References

1. Clarke, E. G. C. and Sowter, S. A. *Nature*, 1964, **202**, 795.
2. Radin, N. S. *J. Lipid Res.*, 1967, **8**, 694.
3. Curtis, P. J. *Chem. Ind. London*, 1966, 247.
4. Zaalishvili, M. M. and Shraibman, F. O. *Biokhimiya*, 1963, **28**, 9.
5. Metzner, H. and Volcsik, H. *Experimentia*, 1964, **20**, 104.
6. Samuels, S. *J. Chromatog.*, 1968, **32**, 751.
7. Vandenheuvel, F. A. *J. Chromatog.*, 1966, 25, 102.

8. Musil, F. and Fosslien, E. *J. Chromatog.*, 1970, **47**, 116.
9. Stahl, E. and Dumont, E. *J. Chromatog.*, 1969, **39**, 157.
10. Scott, T. W. and Beeston, J. W. *J. Lipid Res.*, 1966, **7**, 456.
11. Janák, J. *J. Chromatog.*, 1964, **15**, 15.
12. Janák, J. *J. Chromatog.*, 1965, **18**, 270.
13. Rizescu, I. and Pelloni, V. *Rumanian. Pat.* No. 51,258 (1968).
14. Brodasky, T. F. and Griffith, D. *J. Chromatog.*, 1969, **41**, 494.
15. Powell, R. H. *U.S. Pat.*, No. 3,341,017 (1967).
16. Blondeel, N. J. *U.S. Pat.*, No. 3,474,031 (1969).
17. Snyder, F. and Smith, D. *Sepn. Sci.*, 1966, **1**, 709.
18. Snyder, F. *Anal. Biochem.*, 1965, **11**, 510.
19. Hamman, B. L. and Martin, M. M. *Anal. Biochem.*, 1966, **15**, 305.
20. Khalameizer, M. B. *Fiziol. Rast.*, 1962, **9**, 120.
21. Padley, F. B. *Chem. Ind. London*, 1967, 874.
22. Visser, R. *Anal. Chim. Acta*, 1967, **38**, 157.
23. Pavliček, M., Rosmus, J. and Deyl, Z. *J. Chromatog*, 1963, **10**, 497.
24. Solms, J. *Helv. Chim. Acta*, 1955, **38**, 1127.

Chapter 9

Ion-exchange Chromatography

The development of a wide range of ion-exchange resins has provided the analytical chemist with a powerful separation tool for ions, both organic and inorganic. In particular, ions of closely similar chemical properties may be separated by using ion-exchange columns, as demonstrated by the now classical studies on the amino-acids[1,2], sugars[3], and the lanthanide and actinide elements[4]. For dilute solutions the application of physicochemical principles together with experimentally determined distribution coefficients permits the calculation of operating parameters to achieve the desired separations and also the calculation of separation factors. In addition a wide range of simpler and quicker separations based on ionic charge and size and metal-ion complexing ability have been worked out, using column or batch procedures.

In principle, almost all ion-exchange separations are amenable to some degree of automation, but, not unexpectedly, developments in automatic techniques have centred almost entirely on the time-consuming and exacting separations of amino-acids and sugars, determinations of both of which are important in biochemical and clinical investigations. Two broad levels of automation can be distinguished with respect to equipment so far developed; (*a*) systems where the equipment is designed to control the sequence from loading the sample on the column through to recording of responses for the individual separated components following automatic colorimetric measurement, and (*b*) those which in addition provide for automatic sequential loading of a series of samples following each completed analysis. Examples of each type are described with reference to the classes of compound studied.

9.1 Amino-Acids

The demand for individual amino-acid analyses in physiological fluids, protein hydrolysates, etc. is such that a range of commercial analysers capable of controlling the analytical sequence is now available. The chemical procedure, developed initially by Stein and Moore[1, 2], involves separation of the amino-acids on a cation-exchange resin, with citrate buffers of appropriate pH as eluent. A fraction-collector serves to divide the eluate into a large number of fractions, each of which is colorimetrically analysed with ninhydrin. Amino-acids are detected and determined by virtue of their absorption of 570-nm radiation, with the exception of proline and hydroxyproline for which 440-nm radiation is used.

Although subsequent workers have incorporated many improvements in the ion-exchange technique, the procedure described by Moore, Spackman and Stein[5] provides the basis for most commercial automatic analysers. A two-column technique enables the common amino-acids in protein hydrolysates to be conveniently separated and subsequently determined. Acidic and neutral amino-acids are separated on a cation exchange column 150 cm long which is eluted first with citrate buffer at pH 3.25 followed, after elution of glycine, by citrate at pH 4.25. On completion of the separation the column is regenerated with 0·2*M* sodium hydroxide. Basic amino-acids are separated on a 15-cm cation-exchange column, with sodium citrate solution at pH 5.28; since this column is exhaustively eluted it is used repetitively without regeneration. Both columns are operated at 50° C. The individual fractions are analysed by the colour reaction with ninhydrin. Close adherence to several experimental parameters (resin-quality, flow-rate, pressure-drop) is critical for achieving good repetitive separations; full details are to be found in the original paper.

Before commercial automatic analysers became available, two laboratory-designed automatic systems were described[6, 7], based on the method above. Simmonds[6] built an automatic analyser for discrete analysis of individually-collected fractions and Spackman, Stein and Moore[7] developed an alternative approach in which the column effluent was continuously reacted with ninhydrin and the absorption at 570 or 440 nm was recorded.

Simmonds's equipment provides mechanization of the manual method from the fraction-collection stage through to the absorptiometric measurement; buffer-changing and final evaluation of results remain manual. The column effluent travels a vertical path successively through fraction-collection, pH-adjustment, addition of ninhydrin reagent and photometric measurement. At each stage the flow of solution is controlled by solenoid-operated valves. Appropriate fractions are collected in a vessel containing upper and lower conductivity probes to control the volume. Electrolytic contact then operates a relay which opens the valve at the base of the collector and advances the fraction-collector to the next position. The pH-adjusting diluent and ninhydrin solution are added by means of a capillary-mounted glass hypodermic syringe, rise and fall of the syringe-plunger being solenoid-operated through a piece of iron cemented to the top of the plunger. Colour-development proceeds in a jacketed glass tube in which temperature control is provided by boiling water circulated by a thermosiphon. A cam-timer driven by a synchronous clock-motor lets the colour develop for 1 min, then the valve at the base of the tube opens and allows the solution to fall into the measuring cuvette of a compensating photometer (Sigrist Type UP2LD), the compensating compartment containing distilled water in a sealed cuvette. A single-beam filter photometer proved inadequate, owing to unacceptable base-line drift over the long period of operation. Determination of proline is not possible with the single-filter unit unless the filter is changed manually at the appropriate time and Simmonds preferred to determine proline separately by the method of Chenard[8]. The photometer response is presented on a stable, self-standardizing single-point recorder. Trials of the equipment with synthetic mixtures of amino-acids indicated recoveries of $100 \pm 3\%$ and a threefold improvement in precision in comparison with the manual method is claimed.

The automatic recording apparatus of Spackman, Stein and Moore[7] is depicted in Fig. 9.1. It is a continuous-flow system in which the appropriate buffers are pumped at constant rate through the column into a flow-photometer. On issuing from the column the eluate is mixed with ninhydrin solution separately pumped. The analysis time is reduced by using cation-exchange resin of small particle size (30–70 μm diameter, prepared by hydraulic fractionation) which permits a flow-rate of 30 ml/hr to be used as opposed to 15 ml/hr in earlier studies. The continuous photometer comprises three separate photometer units for measuring absorbance at 570, 440 and 570 nm consecutively, the effective cell path-lengths being respectively 2·2, 2·2 and 0·7 nm. Results are presented on a three-point recorder. Once the eluate and ninhydrin streams have converged it is essential to minimize back-mixing, and to this end all tubing from the point of merging of the two streams through to the photometer is 1 mm or less in diameter. A resolution equivalent to collecting and individually measuring 0·5-ml volumes of eluate is thereby achieved. Special measures are necessary to prevent air-bubbles from separating in the heating-bath and vitiating the subsequent absorbance measurement; the buffer solutions are continuously deaerated before entering the column, a small amount of detergent is included in the buffer solutions and a back-pressure of about 100 mm Hg is applied to the heating coil. The pumps (Milton-Roy Minipump MM1-B-29, 0–320 ml/hr) must be carefully

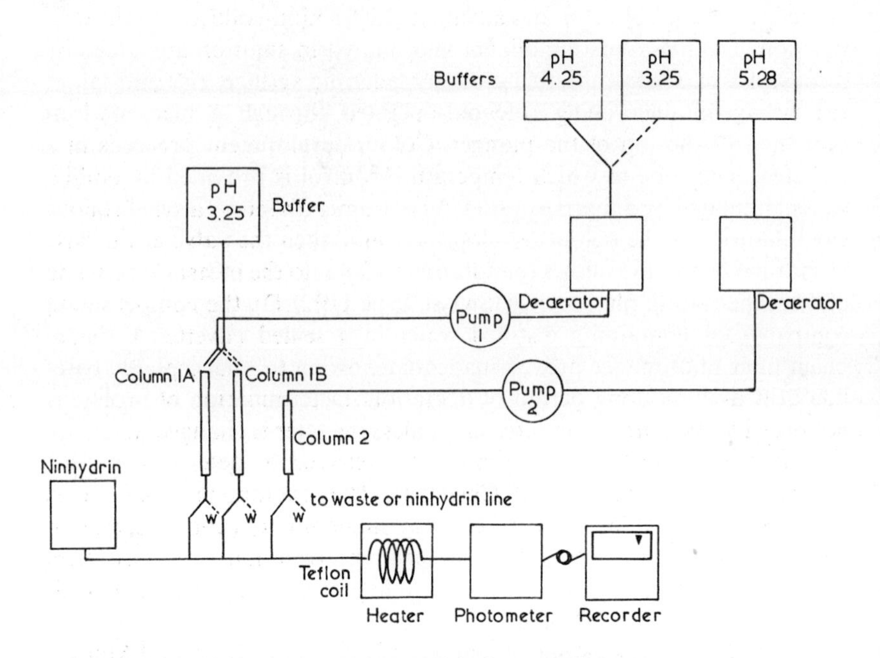

Fig. 9.1 Equipment for automatic ion-exchange separation of amino-acids.
Reproduced with permission from Moore, Spackman and Stein[5] and American Chemical Society.

assembled to ensure absence of entrapped air or leakage. Selection of appropriate buffer or regenerant for the 150-cm column is achieved by a solenoid-operated stopcock manifold, all operations being controlled by a Gra-Lab timer. For column loadings of 0·25 μmole of amino-acid, recoveries of 100 ± 3% resulted. The stability of the photometer base-line (±0·001 absorbance unit immediately before and after a peak) is such that satisfactory results can be obtained with column loadings of 0·06 μmole of amino-acid.

Since the detailed publication of the two-column procedure by Moore, Spackman and Stein[5] many authors have studied means of improving the resolution, sensitivity and speed of the method. In terms of automation of the method two approaches are particularly relevant because they have been incorporated into a number of commercial automatic analysers. The so-called accelerated two-column method due to Spackman[9] achieved an approximately threefold improvement in analysis time by reducing the sizes of the two columns and increasing the flow-rate of eluting buffers.

A single-column procedure developed by Piez and Morris[10] and

Hamilton[11, 12], enables the entire range of acid, neutral and basic amino-acids to be separated on one long column. Hamilton[11] devoted particular attention to resolution and sensitivity. By using a narrow column (0·636 × 125 cm), resin accurately graded in size and cross-linking, and a longer light-path photometer, he achieved a high degree of resolution which enabled him to chart 148 amino-acids and to obtain detection limits of the order of 10^{-10} mole.

9.2 Commercial Amino-Acid Analysers

Automatic instruments for amino-acid analysis are available from a number of commercial suppliers (see for example the *Analytical Chemistry* Buyers' Guide) and for detailed descriptions and specifications the reader is referred to the technical literature of the suppliers.

Facilities for analysis by the two-column technique of Moore, Spackman and Stein[5] are offered by all manufacturers and in several instruments duplicate sets of columns are provided to allow regeneration of one set of columns while separations are in progress on the others. The accelerated system described by Spackman for routine analysis is available from Carla Erba and the single-column approach of Piez and Morris[10] is available from Technicon and Carlo Erba. Although detailed descriptions of all available systems is not attempted here it is pertinent to comment on several aspects where they bear upon the quality and flexibility of the equipment performance.

9.2.1 Cation-Exchange Resin

The quality of the ion-exchange resin affects the separation performance by way of both component resolution and speed of analysis. In earlier studies the resin preparation comprised grinding and sieving to give the required range of particle sizes and elutriation yielded additional size-grading. The resin particles so prepared are irregular in shape and the pressure-drop across columns made from them tends to increase with time and the maintenance of constant flow-rate requires occasional adjustment to the metering pumps. Specially prepared resin particles of uniform size and spherical geometry are now available; these provide optimum ion-exchange separation performance and in packed columns they exhibit virtually constant pressure-drop and enable faster flow-rates to be used. The faster analysis times which can be achieved by using commercial instruments compared with the original work of Moore and Stein[1, 2] stem largely from the use of uniform, spherical resin beads.

9.2.2 METERING PUMPS

Long-term constancy of metering pump performance is critically important. They must be fabricated from corrosion- and wear-resistant materials; sapphire plungers and valves give highly satisfactory performance. Vernier adjustment of the pump-stroke must be provided so that constancy of flow-rate can be maintained regardless of pressure-drop variations across the ion-exchange column.

9.2.3 PROGRAM CONTROL

For routine analysis of similar samples where the program is invariant, control may be exercised by a suitable timer-unit of adequate long-term accuracy. For research studies, where versatility of program selection is an important consideration, a pin-board matrix programmer provides a convenient means of preselection of timing sequence for each stage of the analysis (buffer-selection, column-regeneration, column-switching, etc.).

A novel approach to programming is utilized in the Technicon TSM sequential multisample amino-acid analyser. The instrument comprises essentially the 'AutoAnalyzer' principle for analytical measurement following automatic ion-exchange separation. The entire operation is controlled by a programmed peristaltic valve which is designed to be compatible with the 'AutoAnalyzer' mode of operation. Fig. 9.2 illustrates the principle of operation.

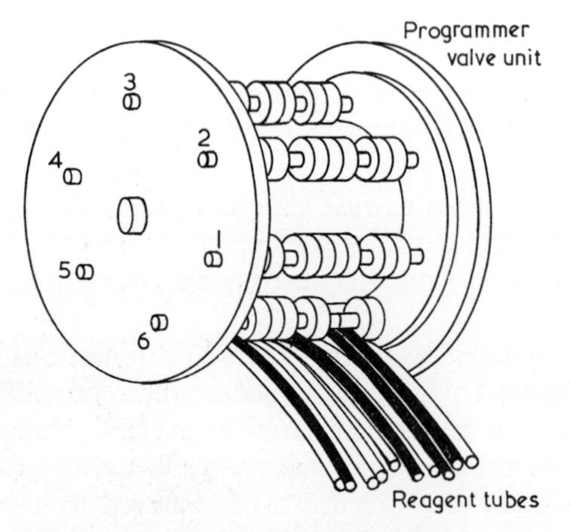

Fig. 9.2 The Technicon peristaltic valve.
Reproduced with permission of Technicon Corporation.

The programmer comprises a closed-loop punched tape that is driven by a synchronous motor past a microswitch which is activated each time a hole is encountered. The analytical cycle time is determined by the length of tape, and the increments of the multiple operating steps are determined by the intervals between the holes. This approach provides extensive versatility since each entire program is determined individually by the operator at the tape-punching stage.

The multichannel valve consists of a spool-shaped mechanism bearing 12 rollers transversely mounted under a platen which has 16 tubes. Each roller carries 16 movable spacers of two different diameters. These are so adjusted that when a roller is positioned across the 16 tubes the tubes that are aligned with the smaller diameter spacers are left open and those

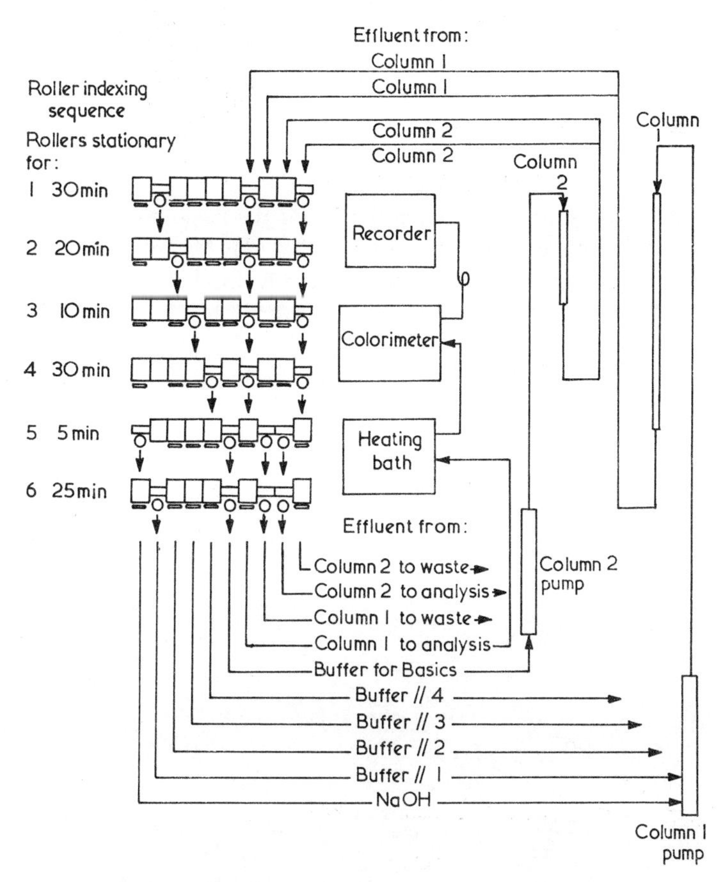

Fig. 9.3 Flow diagram for the Technicon peristaltic valve. *Reproduced with permission of Technicon Corporation.*

aligned with the larger diameter spacers are closed. The arrangement of the spacers on the rollers permits one buffer solution to pass to the column pump. When the tape programmer has advanced to the next hole a microswitch is activated and the valve-unit is indexed to the next roller position; this roller has a different arrangement of spacers which closes the tube carrying the first buffer and opens the tube carrying the second. Suitable arrangement of larger and smaller spacers on each succeeding roller enables any particular flow line to remain open while others are closed. Since the rollers operate over 16 pump tubes the peristatic valve has the capacity to control the flow of reagents to both the column and colorimeter simultaneously, as Fig. 9.3 indicates. In addition, the rotating valve is fitted with a series of microswitches which are driven past large and small diameter spacers and activation of these microswitches controls the operation of pumps, column-switching and the advancement of an automatic sample-injection sequence.

9.2.4 Gradient Elution Techniques

The use of fixed-pH buffer solutions, customarily 3·25 and 4·25 for acidic and neutral amino-acids and 5·28 for the basic ones, suffers from certain disadvantages. The separation time is long because conditions for the resolution of amino-acids which elute in a closely following sequence

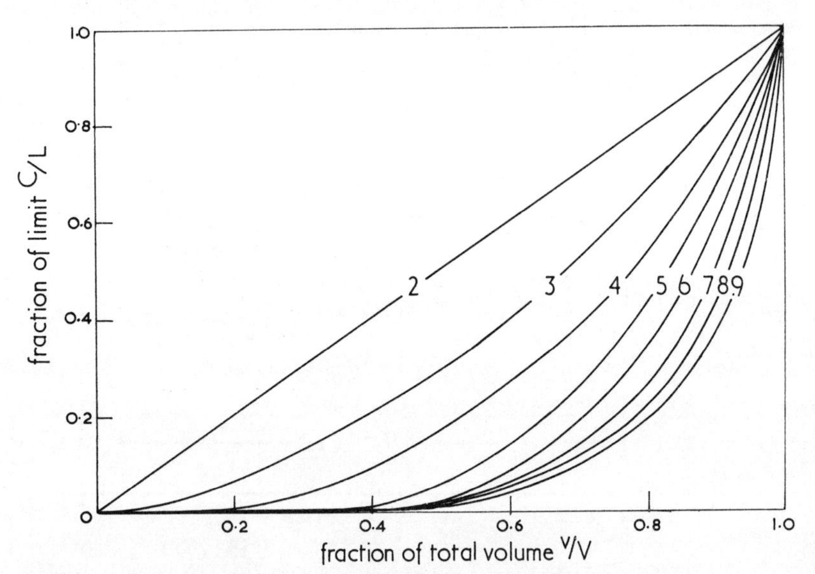

Fig. 9.4 Effect of number of eluent chambers on concentration profile.
Reproduced with permission from Peterson and Sober[13] *and American Chemical Society.*

result in long gaps between certain others. Because both the quality of resolution of adjacent amino-acids and the overall analysis time are, for a fixed column-condition and eluent flow-rate, dependent upon the buffer pH and its variation throughout the separation sequence, attention has been given to methods of generating a predetermined continuously changing pH of the influent buffer solution.

The simplest form of gradient elution is that in which the solution in a vessel dispensing eluate to the chromatographic column is replenished at the same rate from a second vessel containing a different concentration of the same eluting solution. In this way a linear concentration gradient is produced. By extending the number of vessels in series, and varying the initial solution concentrations in each, a whole range of gradient profiles can be produced. Peterson and Sober[13] developed and mathematically

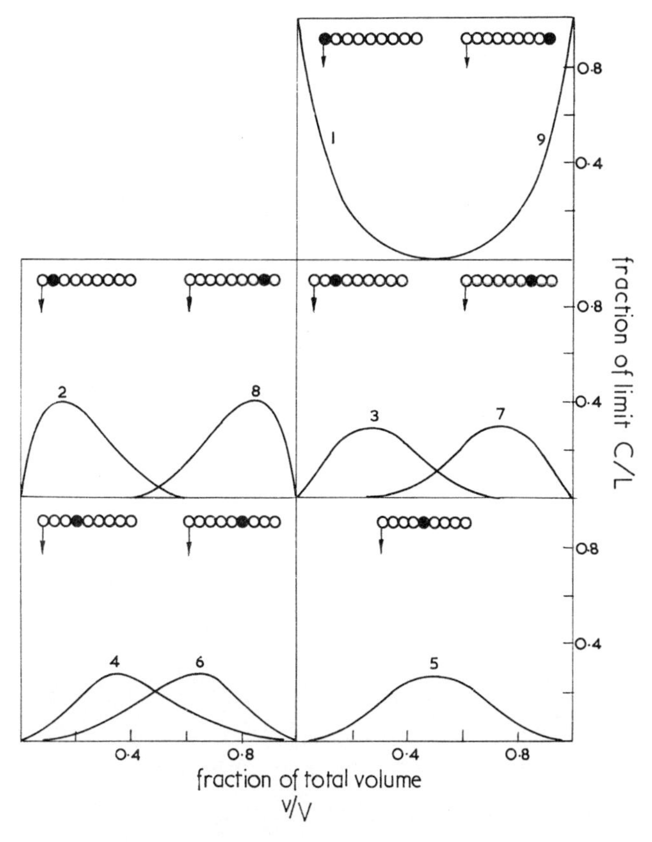

Fig. 9.5 Gradient profiles as a function of location of single eluent with the 9-chamber system.
Reproduced with permission from Peterson and Sober[13] *and American Chemical Society.*

evaluated a nine-chamber system (termed the 'Varigrad') and demonstrated the versatility of such a system. The design is such that the chambers are interconnected and in hydrostatic equilibrium which is continuously re-established as the effluent is drawn off. Each compartment is stirred mechanically from a single motor. Any number of chambers from two to nine could be used by isolating those not required. Equations were derived relating the effluent concentration from the final chamber at any time as a function of individual concentrations in the other chambers.

For gradients changing in concentration in one direction only, the effect of increasing the number of chambers used is illustrated in Fig. 9.4. All chambers contained water except the first, which held eluting solution at the limiting concentration L. The family of curves is a representation of the effluent concentration C, expressed as a fraction of the limiting concentration, as a function of the total volume discharged. Gradient profiles exhibiting a reversal of slope result when the solution of concentration L is placed in any of the chambers other than the first or ninth. Fig. 9.5 depicts the resulting profiles, again with water in the remaining eight chambers. If more than one of the chambers contains a finite concentration of eluting material then the compound gradient profile is the sum of the individual ones for eluting material in one chamber only. Petersen and Sober demonstrate this situation in Fig. 9.6 in which four chambers contain

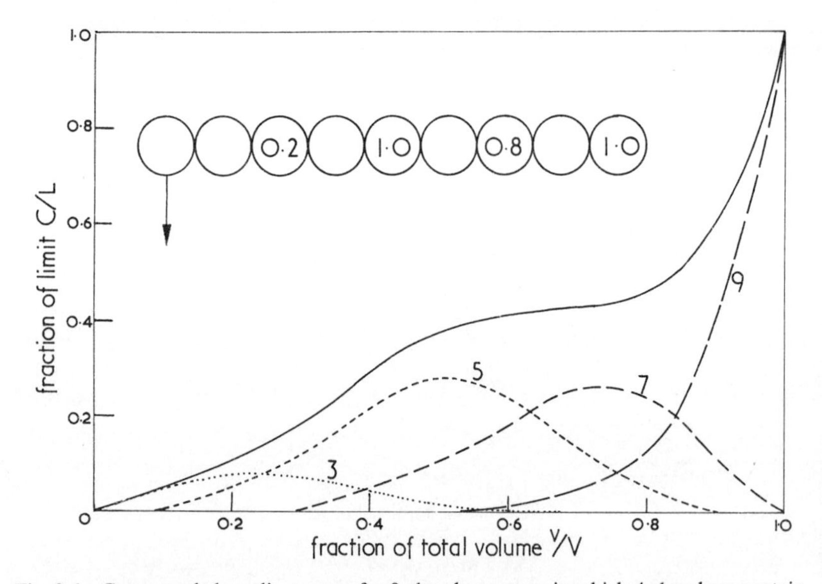

Fig. 9.6 Compounded gradient curve for 9-chamber system in which 4 chambers contain eluent at concentration ratios 0·2, 1·0, 0·8 and 1·0.
Reproduced with permission from Peterson and Sober[13] and American Chemical Society.

eluting material at the concentration ratios shown. Application of these principles enables gradients to be achieved giving optimum resolution and economy for a particular experimental requirement.

The 'Varigrad' approach is obviously well suited to amino-acid analysis. Not only can the overall analysis time be minimized but, where desired, portions of the separation of particular interest can be expanded or other portions of less interest can be compacted. Furthermore the mechanical arrangements for feeding buffer solution to the column are simplified in that only one line, instead of the usual three, is required. Commercial use of the 'Varigrad' principle is found in the 'Grad-o-matic' (Carlo Erba) and the 'Autograd' (Technicon). Fig. 9.7, taken from the

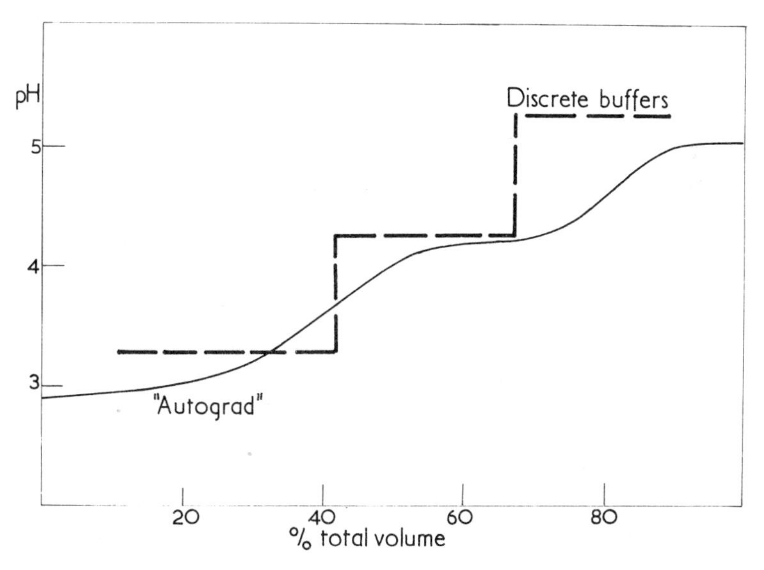

Fig. 9.7 Comparison of 3-stage stepped gradient with gradient produced by 9-chamber 'Autograd'.
Reproduced with permission of Technicon Corporation.

literature of Technicon Instruments Corporation, shows a comparison of a three-stage stepped gradient with the profile obtained with the 'Autograd' prepared with buffers of pH 5·00, 5·00, 4·60, 3·80, 3·10, 2·875, 2·875, 2·875 and 2·875 in the nine compartments.

Keck[14] has described a method of producing concentration gradients by using two solutions only, both of which flow separately into a mixing chamber above the chromatographic column. Access of each solution to the mixing chamber is controlled by a two-position magnetic valve. The relative volumes of the two solutions entering the mixing chamber are

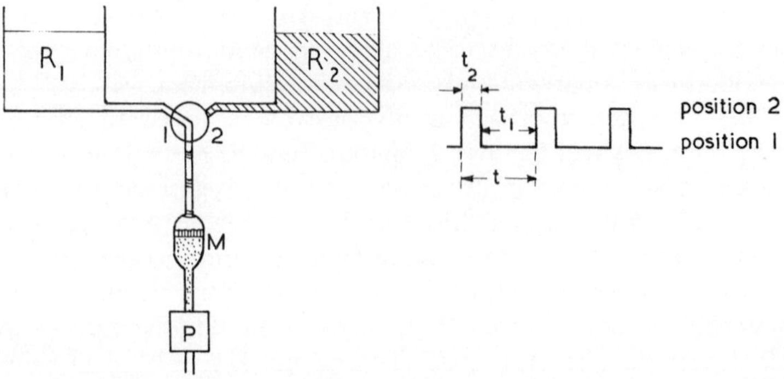

Fig. 9.8 Generation of concentration gradient by means of a programmed magnetic valve R_1, R_2. Reservoirs; M. Mixing vessel; P. Pump.
Reproduced by permission from Keck[14] and Academic Press Inc.

proportional to the length of an electric pulse relative to the total cycle time (Fig. 9.8). The concentration gradient is then described by the on-to-off time ratio as a function of time. To produce the required gradient this ratio is programmed by the gradient as represented by the boundary of a black and white illustration. The curve is scanned by a photocell and the valve position is reversed each time the boundary is crossed. Scanning comprises up and down movement of the photocell while the photocell carriage travels horizontally at constant speed. A potential advantage of this approach is that two gradients, say concentration and pH, can be superimposed by adding a second magnetic valve and reagent reservoir. Alternatively this arrangement could be used to divide a wide-range concentration gradient into two ranges, thereby reducing the error involved in the final concentration.

The use of gradient methods is the subject of a recent monograph[15].

9.2.5 Evaluation of Analytical Data

Although automation of the ion-exchange column operation and of the colorimetric measurement greatly reduced the manpower requirement for amino-acid analysis and also provided increased sample throughput relative to manual procedures, the calculation of results was still performed manually. For each amino-acid peak the area had to be measured and a conversion factor calculated, from standard additions, to enable the concentration of each one to be calculated. For a multicomponent analysis this becomes time-consuming and exacting and consequently the application of automatic data-processing techniques has attracted the attention of several groups of workers.

Graham and Sheldrick[16] described a system for use with the Technicon 'Amino-Acid Analyzer'. It involves manual measurement of the height and width at half-height followed by punching the data on an input tape. A program written in ALGOL and run on an English Electric-Leo Marconi KDF9 computer was used for the calculations. Norleucine is used as an internal standard and the program calculates each peak as a norleucine equivalent, i.e. the number of μmole of any particular amino-acid producing the same colour intensity as 1 μmole of norleucine. The output data comprise (*a*) μmole of amino-acid, (*b*) μg of nitrogen, (*c*) residues/1000 residues for each amino-acid and (*d*) amino-acid nitrogen as per cent of total nitrogen eluted from the column. The accuracy of the computer calculation was shown to be the same as by the manual method.

A system which is capable of operating directly on the analogue response of the photometers is described by Porter and Talley[17]. The signals are digitized by means of a rotating-shaft encoder, the digital output being a function of the angular rotation of the shaft. A programmer-unit subsequently converts the digital information into a decimal code for computer-operation. Input data are punched on to tape in the appropriate format for use with an IBM 1620 computer. An alternative data-acquisition method is that of Krichevsky, Schwartz and Mage[18] in which the photometer outputs are digitized with a digital voltmeter. Necessary analytical information is punched on to paper tape in IBM BCD code together with programming instructions written in FORTRAN. The program is capable of editing the data with regard to noise and compensating for variations of base-line. Peak-areas are calculated over an absorbance range from 0 to 3·000 and in addition unresolved peaks are split and evaluated.

Much of the programming effort in these earlier approaches was concerned with methods of recognizing the start and maximum of a peak, together with techniques for base-line construction and area-calculation. These functions may now be performed with commercially available digital integrators and Starbuck *et al.*[19] have devised a computer program which can be utilized with either an integrator or with the manual calculation approach outlined above. The program is written in FORTRAN and is of general applicability. It is designed to perform the following.

(*a*) Calculation of standard values from separation of known standard quantities of amino-acids, including standard deviations. The data may be averaged over a series of standard analyses or used individually.

(*b*) Calculation of amino-acid compositional data, using the derived standard values for analysis of unknown samples. The output records mole per cent (each amino-acid expressed as a fraction of the total moles of amino-acids recovered), the number of amino-acid residues in the protein, and per cent nitrogen.

(*c*) Calculation of empirical composition of peptides, based upon number of moles of individual amino-acid recovered, ratios of moles of individual amino-acids to number of moles of a given amino-acid or group of acids, and the per cent hydrolysis yield of the peptide, and its purity.

The peak-area ratios for the 440 and 570 nm responses are specific for each amino-acid. Conkerton, Coll and Ory[20] have noted that in the analysis of certain food samples symmetrical but impure peaks can arise which yield different ratios and they have written a supplementary program capable of identifying and recording these anomalous peaks.

Ratios of the 440/570 nm responses are determined for pure amino-acids, by using an integrator, and average deviations are calculated. A FORTRAN program utilizes the data in comparing the peak-ratios for amino-acids separated from samples and yields a print-out of the ratio whenever an anomalous ratio is detected.

9.2.6 Automatic Sequential Sample-Application

Since a full analytical cycle for amino-acid separation and analysis takes several hours, facilities for automatic sequential sample-addition are desirable in order to maximize the equipment utilization and sample throughput. Several systems which achieve this and allow the analyser to operate unattended have been devised. Dus *et al.*[21] put individual samples in narrow-bore tubing held between valves which were rotatable and facilitated sample selection and elution programming. Murdock, Grist and Hirs[22] used coils of Teflon tubing to hold the samples; after alignment of the sample-holder with the column the sample was pumped on to it by displacement. Nylon sample-tubes replaced the Teflon coil in the design of Alonzo and Hirs[23]. Eveleigh and Thomson[24] used a series of very small ion-exchange columns to hold individual samples, each being clamped in turn over the analytical column at the end of each cycle. A needle injection-system with facilities for circulating cooling water round samples awaiting analysis has been described by Dymond[25].

The technique of storing samples on miniature ion-exchange columns has been developed commercially in the Technicon TSM 'Sequential Multisample Amino Acid Analyzer'. The sequential sampling device accommodates up to 40 samples, each contained in a plastic cartridge containing resin which is sandwiched between two thin discs of inert sintered plastic. On completion of the previous analysis the sampler is advanced to the next position whereupon the cartridge is clamped on to the analytical column and becomes an extension of it. Eluting buffers then pass through the cartridge-plus-column assembly. The entire sequence of operations is controlled by the peristaltic programming valve described above. Advantages of this approach are that considerable latitude is allowed for the initial sample-volume and samples can be stored, with minimal deterioration, for long periods in an essentially dry state.

9.2.7 Computer-Controlled Amino-Acid Analysis

The application of computers to process and evaluate data from amino-acid analysers is described above. An amino-acid analyser (Durrum D 500) in which the entire sequence of equipment-operation as well as data-processing is controlled by a digital computer is now available. The computer, a Digital Equipment Corporation PDP 8/L, is programmed either by Teletype or punched paper-tape using coded instructions. The program capacity is 30 steps per full procedure with up to 16 events per step. Once the samples have been loaded in cartridges (up to 80 can be accommodated) the computer initiates and controls the entire sequence of sample-introduction, buffer-selection, column-regeneration and temperature-changes. Computer-operation allows events to be timed to the nearest second. This is almost an order of magnitude improvement over earlier instruments and the reproducibility of performance is thereby enhanced. As each individual amino-acid is eluted from the column the computer calculates the molar quantity of amino-acid and prints out the data, listing the peaks numerically and by name. Unknown peaks, i.e. those with retention times not corresponding to an amino-acid, are listed, as are data points occurring off-scale or below base-line.

The D 500 incorporates several other novel features. The speed of analysis is markedly improved by miniaturizing the chromatographic stage. This has been achieved by designing the system to operate at high pressures, using stainless-steel columns, and flow-line pressures of up to 3000 psig can be accommodated. The single-column procedure is used, the column dimensions being reduced to 43 cm × 1.75 mm, and it is filled with cation-exchange resin beads of diameter 10 ± 2 μm. To make full use of the versatility of operation inherent in a computer-controlled instrument, eight buffer and colour-reagent feed-lines are provided by four hydraulically-driven double-syringe pumps capable of smooth operation at 3000 psig. Colour development with ninhydrin is controlled at 150° C; this obviates the need for a second photometer channel to detect the proline and hydroxyproline peaks, which is essential when temperatures of 50–70° C are used. The photometer is a dual-wavelength a.c. modulated unit with a solid-state photodiode detector yielding linear absorbance output over five ranges; 0·1, 0·2, 0·5, 1·0 and 2·0 absorbance units. A short light-path (0·5 cm) cell is used to minimize overlap of peaks. It is illuminated alternately with radiation of 590 nm (for measurement) and 690 nm (for background); the photodiode responses are fed in parallel to a chart-recorder and the computer-processor which monitors the output 600 times per minute. The stability of the photometer is such that 1 nmole of amino-acid yields a signal-to-noise ratio of 30:1. In consequence initial sample volumes can be as little as 10–20 μl.

The full analysis time for a protein hydrolysate is a function of the operating pressure of the ion-exchange system; figures quoted by the manufacturers are 48 min at 2700 psig and 69 min at 1850 psig.

9.3 Sugars

The resolution and analysis of complex mixtures of sugars is of considerable biochemical and medical importance. Ion-exchange chromatography has proved one of the most effective means of achieving this end. The basic procedure is due to Khym and Zill[3] who utilized the ability of carbohydrates with adjacent *cis* hydroxyl groups to form anionic borate complexes which are separable by ion-exchange. Khym and Zill used borate buffers of stepped pH gradient to elute individual sugars from a column of strongly basic quaternary ammonium type resin. Detection and concentration measurement were based on colour formation with anthrone or cysteine. Subsequently several workers have modified the eluent composition to speed up and increase the resolution of the separation procedure. The most successful approach seems to be that of Catravas[26] who employed a continuous-gradient method in which borate and chloride concentrations and pH were increased linearly throughout the elution.

An alternative method of elution utilizes aqueous ethanol, the anion-exchange resin being used in sulphate form. Samuelson, Larsson and Ramnas[27] described a system in which a column of 95 × 0·6 cm was used and the mixed sugars were eluted with 94% ethanol to separate cellulose components.

The separation and analysis of sugars is procedurally similar to that for amino-acids and identical equipment can be used. Thus any automatic amino-acid analyser can be converted for separation of sugars by using the appropriate resin, eluent, colour-reagent and column-temperatures. For example the Technicon sugar analyser utilizes the borate-complex method; after separation the eluted sugars are determined colorimetrically by reaction with orcinol, using an 'AutoAnalyzer'. The flow-diagram for

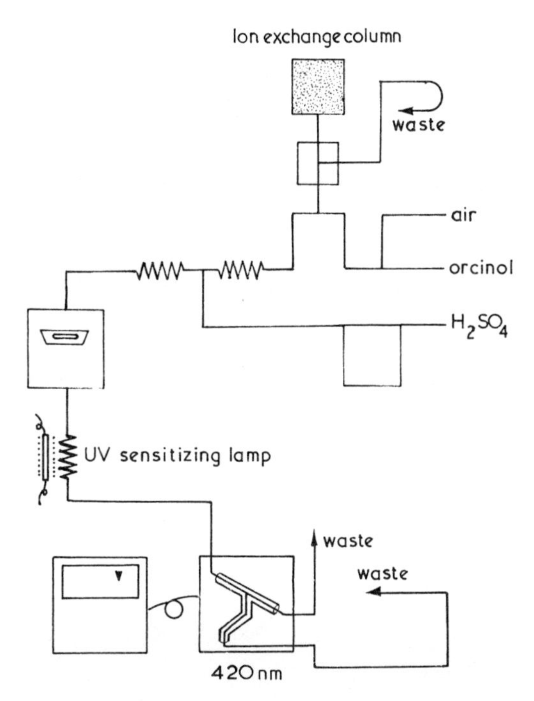

Fig. 9.9 Flow diagram, Technicon sugar analyser.
Reproduced with permission of Technicon Corporation.

the complete method is given in Fig. 9.9. The constant-temperature (45° C) column is packed with a strongly basic anion-exchange resin, Technicon Type S Chromatobeads, which are spherical and graded to give highly uniform bead-diameter. In this way the resolution of the column is maximized and the flow resistance minimized. A 9-chamber 'Autograd' is used to provide a borate–chloride eluent of graded composition which is transferred to the top of the resin column by a micropump. The bulk of the eluate is passed to a fraction-collector while a small portion is pumped directly into the continuous-analysis manifold. Provided internal standards are used, the concentration of individual sugars can be calculated from the peak-areas on the recorder-chart, either manually or by using a digital integrator accessory.

Mopper and Degens[28] draw attention to three shortcomings of standard commercial sugar analysers: the detection limit is of the order of 10^{-8}–10^{-9} mole and better sensitivity is desirable in some instances, the orcinol colour reaction requires the pumping of concentrated sulphuric acid, and the reproducibility of the method is inherently limited by deterioration of the peristaltic-pump tubes. They achieved a sensitivity of 10^{-10} mole by replacing the orcinol reagent by an alkaline solution of Tetrazolium Blue, and since the latter is incompatible with borate/chloride solutions an eluent of 89% ethanol/water was used. This composition represents an optimum balance between peak-resolution and peak-broadening. With an ethanol/water eluate both cation- and anion-exchange resins can be employed; Mopper and Degens preferred Technicon Type S 20-μm Chromatobeads on account of the quality of their separation performance. The automated sugar analyser is shown schematically in Fig. 9.10. A 110-cm resin bed in a nylon column of 0·28 cm bore is housed

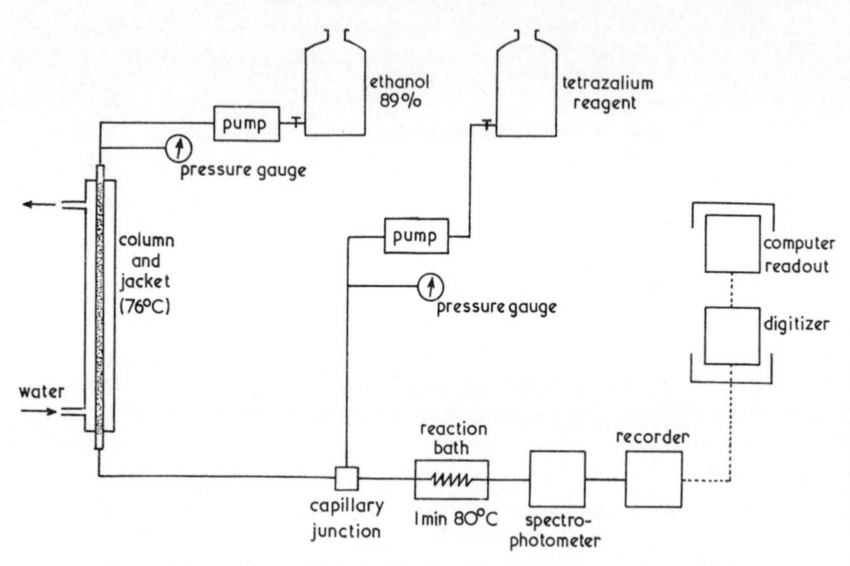

Fig. 9.10 Automatic sugar determination by ion-exchange and colorimetry. *Reproduced with permission from Mopper and Degens*[28] *and Academic Press Inc.*

in a jacket through which water at 76° C is passed. This temperature is higher than that previously used in sugar analysers and has the effect of increasing the rate of diffusion within the resin beads. This results in improved resolution, reduced pressure-drop across the column and reduced analysis-time ($3\frac{1}{2}$ hr for monosaccharides). Both the eluent and Tetrazolium Blue reagent are transported by piston pumps (Beckman 'Accu-Flo'). The two streams merge at a capillary-tubing junction and the colour-forming reaction occurs in a spiral of narrow-bore Teflon tubing maintained at 80° C in a water-bath. A residence time of 1 min is adequate for the reaction to go to completion. Absorbance measurements are made with a Gilford 2000 spectrophotometer fitted with a flow-through cuvette of 5-mm optical path-length. Results are presented as peaks on a chart-recorder and the peak-areas are automatically integrated and digitized by using a standard dual-channel digital integrator (Infotronics Ltd). A FORTRAN II computer program quantitates results from peak-area data from samples and standards.

9.4 Hydroxy-Acids

Anion-exchange chromatography has proved invaluable as a means of separating complex mixtures of hydroxy-acids, especially those derived from sugars. Information about such acids in physiological fluids is of considerable medical interest. The conditions for the anion-exchange separation were developed by Samuelson and co-workers[29, 30]; Dowex 1×8 resin was used, graded to give a particle-size range of 13–17 μm, and elution was performed with either acetic acid or sodium acetate–acetic acid at pH 5·9. In these and subsequent studies[31, 32], automatic techniques were used only at the analytical stage, which depends upon several colorimetric reactions. Automatic control of the ion-exchange column was not included.

The use of several colorimetric detection systems enables maximum information to be obtained regarding the nature of the eluted hydroxy-acids. Fig. 9.11 depicts the system used by Carlsson, Isaksson and Samuelson[31]. A Technicon peristaltic pump was used in conjunction with a

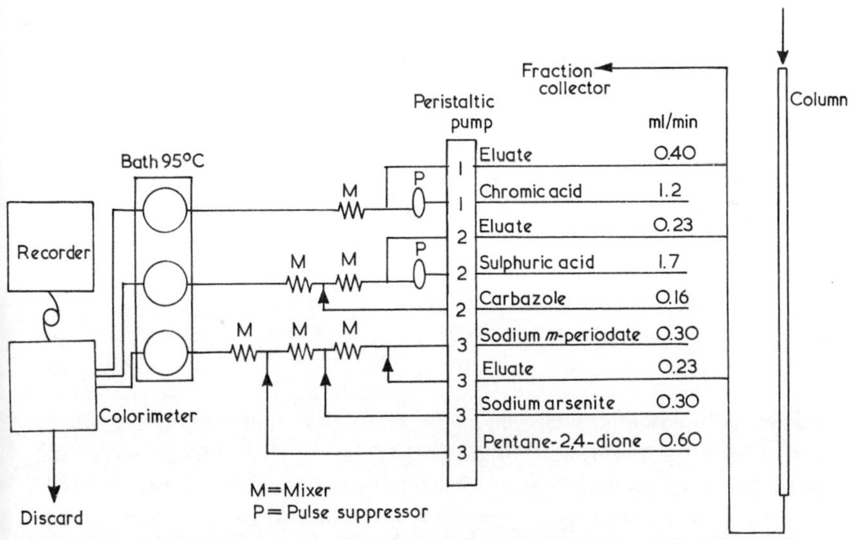

Fig. 9.11 Manifold for automated colorimetric detection of hydroxy-acids following ion-exchange separation.
Reproduced with permission from Carlsson, Isaksson and Samuelson[31] *and Elsevier Publishing Company.*

3-channel colorimeter (LKB Producter, Stockholm). The manifold was designed to enable the eluate stream to be divided into four, one part being diverted to a fraction-collector and the remaining three parts reacted with chromic acid, carbazole and sodium periodate respectively, the resulting colour being measured in the colorimeter in each case. A fourth colorimetric analysis line capable of measuring the amount of sodium periodate consumed in addition to the formaldehyde produced has subsequently been added by Carlsson and Samuelson[32].

The provision of the automatic analytical unit made possible the accumulation of extensive data on the separation and reactions of many hydroxy-acids. Table 9.1, taken from the work of Carlsson and Samuelson[32], summarizes typical data which can be utilized in analysing samples of biological interest.

Table 9.1 DISTRIBUTION COEFFICIENTS, PERIODATE CONSUMPTION AND RELATIVE RESPONSE INDICES OF ALDOBIOURONIC, HEXURONIC AND HEXULOSONIC ACIDS

Acid	*D_v value*			*Moles of periodate per mole of acid*	*Periodate index*	*Formaldehyde index*	*Carbazole index*
	0·5M HAc	*1M HAc*	*0·08M NaAc*				
2-*O*-(α-D-Galactopyranosyluronic acid)-L-rhamnose	8·4		4·16	2·4	0·54	0·01	0·49
4-*O*-(α-D-Galactopyranosyluronic acid)-D-xylose	12·0			2·0	0·48	0·01	0·54
6-*O*-(β-D-Glucopyranosyluronic acid)-D-galactose	15·5			2·2	0·46	0·01	0·47
2-*O*-(4-O-Methyl-α-D-glucopyranosyluronic acid)-D-xylose	13.2			0	0	0	0·57
Cellobiouronic acid	26.0		6·0	1·4	0·25	0·01	0·51
Alluronic		17·1		2.4	0·94	0·04	0·85
Galacturonic		11·0	9·4	2·3	0·90	0·01	1
Glucuronic		22·4	12·8	2.0	0·81	0·01	0·65
Guluronic		12·3		2·1	1·00	0·04	0·50
Mannuronic		18·6	14·2	2·1	0·86	0·04	0·70
Taluronic		14·2		2·3	0·91	0·01	0·94
4-*O*-Methyl-D-glucuronic		18·0	8·0	1·1	0·35	0·01	0·80
arabino-5-Hexulosonic		13·2		1·7	0·67	0·20	1·27
lyxo-5-Hexulosonic		15·4		1·9	0·56	0·27	0·80
ribo-5-Hexulosonic		28·2	17·4	1·5	0·65	0·26	1·30
xylo-4-Hexulosonic		22·3		2·1	0·83	0·90	0·11

Variants of the anion-exchange technique have been developed to analyse complex mixtures such as the hyaluronic acid oligosaccharides[33] and chondroitin sulphate oligosaccharides[34], with formic acid and sulphuric acid respectively as eluents. Fransson, Roden and Spack[35] describe a method using a Technicon automatic sugar chromatography analyser for determining uronic acids and acids containing oligosaccharides obtained from mucopolysaccharides by hydrolysis or enzymic degradation. They established that uronic acids and non-sulphated aldobiouronic acids could be separated by using gradient elution over the range 0·3–0·5*M* formic acid. For separation of the hyaluronic acid oligosaccharides and chondroitin-6-sulphate oligosaccharides gradient elution with lithium chloride solutions was preferred.

9.5 Sodium Monofluorophosphate

Benz and Kelley[36] developed an automatic anion-exchange method for determining the monofluorophosphate content of dentrifices. After elution from the column the monofluorophosphate is hydrolysed with sulphuric acid and reacted with molybdate to form molybdenum blue which is measured colorimetrically. The 'AutoAnalyzer' flow-diagram is

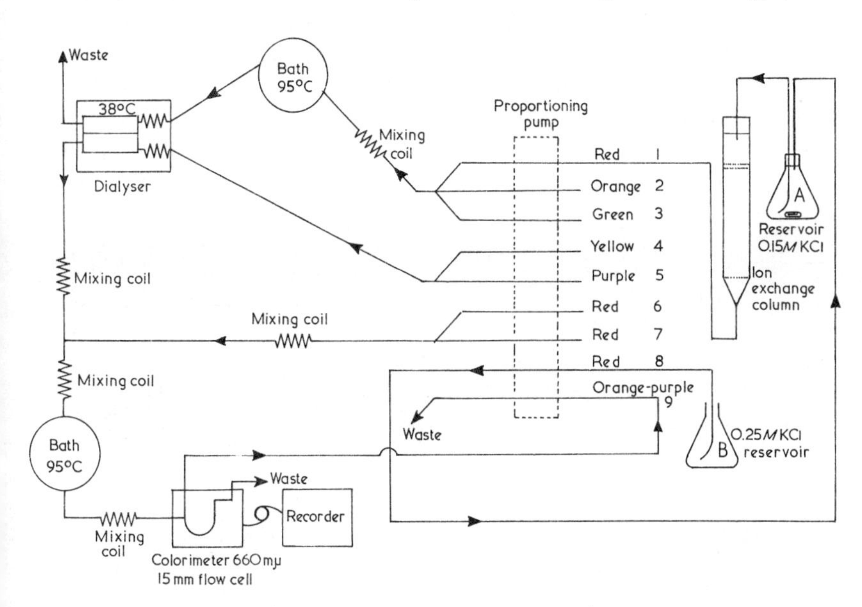

Fig. 9.12 Manifold for automatic determination of monofluorophosphate.
Reproduced with permission from Benz and Kelley[36] and Elsevier Publishing Company.

shown in Fig. 9.12. A gradient elution with potassium chloride solution is achieved by a two-vessel system which is operated from the proportioning pump. The monofluorophosphate concentration is calculated from peak-area ratios relative to an orthophosphate external standard added to the column before the addition of sample. Orthophosphate is eluted first; pyrophosphate is eluted from the column immediately following monofluorophosphate and the column dimensions (15 × 0·8 cm) and potassium chloride gradient concentration were chosen to ensure that no overlap of these two peaks occurred. The method was compared with a standard analytical procedure of precipitating silver monofluorophosphate, distilling the fluoride and measuring it colorimetrically. No bias was detected and the automatic method, total time 70 min, was quicker than the gravimetric one.

9.6 Free Acidity in the Presence of Hydrolysable Ions

Many methods have been developed for determining free acidity in the presence of hydrolysable ions such as Al^{3+}, Th^{4+}, Zr^{4+}, Pu^{4+}, and UO_2^{2+}. Mostly they depend on titrating the acid after complexing or precipitating the hydrolysable cations with fluoride, EDTA, ferrocyanide, etc. Such methods are not readily adaptable to samples where the nature and concentration of the hydrolysable ions are variable. To overcome this problem Gaddy and Dorsett[37] developed a general procedure based on removal of the interfering ions on a cation-exchange column and determining the unchanged acidity in the column effluent by automatic colorimetry. The flow-diagram is shown in Fig. 9.13; the apparatus comprises

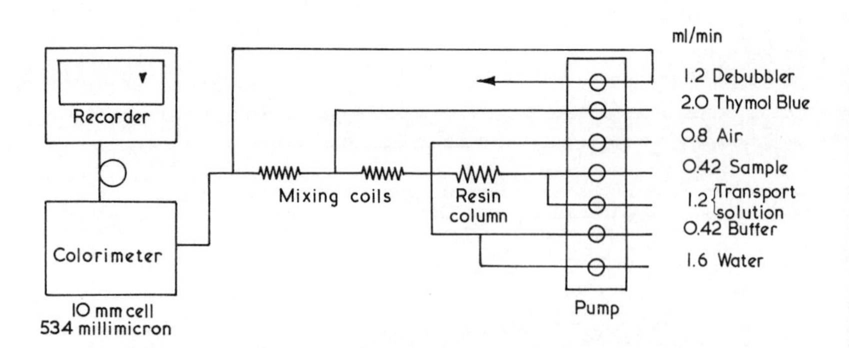

Fig. 9.13 Manifold for automatic determination of free acidity.
Reproduced with permission from Gaddy and Dorsett[37] and American Chemical Society.

a turntable sampling unit, a multichannel proportioning pump, a colorimeter and recorder. The ion-exchange column, contained in coiled Teflon tubing, is an integral part of the pumping system and the elution rate is therefore governed by the diameter of the tubes pumping sample and eluent. Both cation- and anion-exchange were used, depending on the hydrolysable ion present. La^{3+} and Al^{3+} were removed by cation-exchange using an eluent of 1·5*M* sodium chloride; Th^{4+} was taken up on an anion-exchange resin from 5*M* lithium nitrate. In each case the eluted acid was determined by automatic colorimetry at 534 nm through its reaction with

Thymol Blue. The authors consider that the two ion-exchange systems could be combined in a mixed-bed system provided adequate attention is devoted to optimizing the composition of the eluent. Concentrations of these ions up to 0·2*M* were successfully accommodated and produced no bias in the acidity measurement. The limit of determination for the method as described is 0·005 mmole of free acid. For standard solutions of 0·2*M* hydrochloric acid containing 0·2*M* La^{3+} a relative standard deviation of ±1·1% was found. A cycle-time of 9 min between sample analyses was employed, the total time for each analysis from sampling to recorder output being 14·5 min.

9.7 Radionuclide Separations

Samsahl and his co-workers[38–43] have developed a comprehensive separation scheme for a large number of radionuclides formed by neutron activation of biological samples. Some 40 elements may be determined by first separating them individually or in small groups, using 10 small separation columns in series. Four of the columns contain the anion-exchange resin Bio-Rad AG, three contain the chelating resin Chelex 100, two are partition columns utilizing the extracting agent di(2-ethylhexyl)-phosphoric acid (HDEHP) and one contains the inorganic ion-exchanger Bio-Rad KCF1. The full separation scheme showing the disposition of the columns, eluting solutions used and the elements retained on each column is shown in Fig. 9.14.

The ion-exchange separation sequence is designed to operate automatically; the layout is shown diagrammatically in Fig. 9.15. The columns (*I–R*) are coupled in series, with Teflon tubing of 1 mm bore. Teflon is used for all connecting lines. They are fed with eluting reagents from a multichannel piston-pump with each piston connected to the movable plate *H*. A separate pump (*A–F*) is provided for each eluent and one (*G*) serves to wash the sample, contained in a coil ahead of the first column, on to the column. The movable plate is lowered or raised by an electric motor, the downstroke fills the channels with eluate and the upstroke discharges them on to the appropriate column. The pumping rate is adjusted to maintain a flow-rate of 5 ml/min. A five-turn Teflon mixing coil is provided between each column to ensure adequate mixing. On completion of the separation cycle, which can take as little as 5 min, the final effluent is analysed radiometrically and the contents of each column are assayed by gamma-spectrometry *in situ*.

Using the automatic ion-exchange apparatus it is possible to perform a complete analysis of an irradiated sample in 2 hr as against $1\frac{1}{2}$–3 days manually. The manual method requires constant operator-attention

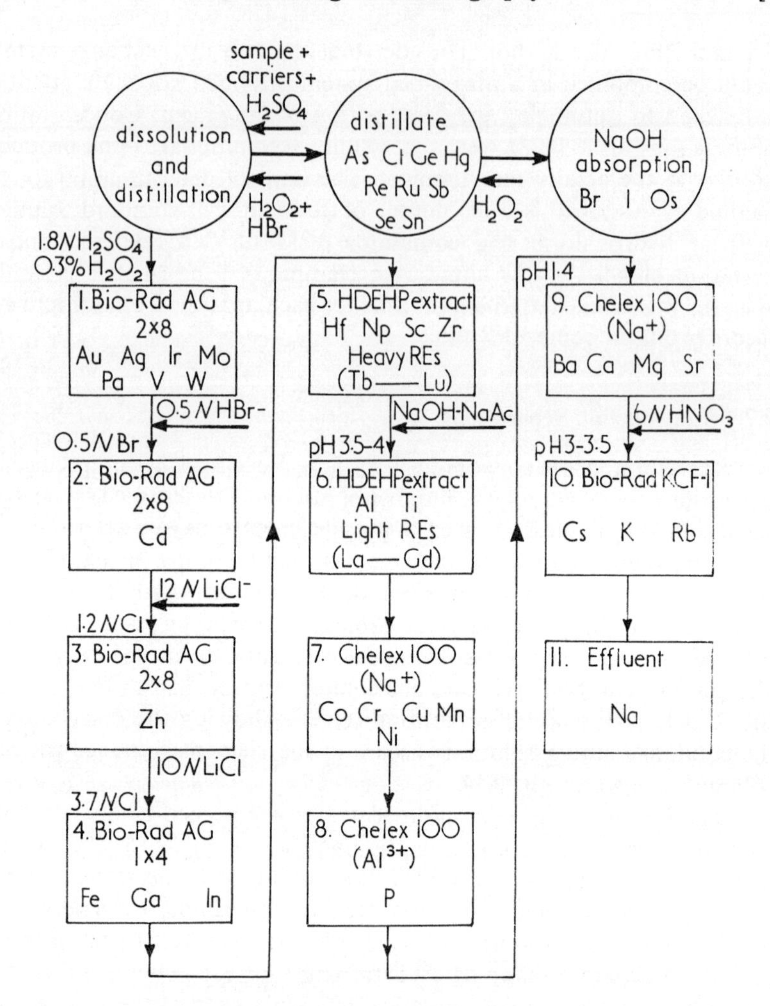

Fig. 9.14 Flow diagram for ion-exchange separation of radionuclides.

whereas the automatic equipment needs little supervision. For almost all trace elements determined, recoveries were demonstrated to be at least 90% and the standard error of a mean result is between 3 and 5%. Undoubtedly such performance depends more upon the detailed derivation of the chemical separation scheme than upon the design of the automatic apparatus.

Baker[44] describes an automated ion-exchange system involving a single column only. It has facilities for admitting a number of eluting solutions in sequence and for collecting the eluates as a series of fractions.

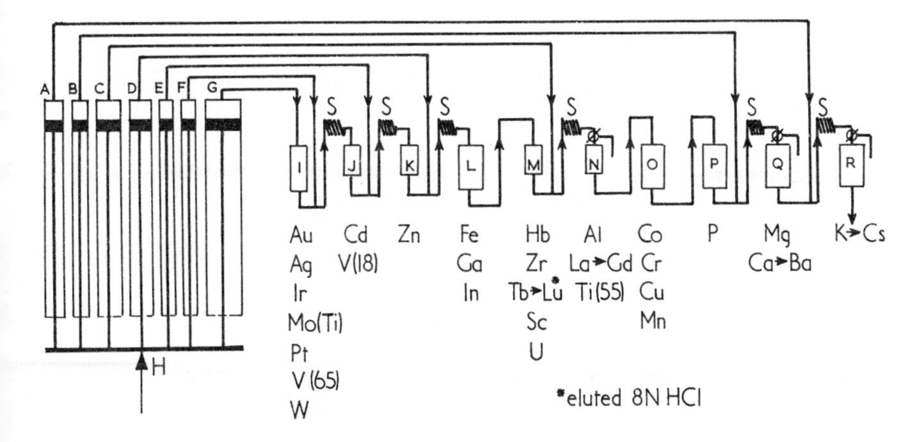

Fig. 9.15 Line diagram of automated ion-exchange system.

It has been designed in response to a need to perform a radionuclide separation procedure lasting several days, but it is capable of general-purpose use. A line-diagram of the apparatus is given in Fig. 9.16. The column is constructed of borosilicate glass (520 × 5 mm) fitted at the top with a reservoir and overflow side-arm to waste and closed at the bottom by a solenoid valve. Collecting beakers are mounted on a turntable beneath the column exit. It is driven by a 1-rpm synchronous motor fitted with an electrically operated brake for instantaneous arrest. Correct positioning of the beakers is achieved by notches in the turntable into which fit microswitch levers. A level-detector comprising a lamp and photodiode is situated at the top of the column just above the top of the resin bed. The signal from the operation of the detector is fed to a relay system which triggers the movement of the turntable to the next position. A bank of 9 reagent reservoirs is mounted above the column, and above each is an electromagnetic pinch-valve connected to an air-manifold and compressor. The pinch-valves are opened sequentially by the relay control circuit. When the first one is opened the contents of the reservoir are discharged pneumatically to the top of the column and eluate is collected until the photodiode detector operates. Then the column-outlet solenoid valve closes, the turntable advances, the next reagent is admitted to the column and the column outlet re-opened.

Automated ion-exchange separation in the determination of fission products in irradiated nuclear fuels is described by Bobleter *et al.*[45] A sample from a bypass line from the fuel dissolver is peristaltically pumped on to the ion-exchange column. The same pump draws eluent from a vessel fitted with a photoelectric or capacitance-type level-detector. This

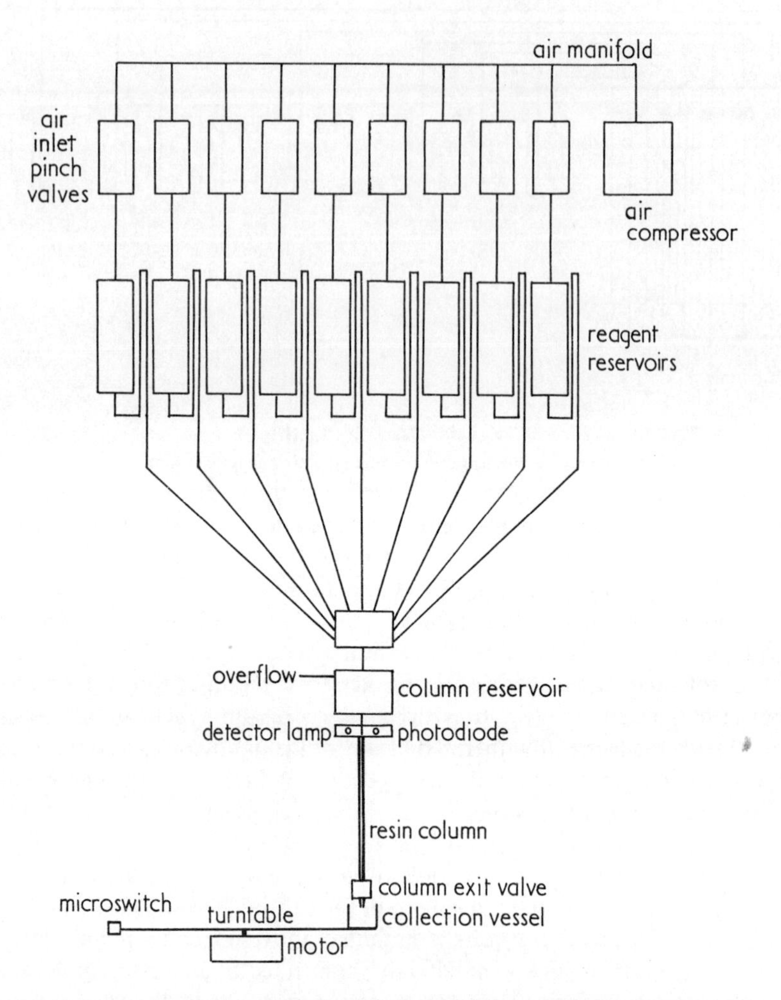

Fig. 9.16 Automatic Sequential elution of a single ion-exchange column.
Reproduced with permission from Baker[44] *and United Trade Press Ltd.*

is used to control the sequence of eluents, and the vessel is connected to a battery of storage bottles through electromagnetic valves. The signal from the operation of the level-detector is fed to a control-unit programmed to open and close each magnetic valve in sequence. Thus a predetermined volume of a series of eluents can be separately admitted to the supply vessel each time the level-detector responds. The eluate from the ion-exchange column is pumped past β- and γ-counting probes and thence either to waste or to collection vessels for further examination.

Comar and Le Poec[46] describe an automatic ion-exchange procedure as part of a system for determining trace elements in ashed biological samples following neutron irradiation. The apparatus is designed so that a column on which sample is adsorbed can be consecutively eluted with three separate reagents. It is based upon 'AutoAnalyzer' components, in particular the 40-position Sampler I and peristaltic pumps. The ion-exchange columns, 15 × 50 mm polyethylene, are attached to a rotatable plate mounted on the same axis as the sampler. Three cones, through which sample and successive eluent can be passed, are mounted in a rack immediately above the columns; they can be raised and lowered sequentially by the operation of an electromagnet; in the lower position the cone is in contact with the top of the column. Column eluates pass into funnels from which they are pumped to waste or to the next analytical stage. A Crouzet 11-position timer is used to program the sequence of events. Sample is pumped on to the top of the column through the first cone and then the turntable holding the columns is rotated to bring the column below the second cone which is lowered and eluting solution passed through it. The procedure is repeated for the third cone and subsequent columns are similarly treated.

Apparatus for simultaneous elution of eight ion-exchange columns for separating radioisotopes formed by neutron irradiation is briefly described by Fourey *et al.*[47] Eluent is delivered to each column by means of a multichannel peristaltic pump. Each line is provided with a separate reagent reservoir and between each reservoir and the pump is situated an electrically-operated valve which can be activated when eluate flow is required. A simple control-circuit, based on pre-set elution times is used to generate and stop the flow; each line is provided with individual control.

References

1. Moore, S. and Stein, W. H. *J. Biol. Chem.*, 1951, **192**, 663.
2. Moore, S. and Stein, W. H. *J. Biol. Chem.*, 1954, **211**, 893.
3. Khym, J. X. and Zill, L. P. *J. Am. Chem. Soc.*, 1952, **74**, 2090.
4. Tompkins, E. R., Khym, J. X. and Cohn, W. E. *J. Am. Chem. Soc.*, 1947, **69**, 2769.
5. Moore, S., Spackman, D. H. and Stein, W. H. *Anal. Chem.*, 1958, **30**, 1185.
6. Simmonds, D. H. *Anal. Chem.*, 1958, **30**, 1043.
7. Spackman, D. H., Moore, S. and Stein, W. H. *Anal. Chem.*, 1958, **30**, 1190.
8. Chenard, F. P. *J. Biol. Chem.*, 1952, **199**, 91.
9. Spackman, D. H. *Federation Proc.*, 1963, **22**, 244.
10. Piez, K. A. and Morris, L. *Anal. Biochem.*, 1960, **1**, 187.

11. Hamilton, P. B. *Anal. Chem.*, 1958, **30**, 914.
12. Hamilton, P. B. *Anal. Chem.*, 1963, **35**, 2055.
13. Peterson, E. A. and Sober, H. A. *Anal. Chem.*, 1959, **31**, 857.
14. Keck, K. *Anal. Biochem.*, 1971, **39**, 288.
15. Liteanu, C. and Gocan, S. *Gradient Liquid Chromatography*, Ellis Horwood, Chichester, 1974.
16. Graham, G. N. and Sheldrick, B. *Biochem. J.*, 1965, **96**, 517.
17. Porter, W. L. and Talley, E. A. *Anal. Chem.*, 1964, **36**, 1692.
18. Krichevsky, M. L., Schwartz, J. and Mage, M. *Anal. Biochem.*, 1965, **12**, 94.
19. Starbuck, W. C., Mauritzen, C. M., McClimans, C. and Busch, H. *Anal. Biochem.*, 1967, **20**, 439.
20. Conkerton, E. J., Coll, E. E. and Ory, R. L. *Anal. Letters*, 1968, **1**, 303.
21. Dus, K., Lindroth, S., Pabst, R. and Smith, R. M. *Anal. Biochem.*, 1966, **14**, 41.
22. Murdock, A. L., Grist, K. L. and Hirs, C. W. H. *Arch. Biochem. Biophys.*, 1966, **114**, 375.
23. Alonzo, N. and Hirs, C. W. H. *Anal. Biochem.*, 1968, **23**, 272.
24. Eveleigh, J. W. and Thomson, A. R. *Biochem. J.*, 1966, **99**, 49P.
25. Dymond, B. *Anal. Chem.*, 1968, **40**, 919.
26. Catravas, G. N. *Technicon Symposium 'Automation in Analytical Chemistry'*, p. 397, New York, 1966.
27. Samuelson, O., Larsson, L. and Ramnas, O. *Technicon Symposium 'Automation in Analytical Chemistry'*, p. 169, 1965.
28. Mopper, K. and Degens, E. T. *Anal. Biochem.*, 1972, **45**, 147.
29. Johnson, S. and Samuelson, O. *Anal. Chim. Acta*, 1966, **36**, 1.
30. Samuelson, O. and Thede, L. *J. Chromatog.*, 1967, **39**, 556.
31. Carlsson, B., Isaksson, T. and Samuelson, O. *Anal. Chim. Acta*, 1968, **43**, 47.
32. Carlsson, B. and Samuelson, O. *Anal. Chim. Acta*, 1970, **49**, 247.
33. Weissman, B., Meyer, K., Sampson, P. and Linker, A. *J. Biol. Chem.*, 1954, **208**, 417.
34. Hoffman, P., Meyer, K. and Linker, A. *J. Biol. Chem.*, 1956, **219**, 653.
35. Fransson, L. A., Roden, L. and Spack, M. L. *Anal. Biochem.*, 1968, **21**, 317.
36. Benz, C. and Kelley, R. H. *Anal. Chim. Acta*, 1969, **46**, 83.
37. Gaddy, R. H. and Dorsett, R. S. *Anal. Chem.*, 1968, **40**, 429.
38. Samsahl, K. *Nucleonik*, 1966, **8**, 252.
39. Samsahl, K. *Analyst*, 1968, **93**, 101.
40. Samsahl, K., Wester, P. O. and Landstrom, O. *Anal. Chem.*, 1968, **40**, 181.
41. Samsahl, K., Brune, D. and Wester, P. O. *Intern. J. Appl. Radiation Isotopes*, 1965, **16**, 273.
42. Samsahl, K. *Anal. Chem.*, 1967, **39**, 1480.
43. Samsahl, K. *Aktiebolaget Atomenergie Report*, AE 389, 1970.
44. Baker, C. W. *Lab. Pract.*, 1972, **25**, 342.
45. Bobleter, O., Forster, H. P. and Reuschel, A. G. *Fourth United Nations Conference on the Peaceful Uses of Atomic Energy*, A/CONF/49/P/211 (1971).
46. Comar, D. and Le Poec, C. *Modern Trends in Activation Analysis*, Proceedings 1965 International Conference, College Station, Texas, p. 351.
47. Fourey, A., Neuburger, M., Garrec, C., Fer, A. and Garrec, J. P. *Modern Trends in Activation Analysis*, eds. J. R. Devoe and P. LaFleur, Vol. 1. p. 160, NBS Special Publication 312, 1969.

Chapter **10**

Separation Techniques (Other than Chromatography)

10.1 Solvent Extraction

Many types of sample cannot be analysed directly but require chemical or physical pretreatment and this part of the analysis is often more time-consuming than the measurement step. Solvent extraction, either for isolating the desired component or for removing interferences, is now established as one of the most useful pretreatment stages and there have been a number of approaches aimed at automating the technique, both for general purpose use and for specific types of determination. In some cases the separation conditions are not critical but in others several stages involving closely controlled conditions of solution chemistry are necessary to yield a satisfactory product for analysis. Where the latter situation prevails the design parameters for an automatic extraction system need to be carefully optimized with particular regard to the automatic control function, and it is essential that the chemistry of the reaction is adequately understood. The main aim should always be towards the design of the simplest system possible, for reasons of economy of capital and development costs, and for ease of maintenance and reliability.

10.1.1 Discrete Analytical Solvent-Extraction Devices

There are various means of obtaining phase separations in discrete analytical systems, and both static and dynamic units can be used. Typically, static units depend on the use of phase-sensors and valve flow-controllers, such as a unit using conductivity-sensing electrodes and electromagnetic pinch-clip valves as described by Trowell[1]. A unit based on these principles is described in section 10.1.4. One problem with the device is in the use of volatile solvents, particularly ether. In such cases there is a potential explosion hazard from the combination of ether vapour and air in the small volume of the vessel, if a spark occurs between the electrodes. A preferred alternative sensor-system also attributed to Trowell[2] utilizes the capacity effect caused by changing the phase in contact with the sensor. A voltage is applied across a pair of copper plates on either side of the extraction vessel so that the liquid in the vessel serves as the determination medium. The change in capacity which results from a change in dielectric constant is used to operate the electromechanical

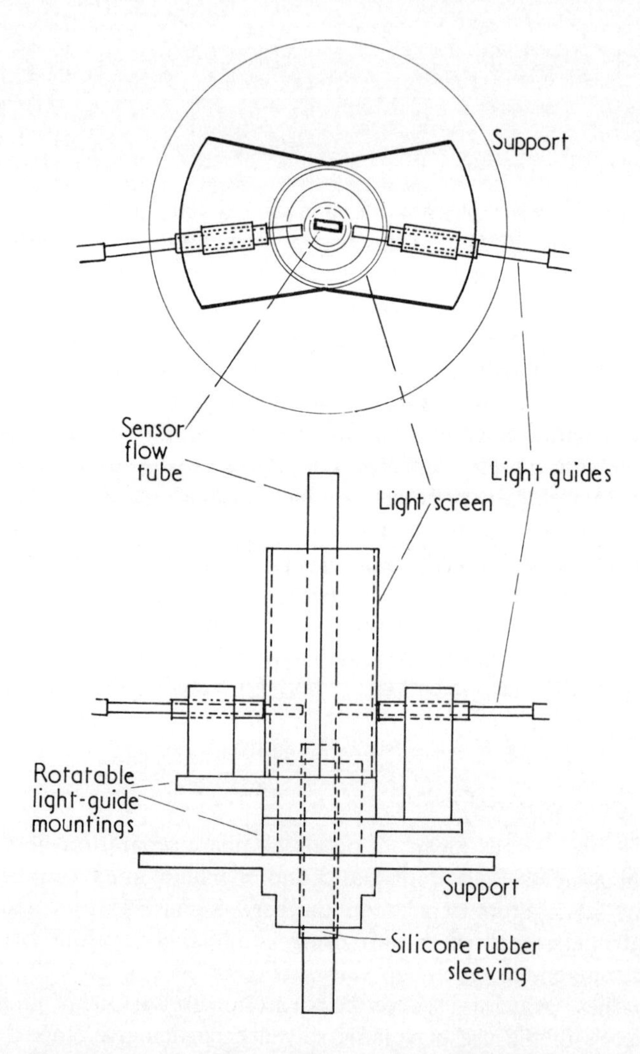

Fig. 10.1 Phase sensor for solvent extraction based on a change of refractive index, designed by Porter[3].

control-valve. The complete device can be encapsulated to provide a cheap, fail–safe sensor system. A commercial version based on a similar principle is marketed by Fisons Scientific Supplies Limited.

An alternative sensing system, which does not involve physical placement of electrodes or other measuring devices in the test-solution, has been reported[3]. This uses the change in refractive index associated with a phase change from organic to aqueous solution. Light from a stable source is directed through a short length of glass tubing with a rectangular cross-section on to a photocell sensor. With the change of refractive index the device forms a simple on/off switch for an associated control-valve. A schematic diagram of this system is shown in Fig. 10.1.

A device which has found a number of applications in solvent extraction is the centrifugal separator designed and developed by Vallis[4], shown in Fig. 10.2. It comprises a cup-shaped vessel mounted on a motorized shaft, complete with a porous lid attached to the lip of the cup. In use, the device is placed inside a collecting vessel and if the porous

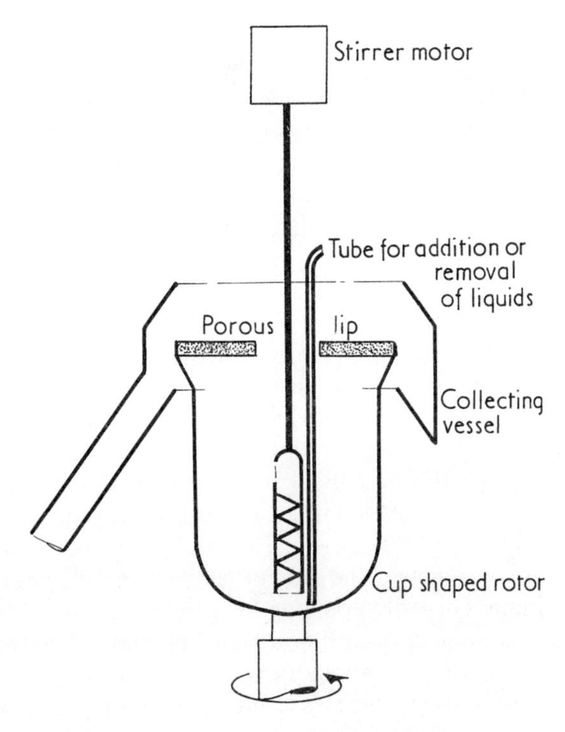

Fig. 10.2 Centrifugal separator.
Reproduced with permission from Vallis[4].

interface is made from a hydrophilic material such as sintered glass, water will pass into the collecting cup at low rotation speeds leaving the organic phase retained in the cup. An increased rotation speed is then used to eject the organic phase. The use of a hydrophobic interface such as sintered PTFE will allow the preferential rejection of the organic phase.

Additional provisions are made in the assembly for the addition of solvents or other solutions and for stirring. Selectively lowering or raising a series of collection-cups in line with a particular membrane enables a complex series of operations to be carried out within a single cup. The 'CentriChem' range of automatic analytical equipment marketed by Joyce Loebl and Co.[5] utilizes the principles mentioned above. A major problem encountered by these systems is the instability of any particular membrane interface. In automatic analysis this either results in continual replacement or requires a regeneration stage. This effectively increases the analysis time and therefore the desired increase in sample throughput may not result.

The apparatus uses a control system developed by Steed[6]. The continuous audiotone programmer utilizes a series of audiotone frequencies which are recorded on a magnetic tape, and each discrete frequency is coupled, through a low-frequency band-pass filter, to an electronic switch-control function. By means of a simple conversion unit and manual control of the switching sequence the filter network can be used to produce the primary source program. A precise record of this sequence can be simply achieved and repeated automatically, offering considerable flexibility and versatility with a particular sequence of commands.

10.1.2 Continuous-Flow Systems

Solvent-extraction problems can also be tackled in the context of continuous-flow systems. Generally these systems make use of a peristaltic pumping device to deliver the liquid streams and a simple separator in which the phases stratify to partition the aqueous from the organic phase. A single mixing coil filled with glass ballotini provides intimate mixing of the two phases to initiate the extraction required. After separation one or other (or both) of the phases can be analysed by a suitable technique such as colorimetry. The ratio of sample to extractant may be varied within the limits normally applying for good operation of the flowing systems. The maximum concentration factor (volume aqueous phase/volume solvent) consistent with good operation normally approximates to 3 : 1. Response-variations in the measuring device can occur owing to globules of the organic phase entering the flow-through cell in the case of the aqueous phase determinations. However, a far more serious restriction is caused

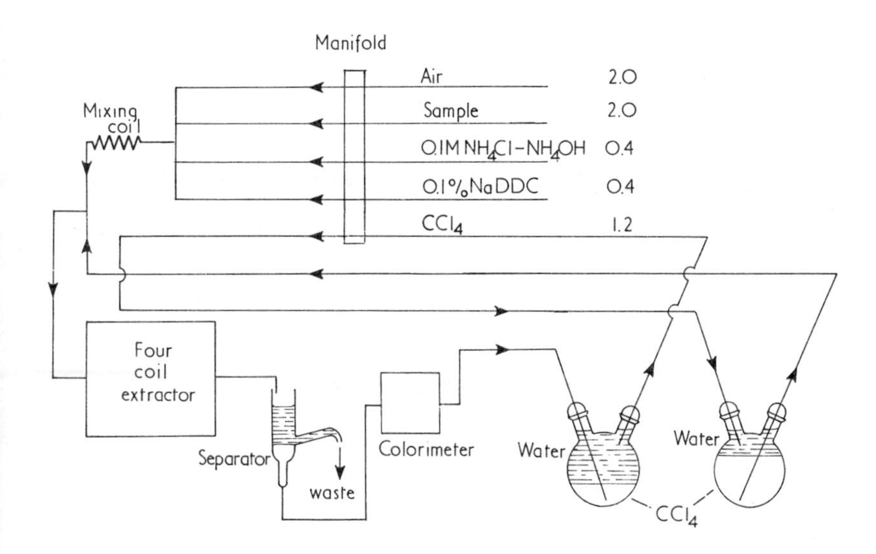

Fig. 10.3 Continuous solvent extraction system (flow rates in ml/min).
Reproduced with permission from Carter and Nickless[7] and Society of Analytical Chemistry.

by the inability of the pump-tubes to deliver consistent aliquots of organic solution for long periods. Despite these drawbacks, the continuous system has found considerable applications in the sphere of 'AutoAnalyzer' work and the flow-diagram, Fig. 10.3, serves to illustrate these uses. The techniques involved, described by Carter and Nickless[7], are designed to overcome the problem of pumping organic liquids, by the use of displacement techniques. Copper is determined by two methods, (*i*) complexation with sodium diethyldithiocarbamate, followed by extraction into carbon tetrachloride and (*ii*) direct extraction with a solution of zinc dibenzyldithiocarbamate in carbon tetrachloride. Vigorous intermixing has not been found necessary to effect either of the phase-mixing operations in the analysis and plain coils have been found adequate. Many other examples using the principles set out in this section can be found in the published accounts of the Annual Technicon Symposia.

10.1.3 Continuous and Industrial Extraction Devices

Many systems have been described in the area of continuous extraction. Butler[8] has adapted a Soxhlet extractor to form a continuous liquid extractor. A Swiss patent to Signer[9] describes separators in which counter-current extraction is used, thin layers of the phases passing over each other in a slowly rotating horizontal cylinder containing many transverse walls with small openings to allow passage of the phases. Two French patents[10,11] also describe counter-current systems for solvent extraction, the solvent flowing within a horizontal tube against the stream of liquid mixture to be separated. Stirring may be done either by a series of paddles or by introduction of compressed air into the extraction medium. These systems serve only as illustrations of the numerous systems that have been proposed for solvent extraction and do not in any way form a comprehensive review.

10.1.4 DETAILS OF SOLVENT EXTRACTION SYSTEMS

10.1.4.1 *Separation of Trace Quantities of Non-ionic Detergents from Water by Batch Extraction*[12]

As a matter of routine, large quantities of water and effluents (i.e. 0·5–3 litres) are extracted with an organic solvent such as chloroform to separate the detergent. Fig. 10.4 shows a schematic diagram of a batch

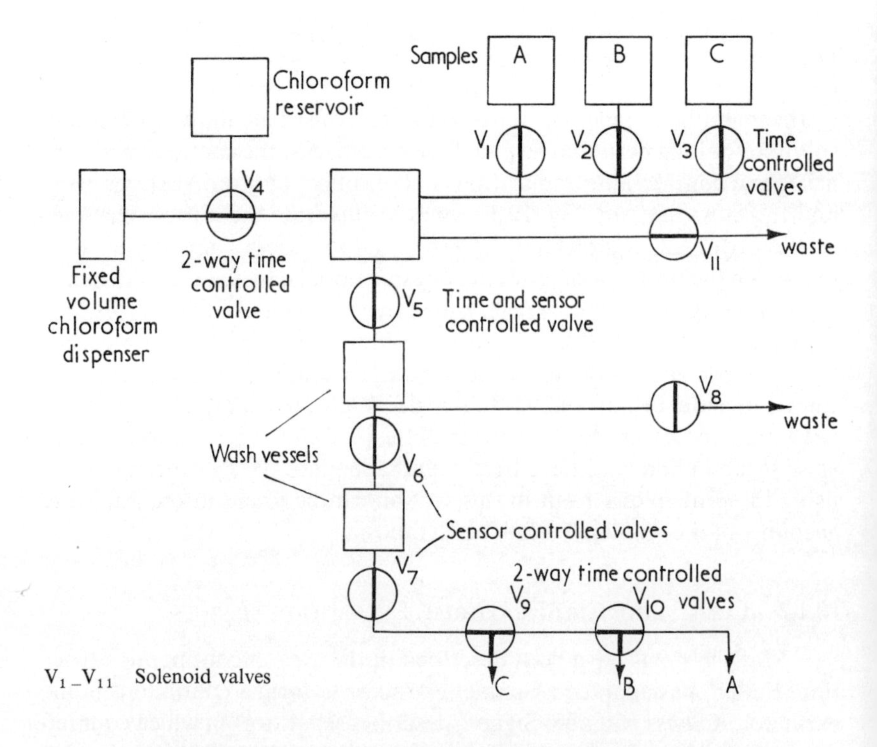

Fig. 10.4 Flow diagram for batch extraction of non-ionic detergent in water. *Reproduced with permission from Sawyer et al.*[12] *and The Society of Analytical Chemistry.*

system for extracting detergent from effluent and concentrating the organic phase by evaporation before further analysis. Effluent mixed with magnesium sulphate is fed into an extraction vessel via a solenoid valve from a sample-manifold, and 50-ml aliquots of chloroform are added from an automatic pipette and the mixture is stirred slowly with a Teflon paddle. For rapid and efficient separation, the dimensions and shape of the extraction vessel and Teflon paddle should be closely related to the chemistry of the particular system in use. For example, should rapid stirring give rise to emulsion formation and consequently prolong separation time on settling, the fault in such a system must be in the design of the vessel and paddle and adjustments will need to be made before really successful automation is achieved.

After stirring, the mixture is allowed to settle and the organic phase drained from the vessel through a solenoid valve using a pair of phase-sensing electrodes as a means of control. In an environment of aqueous ionic solution a current passes between the electrodes; with a change of phase from aqueous to organic the current stops, thus switching the solenoid valve off and on. A gravity-feed, coupled to solenoid valves controlled by phase-sensing electrodes, is used to permit the organic phase to proceed through the apparatus for further treatment in vessels *A* and *B*, *viz.* being washed with acid in vessel *A* and with alkali in vessel *B*. Three separate aliquots of chloroform are used in the extraction procedure.

The organic extract is continuously collected and at a predetermined time in the extraction cycle is evaporated to leave the chloroform-extracted detergents in a pear-shaped flask below the heating mantle. The contents of this flask are dissolved in 0·5 ml of chloroform and subsequently analysed by thin-layer chromatography. A simple cam-timer operating a series of microswitches is used as the sequence control. Manual analysis of the aqueous effluents following automated solvent extraction showed results in the region of 95% extraction efficiency in the automatic procedure described above.

10.1.4.2 *Separation of Non-ionic Detergents from Waters by Continuous Extraction*[13]

As an alternative to the phase-sensing electrodes, an approach using a phase-separator was evaluated. A schematic diagram of a system designed to tackle the problem described in section 10.1.4.1 is shown in Fig. 10.5. An effluent is fed into the lower section of a double vessel from a manifold similar to that described above; the sample is then pumped from this vessel by one head of a Hughes piston-pump. Sodium hydroxide is pumped from a reservoir, via a second head of the piston-pump, so that

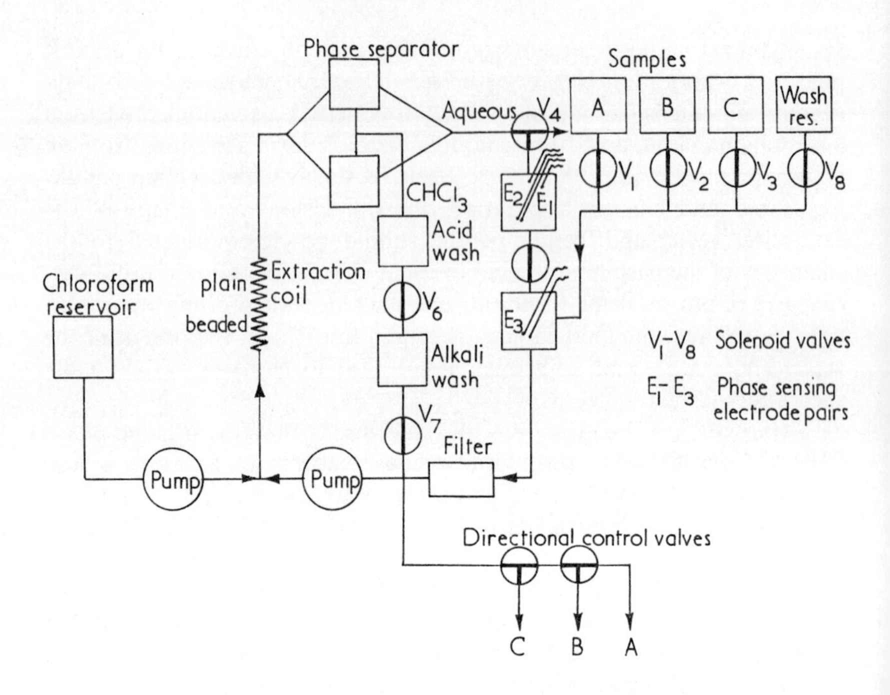

Fig. 10.5 Flow diagram for continuous extraction of non-ionic detergent in water. *Reproduced with permission from Sawyer et al.*[13] *and The Society of Analytical Chemistry.*

the two streams converge at a T-junction at the base of a mixing coil partially filled with glass beads. Forcing the two streams through the beads produces an intimate mixing of the phases. In the upper, unfilled, section of the coil, separation of the two phases begins and separation is completed in the double-weir separator mounted on the top of the coil. The phase separator diverts the organic phase through a series of wash-vessels and into a solvent-evaporator as described previously. Aqueous phase is returned through a two-way solenoid valve to the double-delay vessel for further extraction. A series of phase-sensing electrodes controlling solenoid valves, mounted as shown in the diagram, and programmed by a stepping uniselector, is used to control the sequence of operation. This allows the system to be independent of time and controlled simply by the flow of liquids involved. After three completed extractions the exhausted effluent is diverted to waste. A separation efficiency of 95%, judged as previously by manual analysis of the extracted effluent, was achieved by using this system. An important design consideration in both systems so far described is the construction of the valves and other components such as the pistons of the pumps. Some components were extracted by chloroform from the 'Viton' valve-seats used originally but this was successfully avoided by the use of valves with synthetic ruby seats. The separated extraneous components interfere drastically with the final thin-layer determination. Similar extraction problems were experienced with the piston-pump design and this necessitated the use of ceramic plungers. Both systems described above have been constructed in prototype form and each has a number of points to commend it. Final decisions as to the acceptability of either approach rests almost entirely on the chemistry involved in any particular analysis system.

10.1.4.3 *Extraction of Quinizarin from Hydrocarbon Oils*[14]

A complete analytical procedure that has found use as a routine procedure in the authors' laboratory is a combination of the procedures described in sections 10.1.4.1 and 10.1.4.2, which provides a double-extraction system. Such a system has been specifically designed to extract, identify and estimate the amount of quinizarin in a sample of gas-oil. The original manual methods involve the following steps:

1. to 25 ml of gas-oil add 5 ml of n-butanol;
2. extract with sodium hydroxide;
3. acidify with hydrochloric acid;
4. extract with cyclohexane;
5. measure spectrophotometrically in the region 440–660 nm.

Fig. 10.6 shows the schematic flow-diagram for a system designed to automate the complete manual analysis. Samples with n-butanol added are pumped by one piston-head of Hughes Piston pump, mixed with a stream of sodium hydroxide from a second head and forced through a mixing coil containing one short section of glass ballotini to provide an intimate mixture of the gas-oil phase and the sodium hydroxide. The quinizarin forms an unstable sodium salt soluble in the aqueous alkaline phase which separates partially in the plain section of the coil and is finely completely separated in the double-weir phase-separator itself. The phase-separator returns the gas-oil for further extraction whilst the aqueous phase is passed into the extraction chamber. Before the alkaline extraction an organic solvent and sufficient aliquots of dilute acid to acidify the aqueous extract from this primary extraction, are dispensed into the extraction chamber by automatic dispensing syringes. The quinizarin is immediately liberated as the alkaline phase is neutralized by the acid and is redissolved in the organic phase. The mixture in this extraction vessel is continuously stirred until completion of the extraction. The probe is removed from the sample and the exhausted gas-oil diverted to waste by

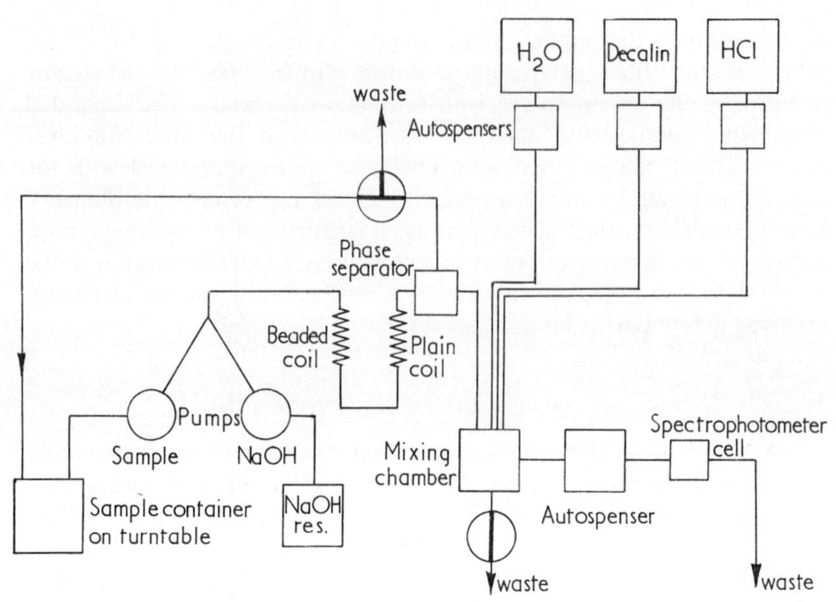

Fig. 10.6 Flow diagram of solvent extraction system for extraction, identification and estimation of quinizarin in gas oil.
Reproduced with permission from Tucker et al.[14] and The Society of Analytical Chemistry.

Fig. 10.7 Photograph of instrument developed in the Authors' laboratory for the automatic determination of quinizarin in gas oil samples.
Reproduced with permission of the Government Chemist.

use of a pinch-clip valve to control the gas-oil flow-path. Each successive sample is then flushed from the extraction coil by continuing the pumping of sodium hydroxide for a further short period. After a final vigorous stirring, the contents of the extraction chamber are allowed to settle and the aqueous phase is discarded by means of a pinch-clip valve controlled by a pair of phase-sensing electrodes. The organic phase can then be sampled by a dispensing syringe into a spectrophotometer equipped with a flow-through cell. The quinizarin extract is then analysed by two successive scans at different attenuations over the range 400–600 nm and by reference to suitable standard solutions. A routine instrument, shown in Fig. 10.7 and incorporating these principles of solvent extraction, has been operated successfully in the authors' laboratory for a number of years and apart from the obvious easing of work-load upon staff a further advantage is obtained. The sodium salt of quinizarin is unstable, so the manual method must rely upon considerable dexterity on the part of the operator, protection from sunlight and rapidity of operation for its success. Almost instantaneous liberation of the quinizarin by continual feeding into the acid and organic solution in the automated system eliminates the problems associated with the manual procedure. Direct automation of the manual method was not possible since the accepted solvent, cyclohexane, has a particularly low flash-point and presents an explosion hazard, particularly in an enclosed system. Simple substitution of decalin (flash-point 100° C) for cyclohexane removed any potential hazard.

10.1.5 Conclusions

Little attention has been paid in the chemical literature to the technique of automatic solvent-extraction, despite the wealth of information available concerning complexing agents. Some of the systems have been discussed in the previous sections and are all capable of development as analytical systems for many applications. Reproducibility of the extraction efficiency is dependent on the chemistry of the system and on the precision in dispensing the reagents. A dispensing system based on timed flow through a precision-bore capillary tube for example, will dispense 10 ml with a precision of $\pm 2\%$[1]. Increased precision can be achieved by using dispensing syringes and enhanced control of the extraction can be attained by the addition of buffered solutions to regulate the pH. (A pH electrode can easily be incorporated into the vessels used for solvent extraction.) The centrifugal separator does present a number of superficial advantages in that it provides facilities for precipitation and/or filtration. Unfortunately the apparent advantages are offset by considerable disadvantages. The interface tends to be unstable, and also hydrophilic interfaces are embarrassed by small quantities of water dissolved in organic phases, and these objections necessitate the inclusion of a regenerating process into the operating programme, to restore the qualities of the interface. Volatile solvents are also prone to evaporate when centrifuged and this must either be prevented entirely or at least taken into account in the overall analytical system. Such restrictions greatly outweigh the useful applications of this device.

With a certain ingenuity on the part of the systems analyst the systems described can be adapted to meet specific analytical requirements. The main criterion for designing any such system is that it should be kept as simple as possible while still achieving the objectives set out for the analysis.

10.2 **Automatic Distillation Techniques**

10.2.1 Distillation Columns

The control of laboratory-scale, pilot and large industrial distillation columns has been the subject of many papers in the literature. Reference is made in this section to the automatic control of distillation columns for short periods of up to a week. The dynamic equilibrium that exists in the distillation column between the liquid entering or leaving it is the overriding control parameter. Three basic methods are used to control the operation of a distillation column: (*i*) optimization of the concentration of a particular component in the column, (*ii*) maintenance of a constant composition that returns to the column (this may be assessed in a number of ways, *viz.* relative to the liquid distilled, the liquid exhausted or the liquid feed), and (*iii*) control of the correlation between boiler heater-load and condenser cooling-load. Complete systems have been developed along these lines, using all or some of these techniques, and these often include completely fail-safe devices[15, 16, 17]. The first of these[15] controls the vapour flow by controlling the pressure at the base of the column, the reflux rate and the volume of the fractions taken. The design incorporates up to 16 data-input channels for system monitoring. The second system[16] incorporates a novel device which volumetrically measures the cuts into self-capping sample-tubes and also controls the temperature of the column. Control of the still-pot heat-input, heat-loss of the condensing column and adjustment of reflux ratio are described in the third system[17]. Numerous safety measures such as identification of cooling-water failure and detection of fires are included in these devices. An alternative approach is to semi-automate the apparatus so that at various points in the operational procedure an alarm attracts the operator's attention. A device has been described by Perry[18] that uses these procedures. The system performs distillation starting from cold. Stable conditions are attained and a stable rate of reflux maintained. Set-points can be recognized or imposed on the column. At cut-points the operator is signalled and the system placed on to stable total reflux control so that the operator can take the cut and reset for the next fraction. The system uses a modular approach and can be used directly in association with standard instrumentation with little operator difficulty.

There are numerous commercially available systems for automatic control of distillation columns, designed primarily as a clean-up procedure for single or multiple fractions; one such system is described in more detail below. The example referred to is shown in Fig. 10.8 and commonly known as the spinning band system; it was designed by Abegg[19]. Separation of liquid mixtures in small-diameter capillary rectifying columns is extremely effective if low through-rates are maintained[20], but the limitation on yield/time factor has restricted the usefulness of such systems. In wider-bore columns the separation efficiency is reduced considerably but by the introduction of mechanical mixing of the vaporized liquid the exchange efficiency can be increased. Lesesne and Lochte[21] used a mechanical strip inside a column of slightly wider bore, and rotating it at 1000–2000 rpm to give the mixing effect, thus allowing vacuum distillation to be carried out efficiently for small quantities of liquid (i.e. between 5 and 50 ml). In contrast the Abegg column (Fig. 10.8) consists of a glass tube inside a silvered Dewar flask, the lower end of the tube being mounted on a distillation flask (13) and the upper end attached to a distillation head incorporating a pendulum (3). Above the head are two separate and independent condensers. At the top end of the uppermost condenser is a suitably housed permanent magnet to which the spinning band is firmly attached. This permanent magnet is rotated by a second magnet attached to the rotor of an a.c. motor. The distillation flask and heater are also insulated in a Dewar vessel.

In operation the condensate refluxes over the glass pendulum and back into the column. An electromagnet controlled by a timer is positioned opposite the pendulum; at the set-point the pendulum is activated, diverting the whole of the distillate into the take-off unit. The required reflux ratio can be preset on the timer and the distillate flows into the collection-vessel through a multipurpose stopcock. This allows the following functions: (*i*) the distillate can be collected at any pressure, (*ii*) the receiver vessel is vented whilst changing over fractions stored under the vacuum with column pressure maintained, (*iii*) the collecting chamber can be pre-evacuated to the pressure set on the manostat, without interfering with the column pressure. A variable heater-control is attached to the jacket of

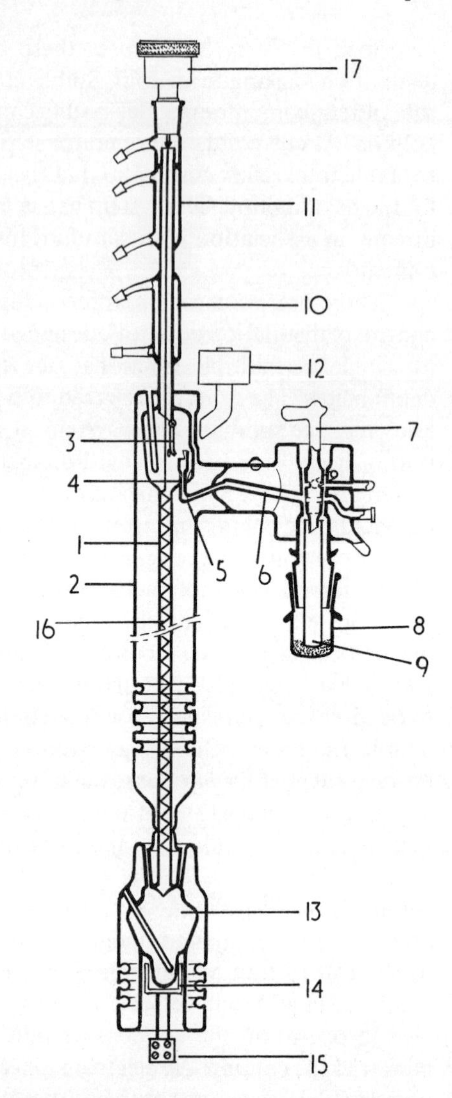

1 Distillation column
2 High vacuum silvered vessel
3 Take-off pendulum
4 Distillate collecting-funnel
5 Vapour trap syphon
6 Distillate outlet
7 Multipurpose stopcock
8 Receiver vessel
9 Collecting tube
10 Lower condenser
11 Top condenser
12 Electromagnet
13 Distillation flask
14 Heater
15 Electrical connection
16 Spinning band
17 Magnetic clutch

Fig. 10.8 Spinning band distillation column designed by Abegg[19].
Reproduced with permission from Orme Scientific Ltd.

the column in order to minimize heat-losses. Semi-automatic control of the column is facilitated by independent timer and heater circuits; operator-alarm signals are easily incorporated to allow easy change-over to another fraction. The system is particularly applicable for the separation of trace impurities from samples of 5–50 ml, either under vacuum or at atmospheric pressure, and also serves as a purification stage for ultrapure fractions as standards in gas chromatography. In the authors' laboratory a similar spinning-band system has found application as the final clean-up procedure for food extracts suspected of containing nitrosamines, before their examination by mass spectrometry[22].

Systems of the type set out above are not readily adapted for sequential sampling of a series of routine samples. Very little effort has in fact been devoted to the use of discrete distillation techniques in automated analysis. Continuous-flow systems discussed in the following section have many interacting variables to control, and it is therefore difficult to effectively improve the precision that they can achieve. Discrete distillation systems in which the precision of each section can be separately assessed could offer a more precise approach than continuous-flow systems.

10.2.2 Flash Distillation in Flow Systems

In the field of continuous sequential analysis, distillation techniques have found numerous applications and many distillation devices have been designed and described in the literature. Mandl *et al.*[23] in an effort to determine the quantity of fluoride in the air and in plant tissues used a standard Technicon 'Digestor' in their preliminary experiments. Encouraged by their preliminary results they designed a microdistillation device, shown in Fig. 10.9, constructed from Teflon tubing mounted on a stainless-steel former (*l*). The distillation coil is placed in a Pyrex glass vessel (*g*) filled with Primol oil (*h*) the temperature of which is maintained at 170° C by an immersion heater controlled by a thermoregulator, and stirring is provided magnetically (*p*). In the particular application considered, sample, sulphuric acid and air enter the distillation coil through inlets (*a*), (*b*) and (*c*). The air-stream carries the acidified sample rapidly through the coil and to the microdistillation columns (*m*). Fluoride and water vapour distilled from the sample are carried out into the microcondenser (*n*) connected to a trap (*o*) where condensate is trapped before resampling to the manifold (*e*); excess of air and uncondensed vapour are removed by a vacuum (*f*). Involatile materials, i.e. acids and ashed solids, are removed to waste by the peristaltic pump from the bottom of the distillation column, cooled by a cold finger. The purified sample-stream is then analysed by conventional 'AutoAnalyzer' techniques involving a colorimetric end-method. The distillation unit shows considerable improvement in precision of results when compared with the original method using the standard digestor module, it provides more rapid analysis and also offers considerable versatility.

Keay and Menagé[24] used a similar approach to that described by Mandl for the determination of ammonia and nitrate in soil-extracts,

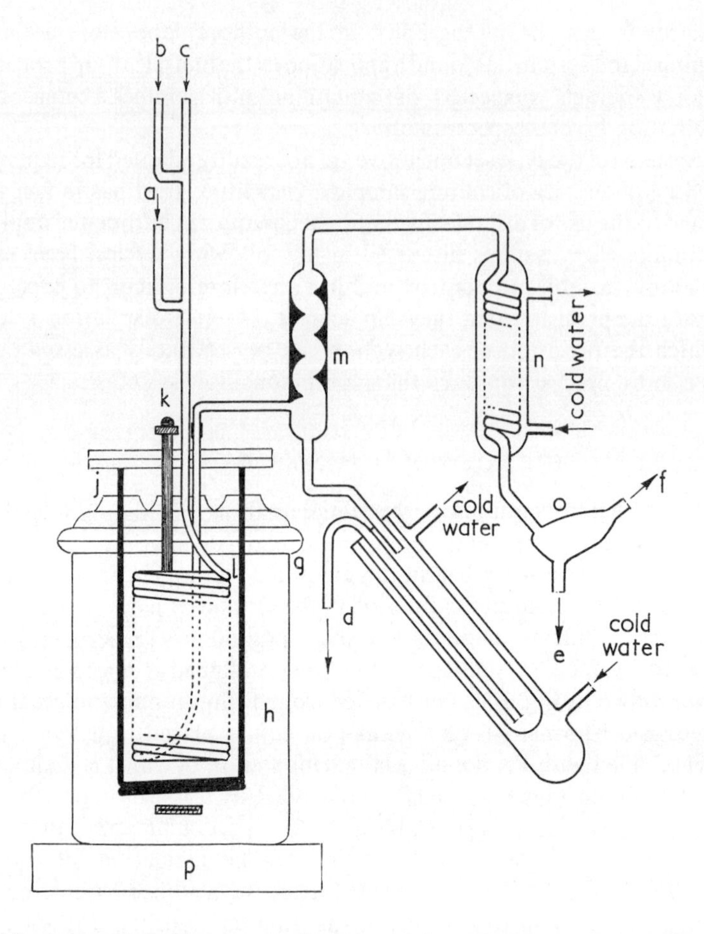

Fig. 10.9 Automatic distillation system for determination of fluoride in air and plant tissues. (Characters explained in text.)
Reproduced with permission from Mandl et al.[23].

involving automatic distillation. A standard Technicon oil-bath was replaced by one with a more flexible design. In the complete distillation assembly, sample is mixed with 10% w/v sodium hydroxide solution and air is added and the mixture pumped through the distillation coil maintained at 116° C in the bath. As the stream emerges from the unit a jet of air is injected to force the hot mixture into a series of splash-heads, trapping the involatile components and pumping them to waste. Steam and ammonia vapour pass through the head to a condenser where a stream of hydrochloric acid is introduced. The condensate is resampled and again analysed by conventional 'AutoAnalyzer' techniques.

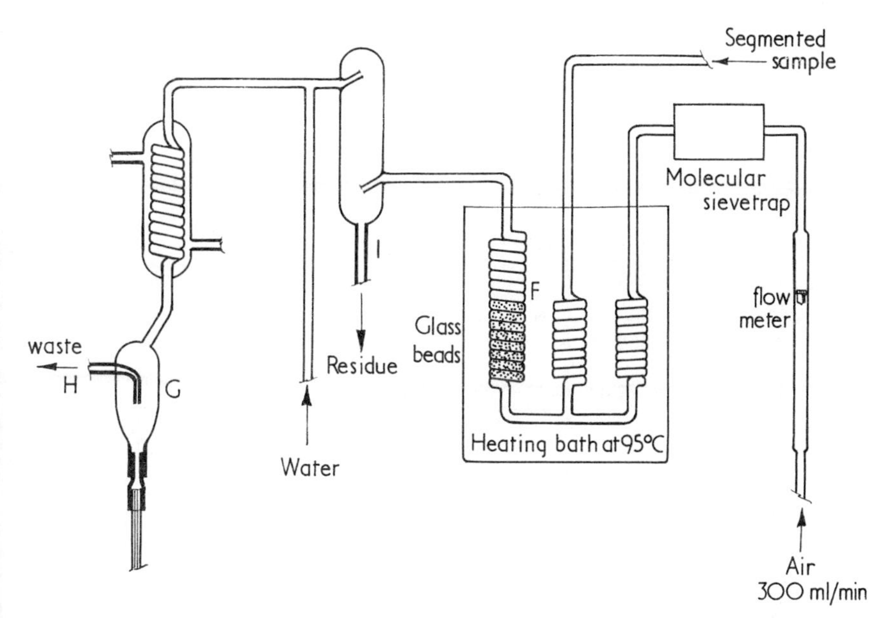

Fig. 10.10*a* Original distillation system for the analysis of beer. (Characters explained in text.)
Reproduced with permission from Sawyer and Dixon[26] *and The Society of Analytical Chemistry.*

Duncombe and Shaw[25] designed an even more flexible distillation unit for determining the concentration of the lower aldehydes and ketones present in experimental cultures of micro-organisms. A similar device, described below, has also been used by Sawyer and Dixon[26] in the analysis of samples of beer and stouts for alcohol and acid content. Further improvements to the design of the distillation assembly enabled considerable improvements to be made to the precision of results, which eventually enabled the beer-system to be modified for analysis of wines. A comparison between the devices described by Duncombe and Shaw and by Lidzey, Sawyer and Stockwell[27] highlights the critical parameters involved in the development of successful distillation techniques.

The main source of variation between replicate analysis of samples, particularly beers and wines, has been found to be directly related to the performance of the distillation device. Figs. 10.10*a* and 10.10*b* show the original distillation system and that developed later. In both of these systems and in the others previously described for continuous distillation, complete volatilization and quantitative recovery are not achieved. The distillation units under consideration for alcohol determination employ a flash distillation from an oil-bath at 95° C, involving the following stages:

(*i*) A stream of air-segmented sample is introduced by a proportioning

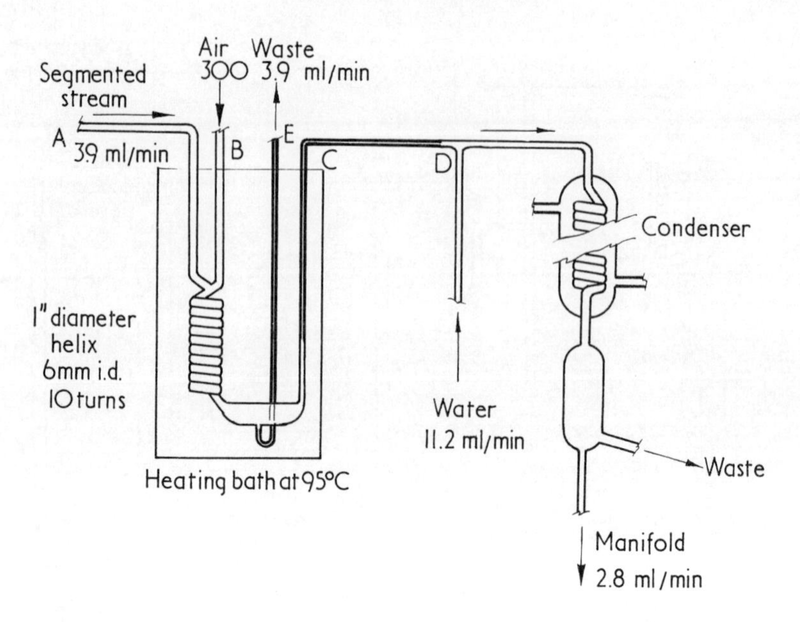

Fig. 10.10*b* Improved distillation system for analysis of beer and wines. (Characters explained in text.)
Reproduced with permission from Lidzey et al.[27] *and United Trade Press Ltd.*

pump into a glass coil immersed in a thermostatically controlled oil-bath.
(*ii*) Sample is vaporized at a nominally constant rate, with air as the separating medium.
(*iii*) The vapour stream is separated from the unevaporated material and washed with water.
(*iv*) The dissolved alcohol is cooled by the condensers and trapped for subsequent resampling to the manifold and hence to a dichromate colorimetric determination.

A critical evaluation of the original unit revealed a number of possible sources of surging of the liquid streams at the inlet and outlet from the unit, resulting in considerable intersample variations. The resistance caused by the glass-bead coil (*F*) and the liquid trap (*G*) requires the air-segmenting stream to be forced against a high back-pressure. The values of these resistances vary depending on the state of the devices, that is to say whether they are full of liquid or not. Air-bubbles in the segmented sample-stream expand rapidly on entering the heated zone and this again produces surging in the liquid-flow lines. These fluctuations in both air- and liquid-lines result in variations in the rate of distillation of any particular sample. Furthermore the separator (*I*) in which the undistilled liquid is separated from the vapour is situated outside the heater-bath and suffers considerable fluctuations of temperature despite close lagging with polystyrene. Similar separators were described by Sorris *et al.*[28] and Keay and Menagé[24], but tests with a thermocouple probe indicated that the vapour phase is still subject to sufficient temperature change to affect separation efficiency. The distillation-unit shown in Fig. 10.10*b* was specifically designed to overcome these disadvantages.

In operation, the air-segmented stream enters the top of the distillation coil at (*A*) and the coil contains no glass beads to restrict the flow through it. At the point of entry the air-bubble used to segment the stream is immediately released and the liquid meets the carrier gas entering at (*B*) and flows in a thin continuous film down the lower surface of a ten-turn vertical coil with air flowing over it. Alcohol and some water are vaporized and transferred to the gas stream. Observations with the unit immersed in a water-bath fitted with observation panels show that the low surface-tension of the solution at elevated temperatures, coupled to the high rate of evaporation, prevents the formation of froth or bubbles. At the bottom of the coil the residual unevaporated liquid is pumped to waste through a vertical tube (*E*) and the carrier-gas, containing water and alcohol vapours, emerges from the coil through the capillary (*C*), to a T-junction (*D*) where wash-water is introduced. The stream of gases and water passes through the vertical condenser and the condensate is collected in a specially designed trap with a wide exit to minimize the back-pressure in the system. A pool of liquid is maintained at a constant level in the trap and is continuously flushed by fresh condensate to ensure accurate representation of each sample in the distillate. The number and dimensions of the helices have been determined experimentally and a ten-turn unit produced approximately 60% distillation efficiency and good sample-discrimination. Further increase in the distillation efficiency serves only to lose sample-discrimination. This unit, used in conjunction with a standard Technicon manifold, showed considerable improvements on the original design. Still further enhancement of performance was obtained by modification of the heating-bath control to provide more stable operation[29]. The device provides rapid and efficient stirring and uses a low thermal-capacity heater controlled by a thermistor-sensor coupled to a latching relay. At equilibrium the unit has a heating-cycle time of only 10 sec and is stable to $\pm 0{\cdot}02°$ C for periods of several days.

A critical evaluation of the distillation-unit was able to prove that attention to minor details in design could effectively improve the accuracy and precision of an automatic system. Coupled to this it is essential to be completely aware of the 'chemistry' of the analytical system. The analysis of beers and wines is an example where this has proved important. Direct comparison of results of beer samples by the automatic distillation and the

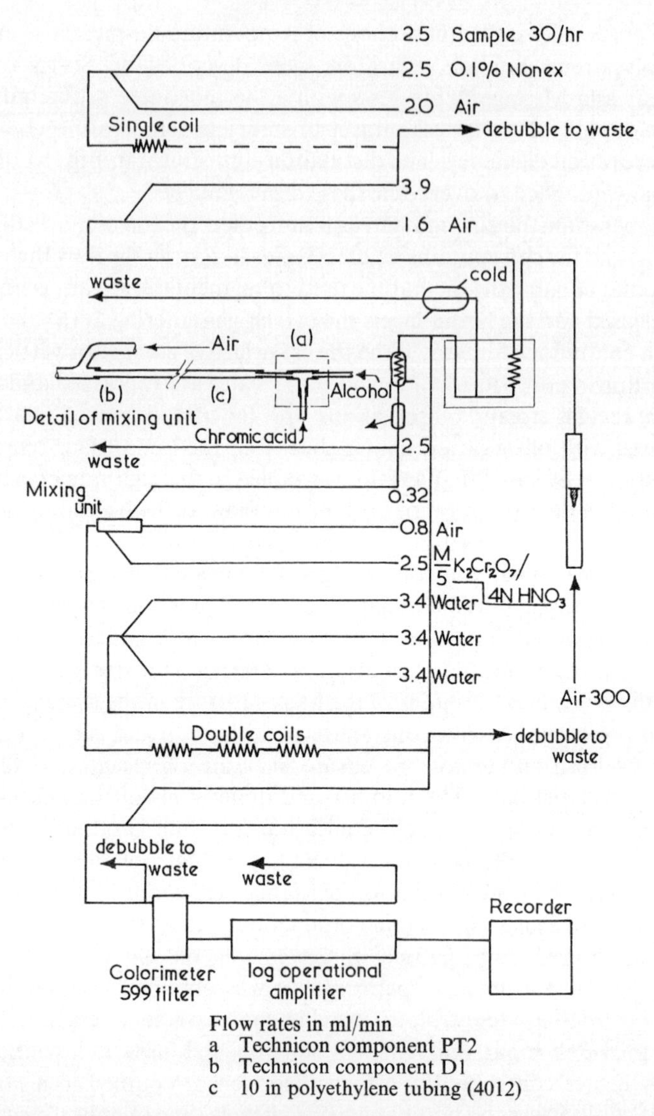

Fig. 10.11 Manifold arrangement for the analysis of beer samples for alcohol content. *Reproduced with permission from Lidzey et al.*[27] *and United Trade Press Ltd.*

statutory manual method showed a number of unexplained discrepancies, whilst results for alcohol/water standards treated by these methods were in good agreement. Fig. 10.11 shows the manifold for automatic analysis of beers. The problem was produced by the proteinaceous material present in the beer and could be reproduced by the addition of 0·5% albumin or malt extract to the standards, whilst glycerin, agar, lactose or other sugars did not modify the results. Addition of 0·1% detergent to the primary diluent stream before its entry into the distillate was sufficient to swamp variations in protein content and results obtained by using this modification showed close correlation with those obtained on beer samples by manual techniques. Table 10.1 shows comparative results by the automatic and manual techniques.

Table 10.1 COMPARISON OF ALCOHOL CONTENT OF BEERS BY MANUAL DISTILLATION AND BY THE AUTOMATIC SYSTEM[27]

Sample type	*Distillation strength* v/v	*Autoanalyser* v/v	
Light ale	2·80	2·84	2·82
Brown ale	2.53	2·59	2·58
Stout	4·55	4·58	4·53
Lager	4·20	4·23	4·24
Strong ale	8·55	8·41	8·43
Export ale	3·98	4·10	4·03
Continental ale	6·58	6·56	6·45
Continental lager	3·17	3·31	3·24
Strong ale	4·97	5·08	4·95
Stout	4·78	4·75	4·72
Strong ale	3.55	3·57	3.55

Further development of the alcohol method for the analysis of wines and fortified wines should present no difficulties, given a good understanding of the chemistry involved. Preliminary results with the beer manifold discussed above produced low and variable results because in addition to alcohol and water many wines contain highly significant quantities of sugars; often up to 30% w/v. Even with a high dilution with water before distillation the efficiency of this operation decreased with increasing

Table 10.2 ANALYSIS OF WINES BY AUTOMATIC DISTILLATION PROCEDURE, VARIATION IN SUGAR CONTENT SHOWN[27]

			Manual		*Autoanalyser*			
Sample Number	*Type*	*% Sucrose* w/v	*Distillation value % proof*		*% Proof mean*	*Number of tests*	*S.D.*	*S.D. overall*
1	Red Wine	8·3	24·93	24·95	25·09	13	0·094	
2	Vermouth	16·1	32·12	32·12	31·91	16	0·168	
3	Vermouth	15·8	39·87	39·97	39·67	14	0·132	0·119
			40·02					
4	White Wine	0·2	23·67	23·64	23·95	16	0·077	
			23·68	23·66				
5	Vermouth	13·1	34·68	34·63	34·38	14	0·078	
			34·59	34·59				
6	Port	10·7	34·0	34·02	33·90	14	0·131	
			33·98	33·95				

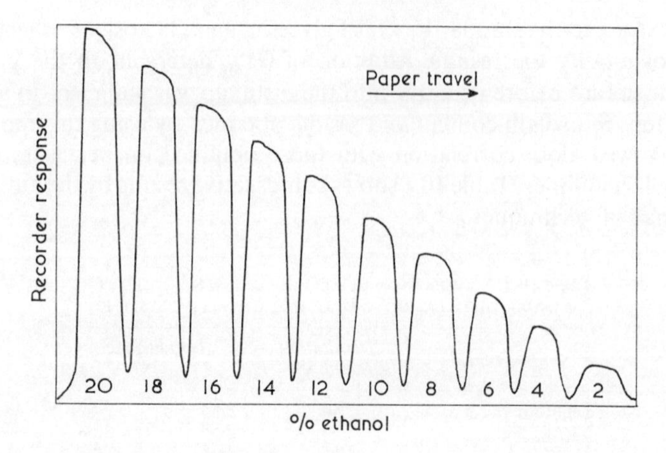

Fig. 10.12 Recorder trace using wine manifold for a range of standards covering the response 0–20% w/v ethanol.
Reproduced with permission from Lidzey et al.[27] *and United Trade Press Ltd.*

sugar concentration, because of the higher boiling point of the sample mixture. A *pro rata* response for alcohol with a wide range of sugar concentrations can be achieved by diluting the sample with a 2% sucrose solution before distillation. The addition of 2% ammonia solution also enhances the performance for two further reasons; it neutralizes any volatile acid present in the wine and produces improved flow characteristics in the distillation unit. Table 10.2 shows comparison of results for alcohol concentrations in wine for the automatic and manual methods and Fig. 10.12 shows a typical recorder response for the colorimetric system, the symmetry and shape of the peaks being further evidence of the controlled and precise analyses performed by the distillation system and manifold.

10.2.3 Conclusions

It is evident from an extensive survey of distillation techniques that a completely reliable and very precise distillation technique is not readily available from manufacturers. Most work on sequential analysis has been carried out with continuous systems and an attempt has been made in this section to highlight points which result in improved analytical results, notably the distillation unit design and chemistry of the system. In the flow-system there are many inter-relating variables and for increased precision discrete systems are likely to offer advantages. At this stage the research worker or routine analyst will have to provide considerable effort on his own account to obtain a working system. He should not be restricted by this consideration since it is evident that only with complete information about the chemistry can automation of the analytical problem be tackled successfully.

10.3 **Automatic Dialysis**

Dialysis, like the other techniques discussed in this chapter, is a separation technique used to provide interference-free analysis, particularly in 'AutoAnalyzer' work. In dialysis the sample-stream flows over a recipient stream and is separated from it by a semi-permeable membrane. The system is so configured that the materials of interest are transferred to the recipient stream by osmotic pressure, leaving interfering ions or molecules in the sample stream. Whether ions or molecules dialyse across the membrane is determined by the nature of the membrane (i.e. pore size and thickness) and the size and nature of the molecule or ions that are diffused through it. Any membrane that is chemically compatible can be used provided it allows the passage of substances to be analysed and eliminates the interferences. In most cases dialysis is used to separate ions and smaller organic molecules from large molecules such as serum proteins. The extent of dialysis efficiency is related to a number of factors; the residence-time in the dialyser module, the operating temperature, the area of contact, the sample concentration and the coefficient of dialysis, which in turn can be influenced by the state of the sample medium. Blood and urine samples often cause difficulties because they adhere to the transferring molecules or ions and prevent transfer into the recipient stream. For an analytical system involving dialysis it is important that samples and standards have the same coefficient of dialysis and for precise analysis the residence-time, the area of contact and the temperature must be maintained constant. Careful control of the flow into and out of the dialyser is also very important. Sample-condition also has a marked effect on analytical precision and the dialysis-rate is dependent on the pH of the solution, and the concentration and the ionic strength of the donor stream. Dialysis is an exponential process and it is almost impossible to bring the procedure even close to completion, especially in continuous-flow systems. As a direct consequence, control of the operating conditions must be stringently maintained for precise analytical results.

10.3.1 Dialyser Design

Many of the automatic analytical methods which involve dialysis are reported in the proceedings of the Technicon Symposia. Whitehead and

Ferrari[30] described a dialyser which was used in the quantitative determination of glucose-levels in blood serum. More precise and reliable results were achieved by the introduction of an air-bubble into the continuous stream before a diluent or reacting liquid was added to it.

The dialyser module is situated in a temperature-controlled water-bath with a round removable cover, which contains four sets of 'Kel-F' tapered nipples to accommodate the inlet and outlet streams from the dialyser-plate assembly. The plates of the dialyser fit over a central threading post, which in turn is attached to four posts on the cover. Glass mixing coils used for delaying and mixing purposes are also arranged to bring the sample and recipient streams to the bath-temperature (usually $37 \pm 0{\cdot}1^\circ$ C). Temperature-control of the bath is achieved by using an immersed stirrer to circulate the water around the assembly comprising the dialyser plate and a heater activated by a mercury contact-thermometer. Modifications to the control-circuitry such as incorporation of thermistor sensors would provide more precise and reliable long-term operation. The plate-assembly accommodates single or multiple sets of dialysis plates. Standard plates are constructed from 'Lucite' and channels are cut out in each plate so that these are mirror images of each other. The plates provide two congruent continuous helical channels which are separated by the semi-permeable membrane. A channel length of 87 in. provides a large area of contact with the membrane. Plates are always supplied as matched pairs and are not interchangeable. If they were not precisely matched, leakage, poor bubble-pattern and loss of dialysis area would result, giving rise to a considerable loss in efficiency. Plates can also be constructed from 'Kel-F', which provides a more chemically resistant system. Correct assembly of the dialysis-module and preparation of the membrane are of prime importance. The dialyser-plates must be thoroughly clean and care taken to avoid scratching or etching the surface. The membrane itself is conditioned in lukewarm water and then stretched over a pair of hoops to remove wrinkles. The membrane is positioned between the two plates and the assembly clamped to form a leak-proof seal, and the excess of membrane is trimmed. The sample and recipient streams are then connected to the upper and lower plates respectively and analyses can be carried out by using either counter-current or concurrent flows of sample and recipient streams. Operating procedures for the dialyser module can be found in the Technicon literature and supplementary information on trouble-shooting and operation can be found in a book by White, Erickson and Stevens[31], a practical manual on automatic analytical techniques. In the context of 'AutoAnalyzer' work concurrent analysis does seem to provide a more reliable approach and leads to less intersample contamination. Numerous methods of analysis, particularly in the field of the life-sciences, are illustrated in the proceedings of the Technicon Symposium from 1966 onwards. Short dialysis-path units have been used in the latest 'AutoAnalyzer' models; these do not produce such a high transfer-rate from sample to recipient streams, but are suitable for many applications.

Counter-current procedures theoretically produce a more efficient dialysis and the designs of many systems can be found in the literature[32–36] Heckly[32] developed a system for precipitation of the toxin of *pseudomonas pseudomallei* from large volumes of the culture filtrate, by counter-current dialysis. Davis[37] describes a dialysis unit, in essence a further development of a unit described in an earlier Swiss patent[38], and this offers more efficient operation than the systems previously described. The dialyser uses the

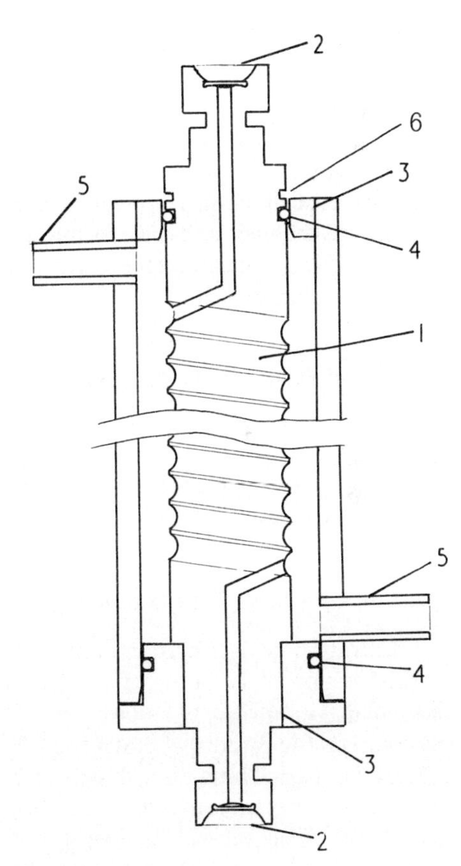

1 Core
2 Ball joints
3 Flanges cemented in place
4 O-rings
5 Diffusate entrance and exit
6 Groove for spring retaining clip

Fig. 10.13 Apparatus for counter current dialysis.
Reproduced with permission from Davis[37] and Academic Press Inc.

inherent efficiency of counter-current processes and allows diffused materials to be collected in the minimum volume and therefore the economic use of special or expensive diffusate solutions. Fig. 10.13 shows a schematic diagram of the dialysis unit constructed from acrylic plastic. The dialyser consists of a central core (*A*) with a helical groove, a regenerated cellulose membrane tightly covering the core, and an outer jacket. In operation with diffusable solids that increase the specific gravity of the solution, the sample stream is pumped upwards through the helical groove inside the membrane and a recipient stream pumped downwards in the outer jacket. O-ring seals and ball-joint connections make the unit simple and quick to dismantle and reassemble. Insertion of the membrane around the core requires particular care, the membrane being first expanded inside a glass tube of slightly larger bore than the core and then shrunk on to the core. Full details of the preparation of the membrane are given[37] and once the dialyser has been assembled, operation is simple.

Stepanishchev[39–41] described a continuous-action dialyser comprised of a number of discrete cells joined together. Each element is a shallow trough carrying the liquid and containing a cell having a meshed bottom provided with a membrane. The liquid to be purified moves through these cells counter-current to the solvent. In the event of membrane-breakdown, cells can be bypassed automatically. Baffles are included which divert the flow of liquid to take a zigzag path through the cells and increase the dialysis efficiency.

10.4 Precipitation and Filtration

Precipitators with applications ranging from microanalysis on the one hand to large scale industrial operation on the other are well established, yet in the field of automatic analyses the literature is sparse and no system enjoys universal acceptance.

10.4.1 Industrial Filtration Units

Filtration devices used in industrial applications are briefly described in this section since some, if not all, could offer possible solutions to particular situations by simple modifications. Rotating-drum filters, plate-filters, the Swedish Landskrona filter and magnetic tape feed filters may all be used. These filter designs are discussed by Bornefalk[42] as well as the critical factors involved in determining their performance. Zievers and Riley[43] also review extensively the automation of filtration equipment. Nickolaus and Dahlstrom[44] discuss the theory and practice of continuous filtration at high pressure where the filtration process becomes more economic, particularly when compressed air or gas is used for agitation during the process.

Automatic filtration processes can broadly be classified in two groups, those designed to recover the filtrate and those to recover the residue. Demeter[45] described a device for filtering liquids and regenerating the filtered material continuously and simultaneously by passing the liquid to be filtered horizontally through several vertical walls made up of filtering-grains depositing continuously. The filtering-grains are returned to the top of the apparatus by compressed air, where the material filtered out is separated from the contaminants before reuse. Moon and Hawkins[46] used a batch system which involved (*i*) the deposition of a filter bed, (*ii*) the filtering of a slurry (particularly of solutions of 1-olefin polymers containing particles of the solid catalyst), (*iii*) washing the filter bed, and (*iv*) removing the filter cake complete with residue before commencing the complete process again.

Zapan[47] described a process which uses the pressure of a hydrostatic column to force the liquid upwards through a filter. At preset intervals the filter cake can be removed, either by dropping under its own weight or by being blown down by compressed air or gas, and the residue recovered periodically. Lopker and Stoltz[48] describe a further continuous process which enables only the well-washed pure portion of a precipitate to be removed whilst the impure inner section is recycled for further purification, thus improving the purity of the desired fraction.

10.4.2 Discrete Analytical Filtration Systems

For quantitative analytical work on the laboratory scale there are few examples of analytical filtration systems. One is a simple modification of the solvent extraction principle described by Vallis[4], section 10.1. Precipitation reactions can be handled by this device provided that either only the supernatant liquid is required or that if the precipitate is required it is readily soluble in a suitable solvent. An alternative approach which readily lends itself to automation is to use a microcentrifuge, an example of which is shown in Fig. 10.14. It consists essentially of a Teflon centrifuge cup mounted on a rotor, which rotates at a fixed speed, and two glass feed-tubes held above the cup. Wash liquid can be fed into the cup through the arm (*a*) and supernatant liquid and precipitate can be extracted through the tube (*b*) which moves in a fixed horizontal plane. Two possible exit paths are available: (*i*) to recover the supernatant liquid and (*ii*) to throw the wash-liquid and precipitate to waste.

The microcentrifuge operates in the following manner. Samples for precipitation are fed into the cup; at the start of a cycle the effluent tube

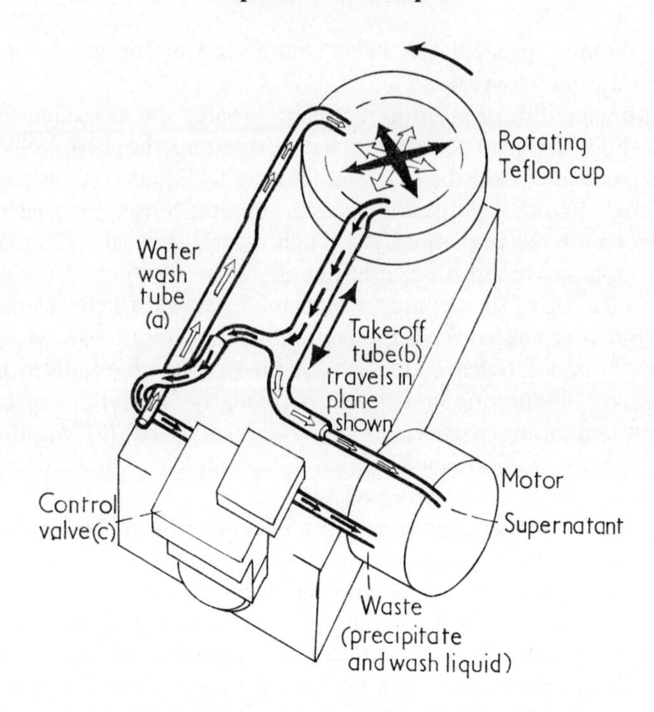

Fig. 10.14 Microcentrifuge.
Reproduced with permission from James A. Jobling and Co. Ltd.

(*b*) is positioned at the centre of the cup. As the rotor is activated the exit-tube moves slowly to the periphery of the cup and supernatant liquid is ejected by centrifugal force and collected for further treatment. At a preset distance from the outside of the cup wash-liquid can be added from a vacuum pump; valve (*c*) is opened and this operation allows the precipitate and wash-liquid to be thrown to waste. The particular unit shown is controlled from a simple cam-timer operating microswitches and is primarily intended as a mechanized aid in precipitation methods. It is a simple and economic device which is easy to incorporate into custom-built total automatic systems where precipitation reactions cannot be avoided.

10.4.3 Continuous Analytical Filtration Systems

For continuous-flow analysis there are two approaches; either a porous filter plug may be placed at a strategic point in the flow-lines, or a Technicon continuous filter, described below, may be used.

The flowing liquid stream from a conventional manifold is dropped

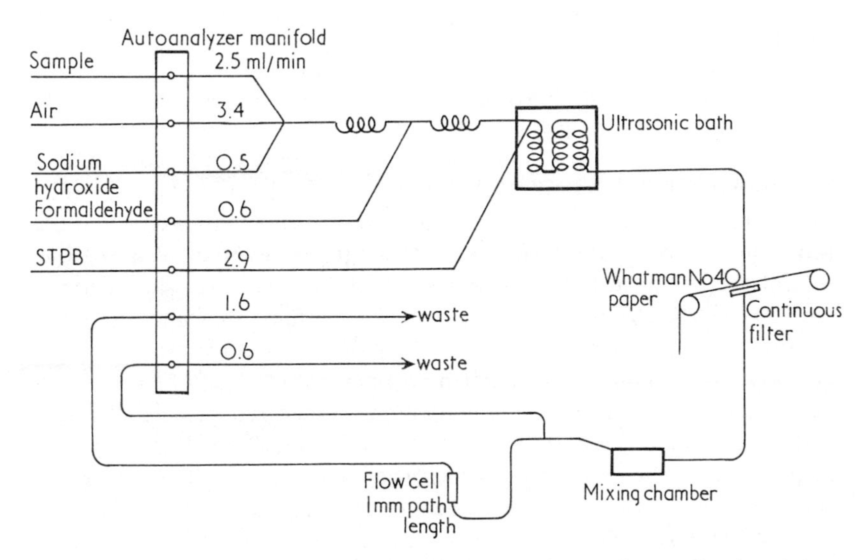

Fig. 10.15 Determination of potassium in fertilizers using continuous filtration. (Flow rates in ml/min.) STPB = Sodium tetraphenyeborate.
Reproduced with permission from Skinner and Docherty[49] and Microforms International Marketing Corporation.

on to a belt of filter paper continuously removing the precipitate. The filtered liquid is then recycled to the pump from the underside of the filter. The system is only suitable for aqueous solutions, since volatile solvents will tend to evaporate, giving variable results. A typical manifold utilizing a continuous filter is shown in Fig. 10.15. This has been described by Skinner and Docherty[49] for the analysis of fertilizers to determine the potassium content. Other systems using similar procedures are well documented in the Technicon series of publications.

Where quantitative results are of importance, filtration is not a suitable procedure to adopt. The system-designer should avoid precipitation reactions whenever possible, either by suitable adjustment of the chemistry of the method or by use of an alternative separation technique such as solvent extraction.

10.5 Automated Digestion

To eliminate the tediousness and possible errors associated with the manual preparation of samples by oxidative digestion for subsequent inorganic chemical analysis two automatic approaches are worthy of discussion; (*i*) a discrete mechanical system, and (*ii*) a continuous analytical approach based on a Technicon 'Digestor' module.

10.5.1 Discrete Automated Digestion

The discrete system described by Hunt[50] automatically performs the sequence of adding reagents, heating the sample and cooling, necessary for digestion. A continuous chain-drive activated by a simple motor moves in a horizontal plane about two chain-wheels, forming two long parallel sides. From each link of the chain a rimmed test-tube is hinged so that the tubes are heated as they pass over the strip-heater mounted along one side of the chain. In the particular application described, plant-samples (approximately 100 mg) are placed in the sample-tubes and sulphuric acid is dispensed into these from a screw-driven syringe and hydrogen peroxide is also added from a solenoid-operated syringe. The sequential operations are controlled by a simple cam-timer and associated relays and microswitches. During the first cycle of operation sulphuric acid is dispensed into each tube from the appropriate dispenser. The heater and cam-timer are activated and the sequence proceeds as follows: eight tubes pass on to the strip heater and hydrogen peroxide is dispensed into them. The tubes are heated for 3 min, after which a further eight tubes are treated in a similar manner. Six passes of the heating and peroxide treatment are required for complete digestion. On completion of the digestion procedure the sample requires only the addition of a fixed volume of water before conventional analysis. The system shows a significant saving in staff-time but is available only in a prototype form.

10.5.2 Continuous Automated Digestion

In the field of continuous-flow analysis the 'AutoAnalyzer' digestor introduced by Ferrari[51] in 1960 has found many widely differing applications. Samples are mixed with the digestion-acid mixtures and then transferred into a rotating helically-grooved glass cylinder, supported horizontally over a bank of three heaters. Sample materials and digestion fluid introduced at one end of the helix are spirally converged to the exit at the other end, the fluid travel speed being governed by an adjustable rotation speed. Faster speeds of rotation allow shorter periods for digestion. The digestion acid is continually transferred to the helix, thus cleaning the helix between samples and preserving the sample identity.

In the preliminary heater-zone the sample/digestion mixture is subjected to a higher temperature than in the other zones, causing water to flash-vaporize from the sample. In the second zone the temperature is lower and oxidation is effected, and in the final zone a diluting water-stream is added to the sample before resampling of the acidic mixture back through the manifold for analysis. Digestion is rapid, showing a considerable time-saving over the manual methods. Sample throughputs of 20/hr are not uncommon with such systems. The reasons for the rapid digestion have been examined by a number of workers[52–54] and are basically that (*a*) the temperature of digestion is higher (since it is no longer limited by the boiling-point of the acid mixture as in the manual procedure), (*b*) a large excess of acid is used, (*c*) there is a thin-film effect whereby breakdown occurs in the layer travelling over the surface of the helix, which is rapidly removed from the source of heat, and (*d*) the products of the digestion procedure, such as CO_2 and H_2O, are continually removed from the system immediately they are formed. Complete digestion cannot often be obtained, but wherever standards of a similar nature to the samples are available this does not limit the field of application. The digestor has been used in the automation of the Kjeldahl nitrogen procedure to digest solid food materials before their examination for the presence of trace metals and pesticide residues. It is important that for a wide range of sample-types the digestion conditions should remain consistent from one sample-type to another. In the authors' laboratory the temperature-controls of the standard digestor have been improved to provide stable heating for up to 10 hr. A prototype automatic digestion system is being evaluated for use in the determination of trace element levels in a variety of foods. The aim is to prepare discrete samples of chelate compounds of the elements following the automatic neutralization of the acid digest by gaseous ammonia; the system includes discrete chelation and solvent-extraction modules.

The 'AutoAnalyzer' digestor system has been used to determine chemical oxygen demand[55] and protein-bound iodine[56] and has been modified for other applications such as distillation[57], solvent extraction[58] and photolysis[59]. It is simple to operate but requires meticulous attention to detail and must be continuously monitored. The helix must not be allowed to run dry when the heaters are on, and the heaters must not be activated without liquid present in the system. Oxidative procedures using an acid mixture produce noxious fumes, and therefore adequate extraction hoods must be provided. The main limitations for digestion procedures are the availability of suitable standards and control of conditions so that for a range of samples of differing type the digestion efficiency is constant. Further practical hints on the operation of the digestor can be found in the Technicon 'AutoAnalyzer' hand-book and in the continuing series of Technicon Symposium reports.

References

1. Trowell, F. *Lab. Practice*, 1969, **18**, 44.
2. Trowell, F. *U.K. Pat. Application*, 17329/67 (1967).
3. Porter, D. G. Private Communication.
4. Vallis, D. G. *U.K. Pat. Application*, 14964/67 (1967).
5. Joyce Loebl 'Centrichem'. Princesway, Team Valley, Gateshead.
6. Steed, K. C. *U.K. Pat. Application*, 1491/67 (1967).
7. Carter, J. M. and Nickless, G. *Analyst*, 1970, **95**, 148.
8. Butler, T. J. *Am. J. Clin. Path.*, 1964, **41**, 663.
9. Signer, R. *Swiss Pat. Application*, 295,646 (1954).
10. Anon. *French Pat.*, 983,981 (1951).
11. Anon. *French Pat.*, 986,639 (1951).
12. Sawyer, R. Stockwell, P. B. and Tucker, K. B. E. *Analyst*, 1970, **95**, 284
13. Sawyer, R., Stockwell, P. B. and Tucker, K. B. E. *Analyst*, 1970, **95**, 879.
14. Tucker, K. B. E., Sawyer, R. and Stockwell, P. B. *Analyst*, 1970, **95**, 730.
15. Huppes, N. and de Jong, J. J. *Chim. Anal. Paris*, 1959, **41**, 436.
16. Galstaun, L. S., Harrison, R. D., Keever, E. R. and Bissol, L. *Am. Chem., Soc., Div. Petroleum. Chem. Symposium*, 1953, **27**, 59.
17. Brandt, H. R. and Langers, F. *Chem.-Ing. Tech.*, 1957, **29**, 86.
18. Perry, J. A. *Am. Chem. Soc., Div. Petrol. Chem., Preprints*, 1964, **9**, 207.
19. Abegg Column (available U.K. Orme Scientific Ltd.).
20. Kuhn, W. and Ryffel, K. *Helv. Chim. Acta*, 1943, **26**, 1693.
21. Lesesne, S. D. and Lochte, H. L. *Ind. Eng. Chem., Anal. Ed.*, 1938, **10**, 450.
22. Collis, C. H., Cook, P. J., Foreman, J. K. and Palframan, J. F. *J. Brit. Soc. of Gastroenterology*, 1971, **12**, 1015.
23. Mandl, R. H., Weinstein, L. H., Jacobson, J. S., McCune, D. C. and Hitchcock, A. E. *Proc. Technicon Symposium, New York, 1965*, 270.
24. Keay, J. and Menagé, P. M. A. *Analyst*, 1970, **95**, 379.
25. Duncombe, R. E. and Shaw, W. H. C. *Automation in Analytical Chemistry, 1966*, Vol. II, p. 15, Mediad Inc., New York, 1967.
26. Sawyer, R. and Dixon, E. J. *Analyst*, 1968, **93**, 680.
27. Lidzey, R. G., Sawyer, R. and Stockwell, P. B. *Lab. Practice*, 1971, **20**, 213.
28. Sorris, J., Morfaux, J. N., Dupy, P. and Hertzog, D. *Ind. Aliment. Agr.*, 1969, **86**, 1241.
29. Bunting, W., Lidzey, R. G., Porter, D.G. and Stockwell, P. B. *Lab. Practice*. 1974, **23**, 179.
30. Whitehead, E. C. and Ferrari, A. *U.S. Patent*, 2,899,280 (1959).
31. White, W. L., Erickson, M. M. and Stevens, S. C. *Practical Automation for the Clinical Laboratory*, Mosby, St. Louis, 1968.
32. Heckly, R. J. *Biochim. et Biophys, Acta*, 1959, **35**, 548.
33. Sastry, S. L. *Current Sci. (India)*, 1949, **18**, 305.
34. Signer, R., Hanni, H., Koestler, W., Rottenberg, W. and von Tavel, P. *Helv. Chim. Acta*, 1946, **29**, 1984.
35. Stauffer, R. E. *Technique of Organic Chemistry*, Vol. III, Part I, 2nd Ed. (ed. A. Weissberger), p. 65, Interscience, New York, 1956.
36. Von Tavel, P. *Helv. Chim. Acta*, 1947, **30**, 334.
37. Davis, J. G. *Anal. Biochem.*, 1966, **15**, 180.
38. Zyma, S. A. *Swiss Pat.*, 291,180 (1953).

39. Stepanishchev, K. P. *U.S.S.R. Patent*, 115,411 (1958).
40. Stepanishchev, K. P. *U.S.S.R. Patent*, 104,861 (1957).
41. Stepanishchev, K. P. *Spirt. Prom.*, 1962, **28**, 22.
42. Bornefalk, J. O. *Tek. Tidskr.*, 1960, **90**, 833.
43. Zievers, J. F. and Riley, C. W. *Chem. Eng. Progr.*, 1958, **54**, 53.
44. Nickolaus, N. and Dahlstrom, D. A. *Chem. Eng. Progr.*, 1956, **52**, 87M.
45. Demeter, L. *Hungarian Patent*, 154,412 (1966).
46. Moon, J. J. and Hawkins, H. M. *U.S. Patent*, 2,993,599 (1961).
47. Zapan, M. *Rev. Chim.* (*Bucharest*), 1956, **7**, 54.
48. Lopker, E. B. and Stoltz, E. M. *U.S. Patent*, 2,839,194 (1958).
49. Skinner, J. M. and Docherty, A. C. *Talanta*, 1967, **14**, 1393.
50. Hunt, J. *Chem. Ind. London*, 1971, 676.
51. Ferrari, A. *Ann. N.Y. Acad. Sci.*, 1960, **87**, 792.
52. Ferrari, A., Catanzaro, E. and Russo-Alesi, F. *Ann. N.Y. Acad. Sci.*, 1965, **130**, 602.
53. Marten, J. F. and Catanzaro, G. *Analyst*, 1966, **91**, 42.
54. Marten, J. F. and Ferrari, A. *Technicon Symp., London*, 1963, 20.
55. Adelman, M. H. *Technicon Symp., New York, 1966*, p. 552.
56. Gambino, S. R., Schreiber, H. and Covolo, G. *Technicon Symp., New York and London, 1965*, p. 363.
57. Gunther, F. A. and Ott, D. E. *Analyst*, 1966, **91**, 475.
58. Mislan, J. P. and Elchuk, S. *Technicon Symp., New York, 1967*, p. 329.
59. Dowd, N. E., Killard, A. M., Pazdera, H. S. and Ferrari, A. *Ann. N.Y. Acad. Sci.*, 1965, **130**, 558.

Chapter **11**

The Role of Digital Computers in Analytical Chemistry

11.1 Introduction

The use of computers for complete control and more simple data-acquisition and reduction will be discussed in this chapter. Only passing reference is made to the use of analogue machines, which have found only limited use in analytical chemistry because of their lack of flexibility, accuracy, storage-capacity and programming ability compared with digital machines. It is important to emphasize that digitization of the signal cannot lead to a more accurate signal than the original. Significant losses in signal can occur if the rate of digitization is not rapid enough and added to this, high-frequency noise may show up strongly with a high speed analogue/digital converter, which was not apparent in the original analogue signal. This necessitates smoothing the data either by software or by hardware. Instruments not specifically designed with computer data-processing in mind often exhibit a high degree of noise and are difficult to incorporate into computer systems. Despite these restrictions the overall flexibility of digital systems has found numerous applications in the fields of analytical chemistry and it has been predicted that by the end of the 1970s all analytical laboratories will include an 'in-house' computer system[1].

11.2 The Digital Computer

Fig. 11.1 shows a schematic block-diagram of a basic configuration of a digital computer which can be considered as four separate units, the memory, the central processing unit (CPU), the arithmetic registers and the input/output devices. Memory is a component capable of storing many thousands of binary-coded bits of information, each package of which is comprised of a number of binary bits and is called a word of information. Each word is identified by a specific address associated with it and the

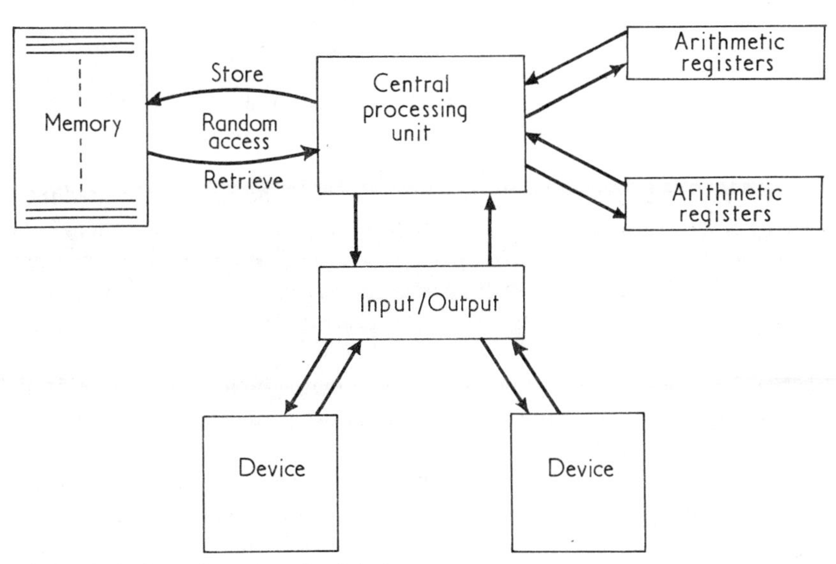

Fig. 11.1 Basic configuration of a digital computer.
Reproduced with permission from Perone[2] and Preston Technical Abstracts Company.

word of information is stored and retrieved by this identity. Speed of transfer into and out of memory is termed the 'memory cycle time'. The working unit of the computer, the CPU, is made up of a series of logic circuits which execute the logical and arithmetic operations which allow the computer to handle complex mathematical and data-processing functions. With the appropriate interfacing of the CPU and other devices a computer can be used to control experiments, acquire data and to calculate and print-out results on the appropriate peripheral device. A sequence of instructions, termed a 'program' (stored as a binary-coded pattern in the memory), controls the computer; the program fetches an instruction from the memory into the CPU, executes it and then proceeds sequentially to the next instruction unless programmed to branch to another instruction. The arithmetic unit is a set of high-speed registers (accumulators) comprising a set of electronic two-state devices which are used to accumulate intermediate results of binary arithmetic. Binary information can easily be transferred to and from the memory and automatic registers by execution of the appropriate programmed machine instructions. The input and output devices allow communication with the CPU in the form of card-readers, paper-tape readers or visual display units (VDU).

11.3 Application of the Digital Computer in Analytical Chemistry

The simplest and most common use of a computer is as a replacement for a desk calculator in long and complex or repetitive calculations. Initially data are transferred manually from a recorder chart by punching on to cards or paper tape; subsequently the data can be incorporated into a computer and analysed or manipulated in the required manner. This procedure is, however, open to operator error and when data are produced in great quantities it is advantageous to present the experimental data automatically in a computer-compatible form, for example by coupling the recording device to an analogue/digital converter connected to a paper-tape punch. Coupling a display device such as a cathode-ray oscilloscope to the computer facilitates on-line manual interaction in processing a set of data.

Complicated calculations which were either impossible or extremely lengthy and laborious before the introduction of high-speed computers are now practical and economically viable. Mathematical models which were often restricted to simple linear functions when manual calculations were used can be made more sophisticated and can include complex functions which can be varied according to the experiment and not limited by the calculations involved. Similarly, mathematical treatment can be used to analyse results from numerous techniques, for example the Newton Raphson[3] method for curve-fitting can be applied to (*a*) ultra-violet spectra[4], (*b*) electron spin resonance[5], (*c*) Mössbauer spectra[6], and (*d*) analysis of experimental kinetic data for a series of consecutive chemical reactions[7].

The most significant development in laboratory automation is the use of the programmable digital computer system which introduces samples, accumulates data and analyses these data, the results being used by the computer to control the laboratory experiment. Results are produced and reports generated automatically and complete files of information can be compiled for future interrogation. This advance has been greatly facilitated by the development of small, powerful and simple-to-operate computers at a relatively low cost, which has in turn resulted in an increasing acceptance of computers in laboratories. Computers are incorporated in analytical instruments, and dedicated to analyse data from a number of similar instruments simultaneously and to produce report sheets of results. A number of systems are currently utilized for nuclear magnetic resonance (nmr) techniques, mass spectrometers, infra-red and ultra-violet spectrometers and electron spin resonance spectrometers. These systems eliminate the tedium of repetitive work, avoid fatigue-induced errors and greatly increase the analytical productivity. In nmr the use of the computer decreases the time necessary for a manual analysis and also extends the range of application of the techniques. The full potential of the combined technique of gas chromatography and mass spectrometry can only be realized by the development of a fully computerized system.

The successful introduction of automatic systems into hospitals and clinics has created an avalanche of data ideally suited for computer manipulation. Many instruments have been successfully introduced which handle routine examinations of blood and serum samples by computer systems. Often the analytical results are made available directly to the particular wards or doctors requesting the tests, but by coupling a small computer to a larger system comprehensive patient-records can be maintained, making patient-histories freely available, on request, to the doctors. By using a store of case-histories it is also possible to compare the symptoms of a particular patient with the medical histories of similar patients to provide a diagnosis. A computer not only allows the clinical laboratory to provide an adequate quality-control system for its analyses but also permits the detection of trends amongst patients. For example, Whitehead[8] as early as 1965 showed that serum potassium levels in men were significantly higher than in women and also that up to the age of menopause women's blood-urea levels are significantly lower than men's, but reach a similar level thereafter.

Further computer applications are possible, for example using remote sensor-systems coupled to a computer to monitor parameters such as a patient's respiration, temperature, blood pressure and electrocardiac activity. Integration of this information with the results of chemical analyses performed in the clinical laboratory will provide the medical researcher with a vast store of information and a complete patient-profile of bodily functions. The use of computers and automation in clinical laboratories has been surveyed by Robinson in a useful monograph[9].

Problems of pollution are currently receiving close attention both academically and politically, and environmental monitoring of a region or area is greatly facilitated by the use of computers. Changes induced by different influences can be rationalized and predictions can therefore be made on the basis of known information. Air-pollution also can be studied by using an instrument specifically designed for the purpose and coupling to a computer allows the degree of pollution and the main constituents to be determined. Correlations between different sampling areas are facilitated and when integrated with standard meteorological information allow the effect of atmospheric pollution on the weather conditions to be investigated.

The X-ray crystallographer was quick to appreciate the uses of a large computer and the calculations have been greatly eased, and the limiting factor in crystallography now involves the quality of the measurements themselves. Improvements can be made by computer-controlled applications of Lorentz and polarization corrections[10], whilst spot-shape, non-linear response and correction for background response can be accomplished by programming. Computer-control of diffractometers enables more precise angular settings to be made. Coupling with an automatic plotting device enables stereoscopic drawings of the structure of the crystals to be produced which show the individual components in their correct spatial relationship.

A few applications of computers have been considered in this section; particular aspects associated with analytical chemistry are further discussed in later sections of this chapter.

11.4 Factors Involved in the Choice of Computer Systems

A laboratory computer system should ideally be chosen with due regard to the role of the laboratory and also to the temperament and skills of the scientists employed by the organization. As far as possible the computer should provide all the facilities required by the scientists, who should be able to feel that they each have the computer at their fingertips. It is possible to purchase from manufacturers a package-deal of hardware and software designed to meet a specific analytical requirement, for example automatic gas chromatography, nuclear magnetic resonance spectrometry, mass spectrometry; such systems, often termed 'turnkey' systems, whilst offering an immediate solution to a particular problem, are inflexible and often not suitable for further 'in house' modification. At this point it is worth discussing the particular skills required in the staff needed to develop and operate a laboratory computer system.

Professional computer programmers are often involved either with programs in high-level languages for mathematical and computational problems or the development of assemblers and compilers. It is extremely difficult for these programmers to write adequate programs for analytical chemistry without at least a flow-chart since they are generally unfamiliar with the problems of analytical instrumentation or chemistry. It is also difficult for computer professionals to communicate with the analytical chemist unless the latter understands the computer jargon. If the scientist is *au fait* with computer usage he will be able to prepare a flow-sheet. In this it is almost inevitable that some function, however small, may be omitted or ill-defined. If a professional programmer designs the program for the computer the scientist will have difficulty in correcting the badly defined function. The software is the most difficult and time-consuming aspect of laboratory automation and for it to be accomplished efficiently the 'motivated' scientist should construct his own programs for the computer.

For example, in the technique of gas chromatography, many programming hours have been spent designing software for dealing with peaks eluted on a solvent tail and unresolved peaks. On the one hand the analyst is unaware of the limitations of the computer and on the other the programmer is unaware of the ability of the analyst to correct for the problems discussed above by relatively simple modifications to his analytical techniques. An analyst with knowledge both of computing and gas chromatography should be able to design the most cost-effective combination of hardware and software effort for these problems.

Automation by computer can take three general courses; (*a*) data-acquisition, (*b*) data-acquisition and on-line processing, and (*c*) on-line data-acquisition and processing coupled with control. With careful programming the use of computerized data-processing leads to greater accuracy and far greater versatility in the overall treatment of results, the accuracy being limited only by instrumentation and the subsequent digitization of the signals; no further loss of accuracy is induced by computerization.

11.5 Off-Line Use of Computers

Data-acquisition, i.e. off-line data-processing can be performed in a number of ways. Information from the analogue devices is usually digitized and presented in a computer-readable form, commonly on paper tape or magnetic tape, but also less frequently on cards. At some later stage the data can be processed in a batch-mode by a computer, which may either be 'in house' or else remote from the analytical devices. Use of a terminal to a large bureau-operated machine offers considerable advantages for a small user; it avoids the major capital expense of a computer, allows particular software packages to be used and also gives access to a far greater computer power than could otherwise be justified economically. Off-line systems, by their method of operation, have one inherent advantage, in that if the computer itself fails, instrumental data can still be produced and processed at a later data. These systems are ideal when extremely lengthy calculations are involved, for example inversion of large matrices, regression fits of mass-spectral data, or the solution of Fourier transforms. Off-line systems offer an introduction to computer techniques and can provide valuable experience before the implementation of an 'in house' computer-system. The availability of off-line processing time is an extremely valuable asset even in on-line control and data-acquisition systems because of the back-up facilities it provides in event of computer-failure. In many instances, for example, for high-precision work in gas chromatography, off-line processing has been found to be more accurate and reliable than systems utilizing on-line processing via a multiplexing system[11]. Advances in design of multiplexers will probably remove this anomaly in the near future. A further advantage is that with the utilization of the large computer, complete programs can readily be written in a high-level language. Fortran and Algol are the most popular.

When a terminal is not available or where the central computer is some distance from the instruments, transport difficulties may arise and

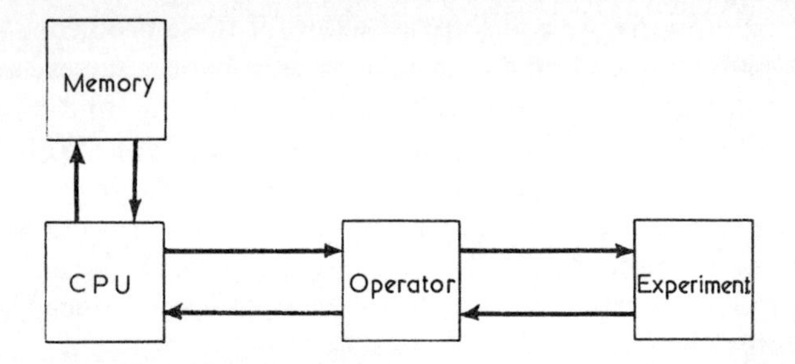

Fig. 11.2 Off-line computer configuration.
Reproduced with permission from Perone[2] and Preston Technical Abstracts Company.

the total turn-around time of a job may become excessive. However, it is impossible to generalize on what constitutes an excessive time delay; this can only be determined by the precise priorities imposed by the work of the laboratory involved. A schematic diagram, Fig. 11.2, shows an off-line system and illustrates the essential role of manual intervention.

11.6 On-Line Use of Computers

There is, however, a real need for versatility of data-acquisition and of control which can easily be achieved by using on-line computers. This is particularly important whenever instantaneous results are required from the instrument either for control of the instruments or for fast reporting. Fig. 11.3 shows schematically the concept of on-line computing; the manual transfer system discussed in the previous section is replaced by an electronic interface, which includes control logic and electronic elements and provides timing synchronization and conversion modules which translate the analogue data into computer-compatible form. Signals from instruments can either be transmitted to the computer

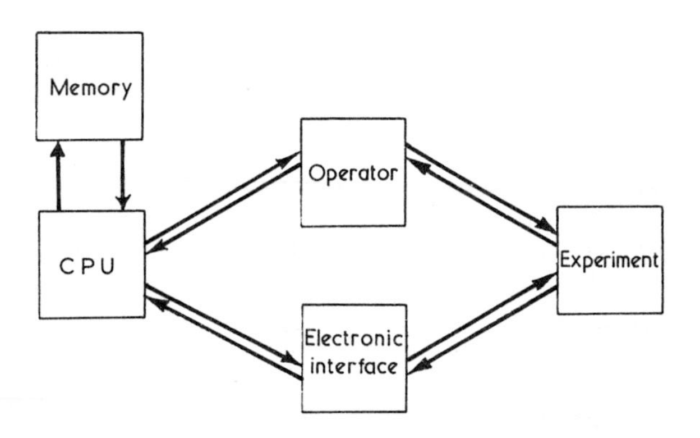

Fig. 11.3 On-line computer configuration.
Reproduced with permission from Perone[2] *and Preston Technical Abstracts Company.*

in the original analogue form and then digitized, or digitized before transmission. The method preferred is generally decided by the rate of sampling required for any particular application, but is often influenced by the precise timetable of development of an on-line system. Where digital voltmeters and integrators are already included in an instrument it is often worthwhile to retain them and transmit the digital signal. Measurements from the instruments are made under direct computer-control or supervision. The program for data-processing is either stored directly in the memory or is directly readable into memory to perform the computation quickly. Results are produced on a teletype, line-printer, or any other suitable output device. Electromechanical or electronic valves or similar devices which can be activated by a voltage-change directly under program control are used to control experimental conditions.

In an on-line computer configuration, a compatible interface takes the role of the manual data collection. This suffers none of the inherent faults of manual operation, accepts data at high speeds and offers the possibility of real-time control. The total resources of the computer's logic and decision-making capabilities can be used for control of a wide range of experiments. Real-time interaction coupled with fast calculating power enables the computer to optimize the experimental conditions during the run. On-line use of the computer also removes the difficulties imposed by off-line techniques where the limiting factor is the speed at which data tapes or cards can be read into the computer before processing.

There are two approaches to on-line computing; (*i*) each experiment has a single small dedicated computer (costing up to £10,000) or (*ii*) several instruments are linked by a time-sharing system (costing between

£100,000 and £150,000). The advantages and disadvantages of both of these approaches are discussed by Perone[12] and are set out in Table 11.1. The importance placed on each of the items shown is open to a considerable discussion and varies with each particular laboratory.

Table 11.1 COMPARISON OF SMALL DEDICATED OPERATION WITH TIME SHARING

Dedicated small computer operation	*Time-shared systems*
Advantages	*Advantages*
I User in complete control of CPU	I Can justify large system
II All CPU power at user's disposal	II Can justify professional staff
III CPU can be shared sequentially	III High-level languages possible for on-line work
IV Small initial investment	IV Efficient use of computer time
V Can justify multiple systems	
Disadvantages	*Disadvantages*
I May be 'computer-bound'	I High initial investment
II Hard to justify large money on expensive peripherals	II Must have professional staff
III Much user programming/ interfacing effort	III CPU's power diluted by time-share 'overhead'
IV Machine language necessity	IV User's applications must fit system
	V Inadvertment user-user interaction
	VI Reliance on single system
	VII User isolated from computer

Time-sharing systems are best implemented where a number of routine control and data-acquisition functions are required; the sampling of the analogue or digital information is usually performed sequentially through a multiplexer. Systems utilizing analogue transmission of signals provide an obvious advantage in that only one analogue-to-digital device is required for the complete system. Provided instruments require similar data-capture rates, sequential sampling in this manner is perfectly adequate, for example, a number of gas chromatographs operating various analyses. Where entirely different data-rates are required the approach can lead to poor control and even loss of the required data. In such instances it is necessary to provide a 'Software Executive Monitor' to handle each instrument function synchronously. This is more expensive, but infinitely more flexible. An ideal time-sharing system should not inhibit any individual operator, and operations on other instruments should not influence him in any way; this is difficult to achieve, since noise generated from one instrument or interface can change a control-function or affect data from other sources. The advantage of increased flexibility and power provided by a high-level language compiler must be offset by the effort and expense of designing a well-organized time-sharing system. (An operating system is often called the Executive.)

Stand-alone systems use smaller computers dedicated to a single task and within the limitations of the power of the computer the instrument user has a complete freedom to experiment, and is not troubled by the intricacies of an Executive, timing errors, or interaction with other user programs. The precise format of his particular program can be tailored specifically to his instrument and its analyses. Results can be serviced as and when required and are not held back by other instruments. At present it is less expensive for a laboratory with a wide range and number of instruments to purchase a time-sharing system; with rapidly changing methods of manufacture and pricing trends this cost-advantage may not always remain.

An obvious conclusion from this is that in future small dedicated computers or specifically designed hardware systems will be used to control and operate individual instruments, whilst these in turn will communicate with a larger computer. The larger machine would be used to perform major computational functions, and offer book-keeping and report-writing facilities. Such an arrangement involves a hierarchical system of computers. Margoshes[12] has examined the benefits both in performance and overall economics of building the computer as part of the instrument, particularly for infra-red transform and nuclear magnetic resonance spectrometers. Computerization is the only practical method to record Fourier transform spectrograms but two further advantages over conventional spectrometers are gained; first the detector senses signals from all wavelengths simultaneously and secondly the design is simpler and individual measurements can be made more rapidly. These effects enable a complete transform spectrometer to record a spectrum in a much shorter time than a conventional instrument. Alternatively use of signal-averaging gives an improved signal-to-noise performance in a given analysis time and hence more precise spectra. A further application of Fourier transform pulsed nmr is also discussed. This technique, coupled to signal-averaging techniques, further extends the scope of nmr such that ^{13}C spectra can be recorded for samples which have not been specifically enriched in the isotope. Using normal as opposed to pulsed technique applications on ^{13}C would require almost continuous operation of the machine for periods of up to a year. This restriction rules out such applications. Margoshes further shows that despite the higher initial expense involved in instruments with dedicated computers the greater throughput achieved reduces the cost per analysis to a more competitive level.

When considering automation and the installation of a computer system a laboratory is faced with the many alternatives discussed above. It may be considered that it is too expensive to use computers, but if computers are justified then three options are available, (*a*) to install small dedicated computers, (*b*) to purchase a turnkey system (accepting the limitations imposed by it or undertaking to modify it), or (*c*) to install a specifically designed computer tailored to the precise requirements of the laboratory. The last approach will be the most expensive, but by definition the precise needs of the laboratory and its staff are fully met.

Time-sharing systems or hybrid systems based on this principle will obviously fulfil requirements for many laboratories and a word of warning is given on the true installation and running costs. A case-history of the installation of a £50,000 computer installed for conventional business use is cited[13] and, although the use and design of the system differ from those considered in this text, various hints can be gained from the experience described in the article. The total overall costs in installation, computer-staff costs, maintenance and other associated expenses for this particular £50,000 computer amounted to £109,000. Considerable effort must be made, before the purchase of a system, in the design and proposed use of the system; too many 'white elephants' have been purchased owing to too little effort at this particular stage.

It is impossible in this chapter to discuss the complete range of computer systems designed and in laboratory use. A few have been chosen from various fields of instrumentation to provide an insight to the reader on the different approaches that have been used. Again, it must be emphasized that the precise computer requirements are governed by the terms of reference of the laboratory as much as by the skills of the staff employed there.

11.7 Laboratory Computer Systems

In this section it is not intended to provide a complete glossary of the computer systems used in analytical chemistry but a few examples have been taken from the range of analytical techniques in order fully to illustrate the possible approaches that are available to the analyst. Generally all the systems described make use of digital data-transmission of mV signals; each is described for a particular application, but their extended use to more general systems is possible. Although this book is not devoted to clinical chemistry as such, the field cannot be ignored because of the scope and volume of 'AutoAnalyzer' work performed in hospital laboratories, and therefore the use of computers in clinical chemistry is considered.

11.7.1 Computer Processing of Data from 'AutoAnalyzers' in Clinical Chemistry

Analysts engaged in work of this nature are strongly recommended to study the invaluable book *Clinical Chemistry and Automation*[9]. Whitby[14]

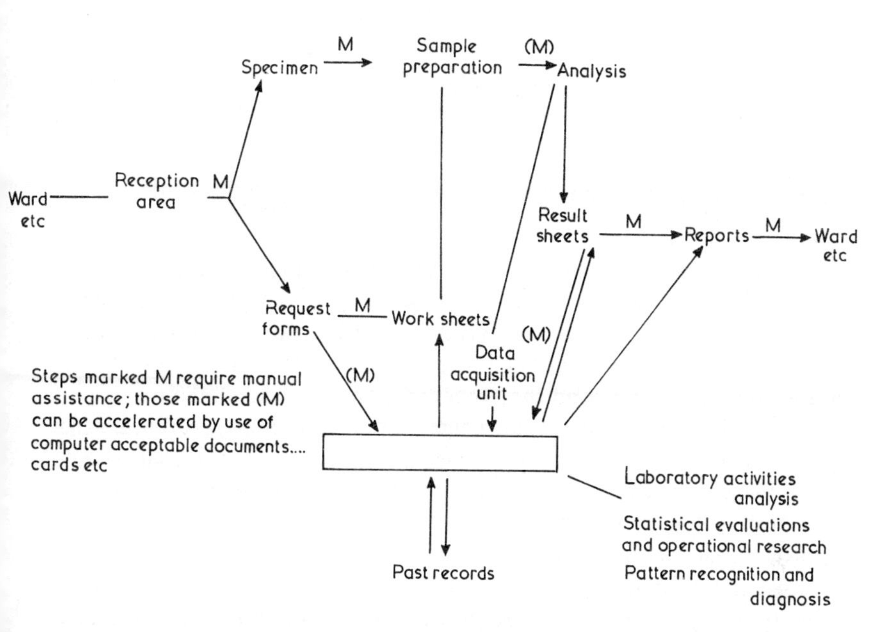

Fig. 11.4 The flow of information in a clinical chemistry laboratory showing the main points where a computer can interact and contribute to the performance of the various processes.
Reproduced with permission from Whitby[14] and United Trade Press Ltd.

has shown (Fig. 11.4) the work-flows in a clinical laboratory and the areas in which a computer can interact. Basically, particular analyses are requested and a work-sheet is prepared, and samples are then analysed along with a set of standard solutions of known concentration. A calibration curve and sample-results are calculated, and returned in the form of a report-sheet to the request-originator. Added to this it is also necessary to update existing patient-files from the new analytical results.

Raw signals can be collected on-line from 'AutoAnalyzers' by one of three methods; (*i*) multiplexing the low level signals coming directly from the detector modules (i.e. colorimeters, flame photometers or fluorimeters), (*ii*) multiplexing the output from individual amplifiers fitted to the detector modules, or (*iii*) multiplexing signals generated from a retransmitting slide-wire fitted to 'AutoAnalyzer' recorders.

Many computer systems devised for on-line monitoring of 'AutoAnalyzer' work have adopted the third approach; briefly stated, the reasons for such an approach are that (*a*) the differential signal from a sample and reference channel of the detecting colorimeter can easily be obtained, (*b*) the recorder damps the high-frequency noise associated with analogue signals, (*c*) retransmitting slide-wires provide signals of a similar level from each 'AutoAnalyzer' channel, and (*d*) recorders are habitually used for monitoring instruments in any case. Whitby, however, in his description of the Elliot ABL system, uses the second approach since he considers it better in principle to take signals directly from the colorimeter detector to the computer. A similar approach to this has been used by Stockwell *et al.*[15] to design a data-handling system for a two-column amino-acid analyser.

Each colorimeter detector in the Automated Biochemical Laboratory, ABL (Elliot 903) system is linked to an amplifier unit situated on the laboratory bench and the output is divided; one component is transmitted via a multiplexer to the analogue-to-digital convertor (ADC) unit situated in the computer room, and the other component is attenuated for display via a normal recorder amplifier which is retained for back-up and error diagnostics. The ADC can only cover a limited range of output signals, so it has been found necessary to decrease the output from the reference photocell below that value in conventional operation. An alternative approach adopted is to stabilize the power supply to the detectors and remove the reference photocell from the circuit[15]. Standard methods have been modified for on-line processing and the methods have been validated by chemical criteria[16]. The ABL system is expensive but the functional aims are extensive and include the computer-dependent preparation of work-sheets, the preparation of cumulative reports on individual patients, on-line monitoring of up to 31 discrete channels and an extensive and informative range of fault diagnostics. Provision of rapid-access storage-devices such as discs or drums and a high-speed output device may be uneconomic but is highly desirable for rapid searches of past records in answer to telephone inquiries and patient-surveys. Discs and drums form standard features of a larger computer installation; it is therefore appropriate for a small dedicated system in the clinical laboratory to be coupled to the larger computer to have direct access to file-manipulation techniques.

Calibration, calibration-drift and specimen-interaction have been discussed[17–19] in some depth by Owen *et al.* Third-order polynominals expressing assay value as a function of peak-height reading on a linear chart-scale are fitted to calibration data. Routine measurement of the exactness of fit is useful in detecting errors in calibration standards or in the calibration procedure. Full correction of calibration requires both appreciation of changes in base-line and of sensitivity. Also the influence that one specimen has on the following specimen depends on the sampling-rate and the sample-to-wash ratio. Carry-over from one specimen to the next is a constant fraction of its assay value and correction of the results for specimen-interaction can be achieved by the subtraction from each assay of a fraction of the previous sample assay-value. The correction procedures are best tackled by an on-line computer. Thiers *et al.*[20] consider that interaction is the principal factor limiting speed of continuous-flow analysis and have designed a program to compensate for interaction and ultimately increase the rate of analysis.

11.7.2 Systems Designed for Gas Chromatography

It has already been stated that a computer system should be tailored to meet the job-requirements of the laboratory and the level of staff involved. The system shown in Fig. 11.5 and described by Deans[21] typifies this requirement. The computer/gas-chromatograph system has been designed to carry out routine off-line analysis for process-control or other analyses where each chromatograph is required to apply several different analytical methods in a random order. Operator-error has been minimized by designing the system to be operated by unskilled process-workers as part of their job of plant-control. Further, the system has been

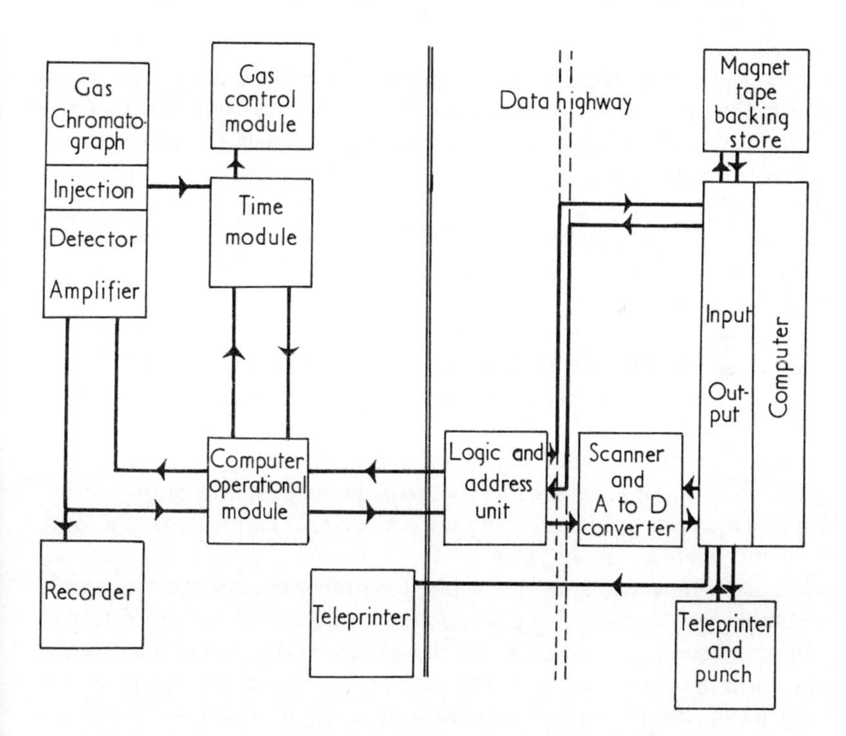

Fig. 11.5 Gas chromatograph/computer configuration.
Reproduced with permission from Deans[21] *and Institute of Petroleum, London.*

designed to function for all essential analyses in the event of the failure of any part of the system, including the computer. This has been achieved without expensive duplication of equipment.

In routine analyses, such as this system has been designed to handle, a considerable amount is known about the analysis; therefore gas chromatographic control-functions such as autozero, change of detector sensitivity, heart-cutting and back-flushing, etc., can be carried out purely as a function of time. Control could be obtained from the computer but interfaces between the computer and the individual items tend to be expensive and since analytical results are vital even in the event of computer-failure, individual timers are used on each chromatograph. A Crouzet timer operated by a plastic program card driven over a series of microswitches has been shown to provide adequate precision of control in this particular environment. The computer-operation module is a plug-in device which interfaces the semi-automatic chromatograph to the computer. In the event of a computer-failure the chromatographic operation remains unaffected. The module contains an amplifier to transmit the chromatograph signal at 0–10 V, an electronic autozero, and facilities to detect the start of analyses and for the selection of appropriate computer programs. For each chromatograph there is required a 'logic and address' unit which allows the operator to interact with the computer system and to select the appropriate computer program. Analogue signals from each of the 16 gas chromatographs are scanned by a multiplexer and fed to the computer system through a 10-bit ADC. The attenuation is controlled from the associated timer and for the analyses performed, all peaks are normally greater than 25% full-scale deflection so that the ADC produces adequate peak discrimination.

The 16,000-word core-store of the Elliot 905 computer is divided into two parts, the first 8000 (8K) words being used to store the main organization program and subroutines used by all gas chromatographs. Modular programming has been used to allow easy interchange and modification of routines. Four levels of priority are designated and the program is automatically called back to the first level of priority every tenth of a second to read the output from each gas chromatograph. More frequent reading can be allowed for particular applications but this reduces the time available for calculations; however, this is not normally a restriction. Program changes and standard data can be added to files whilst the computer is on-line to the gas chromatographs. The second 8K words of core-store are used to hold all the information and constants used in calculations applied to each chromatograph. Each chromatograph may have a number of methods related to it but only the data-card for the method in use is held in core; information for all other methods is stored on the 46K magnetic-tape backing-store and called into core on command from the appropriate program card inserted into the timer module. Calculation programs have been designed to give only the information required by the plant-worker but the more conventional gas chromatography programs based on slope-sensing and peak and trough selection can be used by laboratory staff to develop new methods.

The computer's calculation and comparison ability is used to save the shift-supervisor time and to improve decision-making. When all results are within specification and the gas chromatographs are working correctly, print-out is simple and used for record purposes only. When either the analytical results or the chromatographic parameters are outside limits a special print-out is used to draw the operator's attention to the fault and provide wherever possible specific instructions for immediate corrective action.

Many turnkey computer-systems are available for data-processing in gas-chromatography laboratories; one of the systems with wide applicability is discussed here. The basic concept of the system, the Pye Datacon DP90, is that true computer-compatability in analytical instrumentation implies a two-way conversation between computer and instrument. This allows both control of operation and data-processing of results, thereby achieving a cost-effective computerization of analytical instruments.

The standard configuration is based on a small real-time general-purpose digital computer with a 16K word (16-bit words) memory, and provides data-processing and control for up to 40 instruments on-line with up to 20 operating simultaneously. A unique 'Data Highway' principle is used to connect the computer to the chromatographs, which can be plugged in and out of the system as desired. A local interface at each unit, known as the 'Address Unit', allows the analyst to select and initiate programs and a choice of options. Basically the Address Unit decodes incoming signals from the computer and directs them to the operating function on the gas chromatograph. Address-discriminating elements and logic-gates ensure that only the correct Address Unit obeys the output command given by the computer.

The Address Unit also controls amplification and autoranging of the analogue signal and gives a good signal-to-noise ratio over a dynamic signal range of 2×10^6. The design provides for good resolution by the computer software and also maintains the local recorder trace on scale, which is useful in both method development and in the event of computer-failure.

A specific software language, Anaconda, has been developed which is intended to be a highly flexible and comprehensive system; it comprises a basic Executive and a subroutine library. Analytical data-handling methods for each chromatograph are written in the language. The program represents the sequence of events to be carried out for a particular analysis and also includes constants for calculations; it can also generate signals which operate valves or initiate temperature cycles, etc. Such lists are written after establishing operating conditions for the analysis and briefly studying the analogue record. Programs are stored for subsequent recall by the appropriate signal from the Address Unit. Routines for smoothing data and for area-allocation of peaks are used in most programs; these are designated as common in use and can be called into each program as requested.

The flexibility of Anaconda enables the analyst to write an almost unlimited variety of programs for analytical techniques and calculations, and the modularity of the Datacon approach allows systems to be tailor-made for individual laboratories.

Basically the system provides the analyst with all the computational requirements considered to be standard for gas chromatography. These include raw peak-area, corrected area, peak-heights, normalization, calculation of concentration from internal or external standard, with calibration up-dating, and absolute or relative retention-times for indentification, along with Datacon's control facilities. Each variation is push-button selectable at the Address Unit and easy to operate. Recently an application of this system has been described[22] which illustrates the usefulness of this turnkey approach.

11.7.3 Computer Applications in Nuclear Magnetic Resonance

Fig. 11.6 shows the schematic layout of a simple computer system[23] which both controls the operating parameters of the spectrometer and collects and analyses the spectral data. The diagram relates specifically to the Varian instrument and in this context a 16-bit 4K word computer is used. For simple control and data-reduction on the spectrometer such options as hardware multiply–divide, priority-interrupts and magnetic discs are not required. Simple operation and minimization of core are attained by a trade-off between hardware and software. The control console allows sequential selection of the many different machine variables and entry of parameters for these options; for example, improvement in signal-to-noise ratio can be achieved by selection of the multiple-scanning facility and entry of the number of scans to be performed. The computer controls the nmr spectrometer, performs the tasks requested, such as data acquisition, drives the recorder to supply the raw output data and finally displays the spectrum. Use of the teletype rather than hardware to enter the

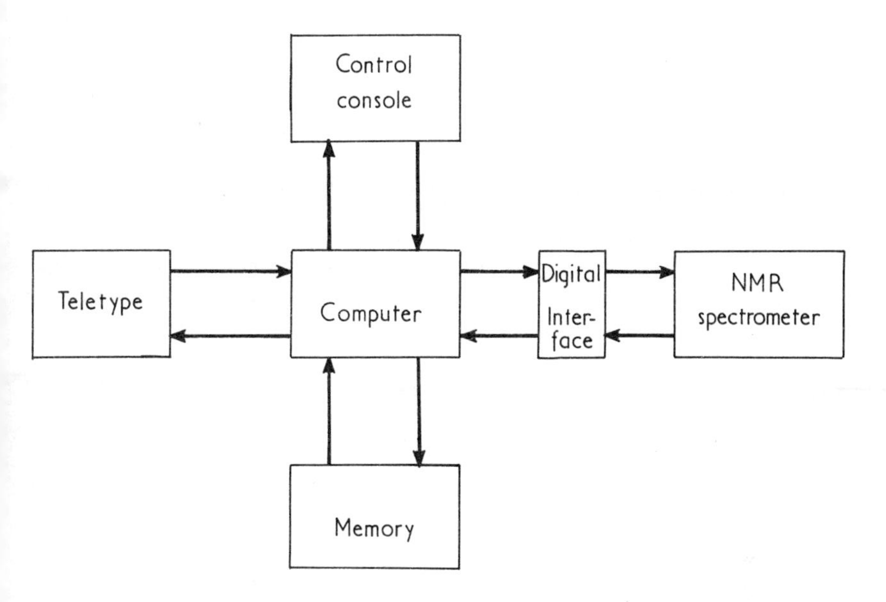

Fig. 11.6 On-line nuclear magnetic resonance spectrometer computer system. *Reproduced with permission from Frazer*[23] *and American Chemical Society.*

necessary options would have required a large section of core memory and would have been slower to operate. The Executive Monitor has only to check if the switch is on or off, and the program then branches to the appropriate subroutine. Control of a number of functions is possible. First the computer sets the current in the *Y*-gradient and the curvature on the shim coils in order to optimize the height of a selected reference peak. This can either be set before each spectrum or at fixed intervals throughout the day. The program accepts starting parameter instructions which denote the start scan-point, the sweep-width and the sweep-time. During a scan, 1024 data points are collected and stored in core, and on request from the operator the stored spectra can be displayed on a recorder. Software allows options for simple scanning, programmed scanning at variable speeds, base-line drift correction, digital signal smoothing, resolution-enhancement, peak-detection and simulation. The instrument integrates data-processing with instrument-control to optimize the spectrometer-performance and provide immediate spectral information. Systems similar to that described above are often considered completely automatic although they do not fulfil the requirements for complete automation set out by Michel, Sauter and Staübli[24], namely:

1. the automatic control of the instrument,
2. the collection and processing of the measured data into a suitable form,

3. the unattended changing of a reasonable number of samples which are under examination.

Commercially-available systems perform the requirements (1) and (2) satisfactorily, but after recording the spectrum the sample must be changed and the recorder initiated for the new spectrum, and the spectrometers cannot be operated outside the normal working hours. However, the system described by Michel *et al.* fulfils all three requirements and is achieved with the minimum alteration to a commercial system. An automatic sampler has been designed, Fig. 11.7, and the conventional

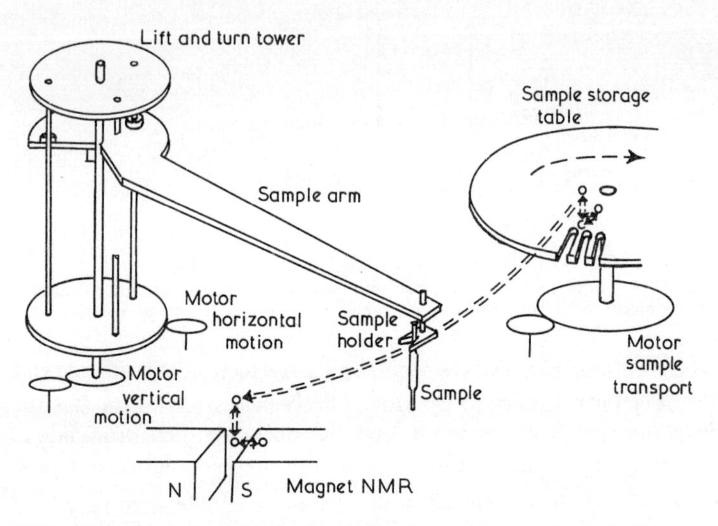

Fig. 11.7 Schematic diagram of nuclear magnetic resonance sample changer. *Reproduced with permission from Michel et al.*[24] *and Institute of Petroleum, London.*

flat-bed recorder replaced by a strip-chart recorder. The sample-changer is operated electronically but receives only the 'start' from the computer-system. The pneumatically-operated changer locates a sample complete with spinning device, picks it up from the turntable and positions it over the sample-holder; as the tube is lowered into position it is also set spinning. Correct alignment of the change-over mechanism is important for unattended operation. The computer checks that the sample is in fact spinning and if not, rejects the sample-tube and prints an error message, before the next cycle is initiated. If the tubes are spinning, the spectrum is recorded and the required options carried out, such as multiscan-averaging, etc. Calculated spectra are plotted whilst the sample is removed and the next sample from the turntable presented for analysis. Full automation is achieved by this device and the sampling-rate increases by a factor of three, provided similar operating conditions are used for a batch of samples.

With the pioneering of Fourier transform spectroscopy, notably by Ernst, nuclear magnetic resonance spectroscopy has taken on a new dimension[25–28]. This would not have been possible without the use of relatively inexpensive computers for data reduction and/or ready access to larger computers. For Fourier transform nmr a pulsed spectrometer is required and the sample in a magnetic field is excited by a radiofrequency pulse of short duration. After the pulse the resonance signal emitted by the sample is recorded as a function of time. The signal is the Fourier transform of the nmr spectrum. Measurement time for this spectrum is short but the signal-to-noise ratio for a single pulse is poor and signal-averaging is essential. A wide variety of pulse-signal sequences may be used, depending on the particular measurement requirements, and computer-control of the instrument allows more flexibility than is offered by hardware alone. Use of the pulsed technique has made possible the recording of ^{13}C spectra of compounds not specially enriched in this isotope. A good ^{13}C spectrum can be obtained by continually recording over a weekend in the pulsed mode; conventional techniques in a non-pulsed mode would require continuous recording for almost a year. Shaw[29] has shown that further information of wider application than coupling-constants and chemical shifts can be provided by study of the relaxation processes. These facts illustrate that the Fourier transform method will undoubtedly increase the usefulness of nmr spectrometry in the next decade.

11.7.4 Computer Applications in Mass Spectrometry

Data-capture rates in mass spectrometry are often of the order of 50,000/sec and this places a large strain on the data-processing equipment. In general, two approaches have been made to solve this problem, namely use of a fairly large computer directly on-line to the mass spectrometer or the use of a preprocessing system before off-line or on-line processing by a larger machine. The systems are typified by the work of Biemann and Carrick respectively and are briefly discussed in this section.

The work of Hites and Biemann[30] relates specifically to gas chromatography–mass spectrometry (gc/ms) computer systems for recording numerous spectra. In early stages of the development of the combined gc/ms with on-line interfacing direct to digital computers, the mass spectra were recorded only during the emergence of a particular gas chromatographic peak, with a few further spectra to obtain a background reference. When multicomponent samples are analysed a large number of spectra are recorded in a relatively short time and each spectrum is converted into an interpretable form. This requires at least identification of mass numbers if not conversion to a mass/intensity plot, which is very time-consuming. Even with a highly critical selection-process the computation and identification of the recorded spectra can hardly keep pace with their accumulation, and minor or unresolved components often escape detection. This risk can only be eliminated if the spectra are continuously recorded and this produces a large and unmanageable number of spectra. For faster processing more automated procedures for the identification of spectra are required and this necessitates data recorded in a digital form, i.e. an *m/e vs.* intensity plot. Such a system was originally described by Hites and Biemann[31] and uses a tape-recording system coupled to off-line data-processing. This off-line approach, however, proved inconvenient and the installation of a medium-sized computer on-line providing real-time data-acquisition, has been shown to be a potentially more useful and flexible approach. Fig. 11.8 shows a schematic diagram of the gas chromatograph–mass spectrometer computer system described by Hites and Biemann. The effluent from the gas chromatograph passes through the Biemann separator[32] and then into the mass spectrometer. The particular mass spectrometer used was a Hitachi Perkin-Elmer RMU6 D, a 90°-sector magnetic scanning instrument which is repetitively started and stopped under computer control. In the gc/ms combination the field

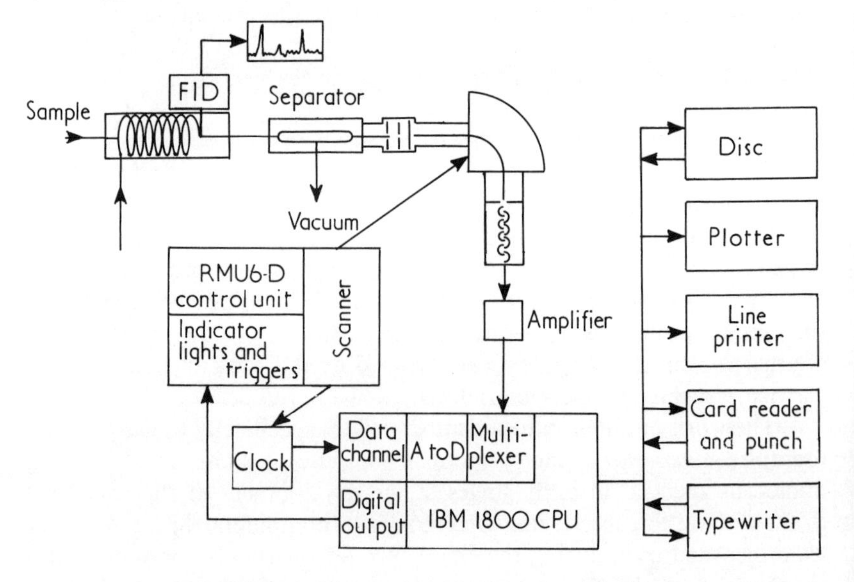

Fig. 11.8 Schematic diagram of gas chromatograph – mass spectrometer – computer system.
Reproduced with permission from Hites and Biemann[30] *and American Chemical Society.*

is swept through from mass 20 to 600 in 3 sec. The computer used is an IBM 1800 designed for fast data-acquisition and process-control; the configuration used in this system has 32K words of core-storage (16-bit words), complete with several input/output peripherals including three magnetic discs. The interface to the mass spectrometer is a pulse-generator set for 350-μsec intervals; this rate is chosen to produce about 3000 digital samples/sec. The pulses are sensed by the computer through a data channel. Each pulse causes the analogue input of the computer to carry out the following sequence; it determines which multiplexer point to read, where in core-storage to place the result, and then converts the analogue signal into digital form. Precise details of the programming are given in the original paper. In operation, the computer software is able to sort the spectra during periods when no new data are being collected. Each spectrum is spooled out on to disc and since the discs are removable, one can be changed while another is receiving data; the system permits an unlimited number of spectra to be recorded.

Generating the spectra in the computer-compatible format opens the way for many types of further processing such as searching files of authentic spectra for matching purposes, the resolution of spectra of mixtures into individual components and correction for background interferences[33].

Capture of mass-spectral data is difficult because peaks appear randomly and because extremely high data-rates are required to identify each peak. A software-controlled ADC computer-system such as described above spends much of the scan-time in rejecting base-line information, and during scan of a peak the computer is involved in lengthy calculations in real time to establish the position and intensity of a peak. Even a fairly

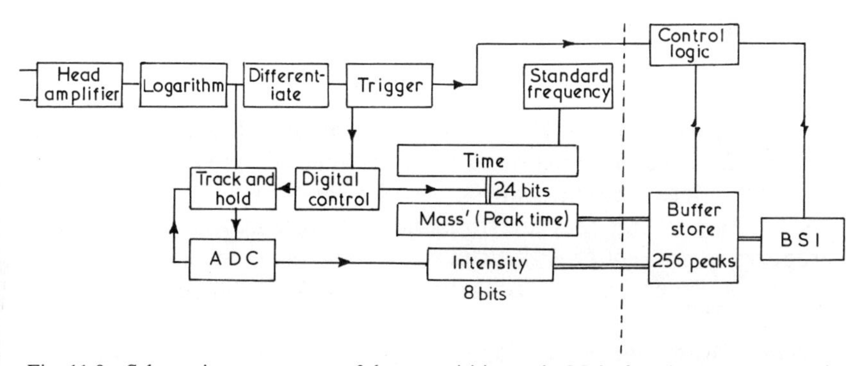

Fig. 11.9 Schematic arrangement of data acquisition unit. Main functions are separated by dotted line into analogue interface and buffer store.
Reproduced with permission from Carrick[34] and Institute of Petroleum, London.

large computer coupled to a mass spectrometer will have little capacity to handle other equipment or work on-line. It is therefore often advantageous to provide a system that matches the random characteristics of the spectra and which works under hardware-control as a peak emerges and calculates the peak-position and intensity directly. This approach has been described by Carrick[34–35] and requires a smaller, slower and less costly computer. A schematic diagram of the system is shown in Fig. 11.9. The system specifies peaks covering four decades of time widths (between 200 ms and 16μs) and three decades of peak-height and uses a combination of analogue and digital techniques. Information on peak-position and intensity is held in a fast-access buffer store (256 words); such information can be recorded continuously. The recorded data can be processed off-line either by producing paper-tape output or by using a 'block' transfer to a larger computer through an interface such as the British Standard interface (B.S. 4421). The peak maxima are detected by using a fast analogue computation and peaks referenced to a free running-time base giving a close approximation to true real-time measurements. The precision is limited only by the time-base and typically a clock frequency of 10 MHz produces an error of less than 1 ppm with scanning speeds of 0·1 sec per mass decade. The limiting factors of this equipment are generally the inherent errors associated with the mass spectrometer itself; the electronics of the data-capture equipment do not introduce any signal-noise or distortion.

A computer-search of a library of mass-spectral data can be used for positive identification of an unknown compound, but relies on the compound being in the library. Consequently, a library needs continuous attention to be up to date. Several workers have designed programs to mimic the technique that the organic mass spectroscopist uses for interpretation of mass spectra. Reasonably successful results have been obtained for specific classes of chemicals[36–40]. McLafferty[41] discusses the problems of adopting this approach, using software. Morrison and Crawford[42] also describe a software system that determines directly from the mass spectrum the molecular mass, the presence of functional groups, groups adjacent to the functional group, and finally the molecular skeleton. The conclusions are repeatedly checked for consistency. As the details of the structure emerge they are stored in a structure matrix. The program produces a conventional structural diagram on completion. Systems devised by Lederberg and co-workers[43, 44] work on different principles; the molecular weight is first determined and all possible structures are generated before an elimination procedure is used to reject the incorrect formulae. For small structures the approach is very successful, but the computing time required increases rapidly with molecular weight. More recent approaches rely on a rapid crude-sorting procedure using the computer and a final manual search using microfilm spectra displayed under computer control[45].

A far more novel approach to the interpretation of data generally and mass spectra in particular has been proposed by Isenhour and Jurs[46] in which programs are designed which learn to classify data on the basis of an inherent pattern structure rather than on known chemical theories. A learning pattern is established from a series of known compounds and unknown spectra can be interpreted by the computer on the basis of this learning pattern.

11.7.5 Complete Laboratory Computer Systems

Sections 11.7.1–11.7.4 discussed computer-applications for single analytical techniques; laboratories employing many of these techniques often benefit from the implementation of a larger computer servicing all the instruments. Design of such a system must pay particular attention to (*a*) the instrumentation hardware and software immediately suitable for on-line applications and (*b*) further instruments likely to be handled on-line. Several systems that have been formulated are discussed in this section and finally a description of the use of an off-line terminal to a bureau is considered.

The use of satellite computer systems has been described by Grohé *et al.*[47], an IBM 1810 being used as the satellite computer to a host IBM 360/65 central computer-system. In such a system, shown in detail in Fig. 11.10, the satellite collects the analogue signal from a number of different analytical instruments, processes them, prepares them in the form of complete blocks of information, transmits the information to the central computer, receives from the latter the evaluated results and finally prints results on the typewriter in the laboratory of origin. Certain functions such as counting peaks or column-switching can be effected in real time directly by the satellite computer without recourse to the central computer. The satellite computer used for data acquisition is small and acts virtually as a front-end processor. The 'host' computer performs batch-entry jobs in the background as well as evaluating preprocessed data from the satellite machine. The system accepts high data-rates for processing and in the unlikely event of a failure of the host-machine, facilities are provided to spool results to disc from the satellite machine. If the satellite computer fails, the analytical instruments cannot be used. The system, by ensuring adequate data-capture for mass spectrometers and nuclear magnetic resonance spectrometers, collects data far too rapidly for gas chromatography. In gas chromatographic applications

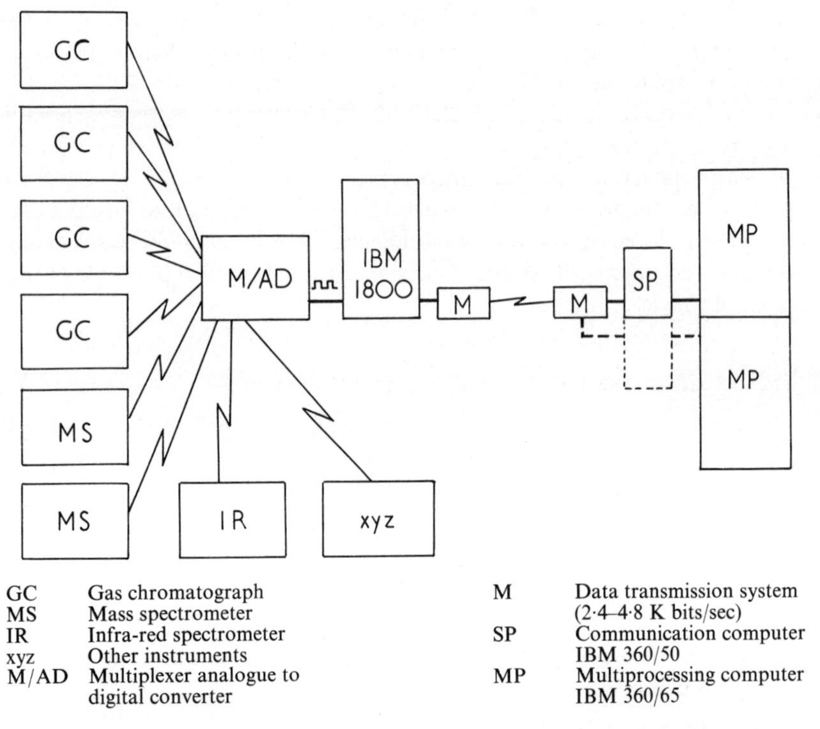

Fig. 11.10 Universal Satellite IBM 1800 on-line to central computer system.
Reproduced with permission from Grohé et al.[47] *and Institute of Petroleum, London.*

many of the data collected by using a fast ADC are automatically discarded; this disadvantage can be avoided by the use of two separate ADCs, one for fast information and the other for slow inputs. Such a system has been described by Ziegler, Henneberg and Schomburg[48]. A schematic diagram of the hardware based on a PDP10 central processing-unit is shown in Fig. 11.11. The unit comprises relocation and bit registers, floating-point and bit-handling instructions and seven levels of priority-nested interrupts. Core-memory consists of 32K 16-bit words and this is backed by a 500K fast-access disc. User programs are transferred into and out of core from disc as required and the disc also contains the source programs, compilers and assemblers as well as some data-storage. Several special devices are interfaced to the PDP10 for real-time data-acquisition through two separated ADCs. Thirty slow instruments running simultaneously are connected by a multiplexer to one ADC. The specification of this ADC provides 13-bit resolution, a dynamic range of 10^6 and eleven automatic gain-changes. Software selection of range-changes provides data-capture rates of 8000/sec, whereas hardware-control allows only 3000–5000/sec

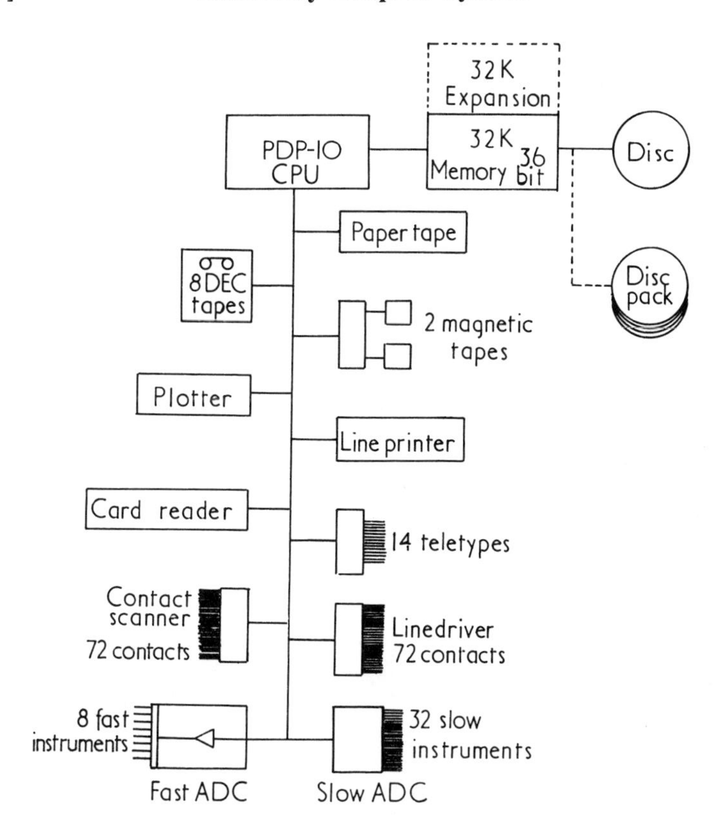

Fig. 11.11 Hardware configuration of Mulheim computer system.
Reproduced with permission from Ziegler et al.[48] *and American Chemical Society.*

for automatic range-selection. For fast instruments, particularly fast-scan mass spectrometers, a second ADC with data-rates between 1250 and 20,000 per sec is provided; this ADC has 10-bit resolution and dynamic range of $2{\cdot}5 \times 10^5$. A multiplexer connects eight data-lines to this ADC but only one fast instrument is allowed to transmit data at any given time; this avoids any problem of missing or wrongly assigning data. Each instrument is activated for only short periods, typically 2–3 sec, and the fact that only one can be used at a time does not restrict any instrument in practice.

Applications of this particular computer installation have been discussed by Schomburg[49], and fall into three categories, shown in Table 11.2.

In operation, the time-slice principle allocates within a fixed time interval a fraction of that interval to servicing data from a particular

Table 11.2 APPLICATIONS OF MULHEIM COMPUTER SYSTEM[48]

Off-line batch operation	*Real-time operation*	
	Slow ADC	*Fast ADC*
X-ray diffraction calculations	Up to 30 gas chromatography	Fast scan
Molecular orbital calculations	Low-resolution mass spectrometry	Mass spectrometer
Chemical kinetics		Pulsed
Spectral simulations	2 or 3 nmr spectrometers	Fourier transform
Documentation and searching spectral data	Infra-red spectrometers	Nmr spectrometer
Information retrieval	Raman spectrometers	
Administration	Esr spectrometers	

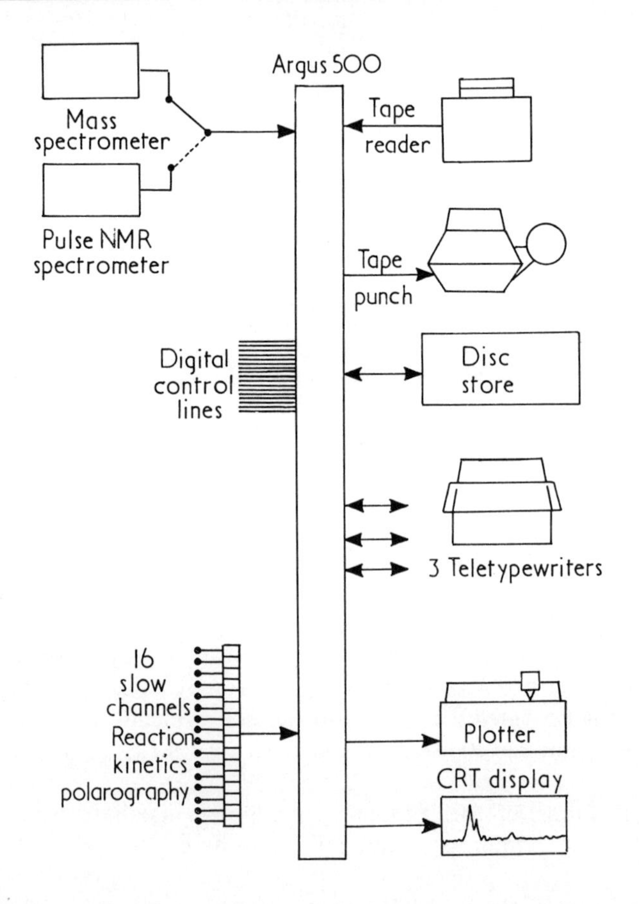

Fig. 11.12 Schematic diagram of time-shared computer system for laboratory automation. *Reproduced with permission from Hallett et al.*[50] *and Institute of Petroleum, London.*

instrument. At the end of the time allocation the computer suspends calculations until the equivalent period in the next time interval.

A further approach using a time-slice principle has been described[50] by Hallett *et al.* This is shown schematically in Fig. 11.12. The system again suffers from the limitation that data can only be accepted from one fast instrument at a time, and this for the nmr user can extend the scan-time for pulsed nmr, typically 15 min, by a factor of two. The system again uses two discrete input routes but the fast access is provided direct into core by a procedure known as cycle-stealing direct store-access. Information from the instrument is passed to one half of a specified buffer area whilst the other half of the store is transferred on to disc. The data-rates necessary specify the size of buffer-area required. Slow instruments are multiplexed by direct store-access into a 16-word buffer at sampling rates of 100/sec. Software is used to obtain sampling-rates normally required for slow instruments and samples are read from the 16-word buffer only when they are required. Off-line applications can also be handled in a manner similar to that[48] proposed by Ziegler *et al.*

Many laboratories will have neither the work nor resources to justify an 'in house' computer-system; at the authors' laboratory computer techniques have been developed by the application of an off-line terminal. Originally, the work, which has been discussed by Stockwell and Telford[51], involved using the London University Computer Services' (LUCS) Atlas computer, remotely through a terminal. The terminal comprised a paper-tape reader/punch, a modem link to the computer and a Friden 'Flexowriter' for paper-tape preparation. A number of automatic analytical systems, such as an automatic gas chromatograph, an amino-acid analyser, an automatic beer-analyser and an automatic specific-gravity instrument were designed and built with paper-tape output suitable for processing over the terminal. Software systems have been written in Fortran IV to analyse these tapes and produce results in the required format, often in the form of printed reports.

These systems have been written almost entirely 'in house' by chemists and not computer specialists. Recently the Atlas computer has been phased out and a Control Data Corporation 6500 computer installed at the LUCS bureau. As a consequence of this a DCT (132) terminal (Electronics Associates Limited) comprising card-reader, paper-tape reader, video-display unit, line-printer and controller has been installed. This is coupled by a modem and private line to the CDC 6500. This instrumentation has been able to increase flow of computing and to widen the range of applications so much that plans have been made for the provision of an 'in house' computer-system. Valuable experience of computer applications has been gained without recourse to large capital cost, and this has simplified the design problem involved in configuration of the 'in house' computer.

11.8 Information Retrieval

Although this chapter is primarily aimed at reviewing the applications of computers to analytical techniques and instruments it is also worthwhile to add a few words on retrieval of chemical information by computer. There is an almost embarrassing variety of chemical compounds known, and this coupled to the rapid growth of chemical publications has produced a major problem of communication. The rate of growth of publication is highlighted by statistics concerning the rate of publication of successive millions of Chemical Abstracts. This Journal began publication in 1907 and the millionth abstract was published in 1939. The second, third and fourth millionth abstracts were published in 18, 8, and 4 years respectively. That is to say, the total number of papers, books and patents doubles approximately every 12 years. The sheer volume of literature caused an inevitable time-delay in the production of abstracts and indexes and in 1961 a computer-produced current-awareness journal, Chemical Titles (CT) was published by Chemical Abstract Services (CAS). In this publication chemical journals were listed and manipulated by computer to form a keyword-in-context index. These titles were available on magnetic tape and it was an obvious step to consider the searching of these tapes by computer rather than relying on manual procedures. Other publications followed suit, such as Chemical Biological Abstracts (CBAC) in 1965, and for this publication, in addition to titles, author names and biographic information, an extensive digest was also provided on tape. This digest is usually 50–100 words long and as well as text contains the molecular formulation. In parallel with mechanical services developed by CAS, other services became available; the MEDLARS service provided by National Library of Medicine, Washington, U.S.A. (Medical Literature Analysis and Retrieval Services) and the Science Citation Index produced by the Institute of Science and Technical Information in Philadelphia. In 1966 the Chemical Society, London, together with OSTI (the Office of Scientific and Technical Information) set up an experimental information service at Nottingham University, and the work of this unit has been described by Kent in 1968[52]; work of a similar group in Pittsburg has also been reviewed[53].

Veal[54] has recently discussed the application of computer techniques to literature information retrieval in four distinct categories.

1. Textual retrieval systems based entirely on computer-tapes available from original publications, for example chemical titles.

2. Numerical data banks containing analytical data on a library of compounds. Information on unknown compounds is then matched to those in the reference library.
3. Applications in the context of mass spectra data (discussed in section 11.7.4).
4. Chemical structure files based on some line notation system for representing the spatial configuration of the molecule. (Lynch *et al.*[55] have discussed in outline notation systems, of which the Wiswesser technique is probably the most widely used.)

Many complete integral systems using varied combinations of these information bases have been devised and the use of integral systems, although presenting a large initial cost for preparation will undoubtedly produce many untold benefits to large organizations. Careful consideration must be given before the installation of a retrieval system because the system will only produce the information that it was designed to produce. Too little effort spent at the design stage of the system will often negate any useful benefits of the retrieval system. This requires a close co-operation between the computer-system designers and the ultimate users of the retrieval system.

11.9 Conclusions

This chapter serves to illustrate some of the advantages of computer applications in analytical chemistry. Several systems for both complete laboratory installation and simple dedicated systems are discussed. The survey is by no means complete but many of the major techniques are covered, but it is probably worthwhile to draw attention to the application of computers to electrochemical processes and in particular to the work of Perone and co-workers[56–57].

It should be concluded that whilst computers provide an extremely sophisticated tool to aid the chemical analyst, there can be no substitute for precise and accurately controlled chemistry. Considerable effort must be made in the design and development stage, since errors in configuration can cause a great deal of embarrassment once the system is installed. Which tasks the computer undertakes must also be carefully considered. Computers are often used as clocks, a large portion of computer-time being spent in controlling time-cycles which could be controlled with adequate precision by simple electric timers at a fraction of the cost.

References

1. Kramer, E. *Am. Laboratory*, 1970, Feb., 32.
2. Perone, S. P. *J. Chromat. Sci.*, 1969, **7**, 714.
3. Nielson, K. J. *Methods of Numerical Analysis*, p. 200, Macmillan, New York, 1956.
4. Roos, B. *Acta Chem. Scand.*, 1964, **18**, 2186.
5. Marquardt, D. W., Bennett, R. G. and Burrell, E. J. *J. Mol. Spectry.*, 1961, **7**, 269.
6. Marshall, S. W., Nelson, J. A. and Wilenzick, R. M. *Comm. Assoc. Computing Machinery*, 1965, **8**, 313.
7. Atkinson, B. and Stockwell, P. B. *J. Chem. Soc. (Phys. Organic)*, 1966, 984.
8. Whitehead, T. P. *Progress in Medical Computing*, p. 52, Blackwell, Oxford, 1965.
9. Robinson, R. *Clinical Chemistry and Automation*, Griffin, London, 1971.
10. Abrahamsson, S. and Larsson, K. *Arkiv Kemi*, 1965, **24**, 383.
11. Guiochon, G. *Techmation Informal Symposium*, 1969.
12. Margoshes, M. *Anal. Chem.*, 1971, **43**, No. 4, 101A.
13. Anon. *Data Management*, 1971, **3**, 7.
14. Whitby, L. G. *Lab. Practice*, 1970, **19**, 170.
15. Stockwell, P. B., Bunting, W., Morley, F. and Telford, I. *Lab. Practice* (in press).
16. Whitby, L. G. and Simpson, D. *J. Clin. Path.*, 1969, **3**, 107.
17. Gray, P. and Owen, J. A. *Clin. Chim. Acta*, 1969, **24**, 389.
18. Bennett, A. Gartelmann, D., Mason, J. and Owen, J. A. *Clin. Chim. Acta*, 1970, **29**, 161.
19. Abernethy, M. H., Bentley, G. I., Gartelman, D., Gray, P., Owen, J. A. and Quan Sing, G. D. *Clin. Chim. Acta*, 1970, **30**, 463.
20. Thiers, R. E., Meyn, J. and Wildermann, R. F. *Clin. Chem.*, 1970, **16**, 832.
21. Deans, D. *Gas Chromatography 1970*, p. 292, Institute of Petroleum, London, 1971.
22. Lancaster, C. B., Mitchell, P. and Moss, A. R. L. *The Applications of Computer Techniques in Chemical Research*, p. 112, Institute of Petroleum, London, 1972.
23. Frazer, J. W. *Anal. Chem.*, 1968, **40**, No. 8, 26A.
24. Michel, G., Sauter, H. P. and Stäubli, A. *The Applications of Computer Techniques in Chemical Research*, p. 101, Institute of Petroleum, 1972.
25. Ernst, R. R. *The Applications of Computer Techniques in Chemical Research*, p. 61, Institute of Petroleum, London, 1972.
26. Ernst, R. R. *Rev. Sci. Inst.*, 1966, **36**, 1689.
27. Ernst, R. R. and Anderson, W. A. *Rev. Sci. Inst.*, 1963, **34**, 754.
28. Ernst, R. R. *J. Mag. Res.*, 1969, **1**, 7.
29. Shaw, D. *The Applications of Computer Techniques in Chemical Research*, p. 76, Institute of Petroleum, London, 1972.
30. Hites, R. A. and Biemann, K. *Anal. Chem.*, 1968, **40**, 1217.
31. Hites, R. A. and Biemann, K. *Anal. Chem.*, 1967, **39**, 965.
32. Watson, J. T. and Biemann, K. *Anal. Chem.*, 1965, **37**, 844.
33. Biemann, K. *The Application of Computer Techniques in Chemical Research*, p. 5, Institute of Petroleum, London, 1972.
34. Carrick, A. *Advances in Mass Spectrometry*, Vol. 5, p. 330, Institute of Petroleum, London, 197.
35. Carrick, A. *U.K. Pat. Application*, 35894/68 (1968).

38. Duffield, A. M., Robertson, A. V., Djerassi, C., Buchanan, B. G., Sutherland, G. L., Feigenbaum, E. A. and Lederberg, J. *J. Am. Chem. Soc.*, 1969, **91**, 2977.
37. Schroll, G., Duffield, A. M., Djerassi, C., Buchanan, B. G., Sutherland, G. L., Feigenbaum, E. A. and Lederberg, J. *J. Am. Chem. Soc.*, 1969, **91**, 7440.
38. Barber, M., Powers, P., Wallington, P. and Wolstenholme, M. J. *Nature*, 1966, **212**, 784.
39. Biemann, K., Cone, C., Webster, B. R. and Arsenault, G. P. *J. Am. Chem. Soc.*, 1966, **88**, 5598.
40. Hertz, H. S., Hites, R. A. and Biemann, K. *Anal. Chem.*, 1971, **43**, 681.
41. McLafferty, F. W., *Mass Spectral Correlations*, American Chem. Soc., Washington, 1963.
42. Crawford, L. R. and Morrison, J. D. *Anal. Chem.*, 1971, **43**, 1790.
43. Lederberg, J., Sutherland, G. L., Buchanan, B. G., Feigenbaum, E. A., Robertson, A. V., Duffield, A. M. and Djerassi, C. *J. Am. Chem. Soc.*, 1969, **91**, 2973.
44. Lederberg, J. and Wightman, M. *Anal. Chem.*, 1964, **36**, 2365.
45. Biller, J. E., *Ph.D. Thesis*. Massachusetts Inst. of Tech., 1972.
46. Isenhour, T. L. and Jurs, P. C. *The Applications of Computers in Chemical Research*, p. 189. Institute of Petroleum, London, 1972.
47. Grohé, F., Hesse, W., Kaiser, R. and Schneckenburger, K. H. *Gas Chromatography 1970*, (ed. Stock, R.), p. 247, Institute of Petroleum, London, 1971.
48. Ziegler, E., Henneberg, D. and Schomburg, G. *Anal. Chem.*, 1970, **42**, No. 9, 51A.
49. Schomburg, G., Weeke, F., Weimann, B. and Ziegler, E. *Gas Chromatography 1970*, (ed. Stock, R.), p. 280, Institute of Petroleum, London, 1971.
50. Hallett, J. G., Lawson, P. A., Richards, D. and Stanier, H. M. *The Applications of Computers in Chemical Research*, p. 270, Institute of Petroleum, London, 1972.
51. Stockwell, P. B. and Telford, I. *Lab. Equip. Digest* (in the press).
52. Kent, A. K. *Chem. Ind., London*, 1968, 1214.
53. Arnett, E. M. *Science*, 1970, **170**, 1370.
54. Veal, D. C. *The Applications of Computers in Chemical Research*, p. 129, Institute of Petroleum, London, 1972.
55. Lynch, M. F., Harrison, J. M., Town, W. G. and Ash, J. E. *Computer Handling of Chemical Structure Information*, MacDonald, London, 1971.
56. Perone, S. P., Harrer, J. E., Stephens, F. B. and Anderson, R. E. *Anal. Chem.*, 1968, **40**, 899.
57. Perone, S. P., Jones, D. O. and Gutknecht, W. F. *Anal. Chem.*, 1969, **41**, 1154.

Index